Selected Titles in This Series

12 **W. J. Kaczor and M. T. Nowak,** Prob. II: Continuity and differentiation, 2001

11 **Michael Mesterton-Gibbons,** An introduction to game-theoretic modelling, 2000

10 **John Oprea,** The mathematics of soap films: Explorations with Maple®, 2000

9 **David E. Blair,** Inversion theory and conformal mapping, 2000

8 **Edward B. Burger,** Exploring the number jungle: A journey into diophantine analysis, 2000

7 **Judy L. Walker,** Codes and curves, 2000

6 **Gérald Tenenbaum and Michel Mendès France,** The prime numbers and their distribution, 2000

5 **Alexander Mehlmann,** The game's afoot! Game theory in myth and paradox, 2000

4 **W. J. Kaczor and M. T. Nowak,** Problems in mathematical analysis I: Real numbers, sequences and series, 2000

3 **Roger Knobel,** An introduction to the mathematical theory of waves, 2000

2 **Gregory F. Lawler and Lester N. Coyle,** Lectures on contemporary probability, 1999

1 **Charles Radin,** Miles of tiles, 1999

Problems in Mathematical Analysis II

Continuity and Differentiation

STUDENT MATHEMATICAL LIBRARY
Volume 12

Problems in Mathematical Analysis II
Continuity and Differentiation

W. J. Kaczor
M. T. Nowak

AMS
AMERICAN MATHEMATICAL SOCIETY

Editorial Board

David Bressoud   Carl Pomerance
Robert Devaney, Chair   Hung-Hsi Wu

Originally published in Polish, as
Zadania z Analizy Matematycznej. Część Druga
Funkcje Jednej Zmiennej—Rachunek Różniczowy

© 1998, Wydawnictwo Uniwersytetu Marii Curie–Skłodowskiej, Lublin.

Translated, revised and augmented by the authors.

2000 Mathematics Subject Classification. Primary 00A07;
Secondary 26A06, 26A15, 26A24.

Library of Congress Cataloging-in-Publication Data
Kaczor, W. J. (Wieslawa J.), 1949–
  [Zadania z analizy matematycznej. English]
  Problems in mathematical analysis. I. Real numbers, sequences and series /
W. J. Kaczor, M. T. Nowak.
    p. cm. — (Student mathematical library, ISSN 1520-9121 ; v. 4)
  Includes bibliographical references.
  ISBN 0-8218-2050-8 (softcover ; alk. paper)
  1. Mathematical analysis I. Nowak, M. T. (Maria T.), 1951– II. Title.
III. Series.
QA300K32513   2000
515'.076–dc21   99-087039

Copying and reprinting. Individual readers of this publication, and nonprofit libraries acting for them, are permitted to make fair use of the material, such as to copy a chapter for use in teaching or research. Permission is granted to quote brief passages from this publication in reviews, provided the customary acknowledgment of the source is given.

Republication, systematic copying, or multiple reproduction of any material in this publication is permitted only under license from the American Mathematical Society. Requests for such permission should be addressed to the Assistant to the Publisher, American Mathematical Society, P. O. Box 6248, Providence, Rhode Island 02940-6248. Requests can also be made by e-mail to reprint-permission@ams.org.

© 2001 by the American Mathematical Society. All rights reserved.
The American Mathematical Society retains all rights
except those granted to the United States Government.
Printed in the United States of America.

∞ The paper used in this book is acid-free and falls within the guidelines
established to ensure permanence and durability.
Visit the AMS home page at URL: http://www.ams.org/

10 9 8 7 6 5 4 3 2 1   06 05 04 03 02 01

Contents

Preface — xi

Notation and Terminology — xiii

Problems

Chapter 1. Limits and Continuity
- 1.1. The Limit of a Function — 3
- 1.2. Properties of Continuous Functions — 9
- 1.3. Intermediate Value Property — 14
- 1.4. Semicontinuous Functions — 18
- 1.5. Uniform Continuity — 24
- 1.6. Functional Equations — 27
- 1.7. Continuous Functions in Metric Spaces — 32

Chapter 2. Differentiation
- 2.1. The Derivative of a Real Function — 37
- 2.2. Mean Value Theorems — 45
- 2.3. Taylor's Formula and L'Hospital's Rule — 52
- 2.4. Convex Functions — 61

2.5. Applications of Derivatives	68
2.6. Strong Differentiability and Schwarz Differentiability	77

Chapter 3. Sequences and Series of Functions

3.1. Sequences of Functions, Uniform Convergence	81
3.2. Series of Functions, Uniform Convergence	87
3.3. Power Series	96
3.4. Taylor Series	102

Solutions

Chapter 1. Limits and Continuity

1.1. The Limit of a Function	111
1.2. Properties of Continuous Functions	129
1.3. Intermediate Value Property	146
1.4. Semicontinuous Functions	160
1.5. Uniform Continuity	171
1.6. Functional Equations	181
1.7. Continuous Functions in Metric Spaces	198

Chapter 2. Differentiation

2.1. The Derivative of a Real Function	211
2.2. Mean Value Theorems	233
2.3. Taylor's Formula and L'Hospital's Rule	245
2.4. Convex Functions	267
2.5. Applications of Derivatives	285
2.6. Strong Differentiability and Schwarz Differentiability	310

Chapter 3. Sequences and Series of Functions

3.1. Sequences of Functions, Uniform Convergence	317
3.2. Series of Functions, Uniform Convergence	336
3.3. Power Series	355

Contents

    3.4. Taylor Series   372

Bibliography - Books   393

Index   397

Preface

This is the second volume of a planned series of books of problems in mathematical analysis. The book deals with real functions of one real variable, except for Section 1.7 where functions in metric spaces are discussed. Like the first volume, *Problems in Mathematical Analysis I, Real Numbers, Sequences and Series*, the book is divided into two parts. The first part is a collection of exercises and problems, and the second contains their solutions. Although often various solutions of a given problem are possible, we present here only one. Moreover, problems are divided into sections according to the methods of their solutions. For example, if a problem is in the section *Convex Functions* it means that in its solution properties of convex functions are used. While each section begins with relatively simple exercises, one can still find quite challenging problems, some of which are actually theorems. Although the book is intended mainly for mathematics students, it covers material that can be incorporated by teachers into their lectures or be used for seminar discussions. For example, following Steven Roman (Amer. Math. Monthly, 87 (1980), pp. 805-809), we present a proof of the well known Faà di Bruno formula for the nth derivative of the composition of two functions. Applications of this formula to real analytic functions given in Chapter 3 are mainly borrowed from the book *A Primer of Real Analytic Functions* by Steven G. Krantz and Harold R. Parks. In fact, we found this book so stimulating that we could not resist borrowing a few theorems from it. We

xi

would like also to mention here a generalization of Tauber's theorem due to Hardy and Littlewood. The proof of this result that we give is based on Karamata's paper (Math. Zeitschrift, 2 (1918)).

Many problems have been borrowed freely from problem sections of journals like the American Mathematical Monthly, Mathematics Today (Russian) and Delta (Polish), and from many textbooks and problem books. The complete list of books is given in the bibliography. As in the first volume, it was beyond our scope to trace all original sources, and we offer our sincere apologies if we have overlooked some contributions.

All the notations and definitions used in this volume are standard. However, in an effort to make the book consistent and to avoid ambiguity, we have included a list of notations and definitions. In making references we write, for example, 1.2.33 or I, 1.2.33, which denotes the number of the problem in this volume or in Volume I, respectively.

We owe much to many friends and colleagues with whom we have had many fruitful discussions. Special mention should be made, however, of Tadeusz Kuczumow for suggestions of several problems and solutions, and of Witold Rzymowski for making his manuscript [28] available to us. We are very grateful to Armen Grigoryan, Małgorzata Koter-Mórgowska, Stanisław Prus and Jadwiga Zygmunt for drawing the figures and for their help with incorporating them into the text. We are deeply indebted to Professor Richard J. Libera, University of Delaware, for his unceasing help with the English translation and for his valuable suggestions and corrections which we feel have greatly improved both the form and the content of the two volumes. Finally, we would like to thank the staff at the AMS for their dedicated assistance (via e-mail) in bringing our work to fruition.

W. J. Kaczor, M. T. Nowak

Notation and Terminology

This is a supplement to the notation and terminology of *Problems in Mathematical Analysis I, Real Numbers, Sequences and Series.*

If $(\mathbf{X}, d)$ is a metric space, $x \in \mathbf{X}$ and $\mathbf{A}$ is a nonempty subset of $\mathbf{X}$, then

- $\mathbf{A}^c = \mathbf{X} \setminus \mathbf{A}$ is the complement of the set $\mathbf{A}$,
- $\mathbf{B}_{\mathbf{X}}(x, r)$, $\overline{\mathbf{B}}_{\mathbf{X}}(x, r)$ denote the open and the closed ball centered at x and of radius $r > 0$, respectively. If $\mathbf{X}$ is fixed we omit the index and simply write $\mathbf{B}(x, r)$ or $\overline{\mathbf{B}}(x, r)$,
- $\mathbf{A}^\circ$ is the interior of $\mathbf{A}$ in the metric space $(\mathbf{X}, d)$,
- $\overline{\mathbf{A}}$ denotes the closure of $\mathbf{A}$ in the metric space,
- $\partial \mathbf{A} = \overline{\mathbf{A}} \cap \overline{\mathbf{X} \setminus \mathbf{A}}$ is the boundary of $\mathbf{A}$,
- $\mathrm{diam}(\mathbf{A}) = \sup\{d(x, y) : x, y \in \mathbf{A}\}$ denotes the diameter of the set $\mathbf{A}$,
- $\mathrm{dist}\,(x, \mathbf{A}) = \inf\{d(x, y) : y \in \mathbf{A}\}$ denotes the distance between x and the set $\mathbf{A}$,
- $\mathbf{A}$ is of type $\mathcal{F}_\sigma$ if it is a union of countably many sets which are closed in $(\mathbf{X}, d)$,
- $\mathbf{A}$ is of type $\mathcal{G}_\delta$ if it is an intersection of countably many sets which are open in $(\mathbf{X}, d)$,

- **X** is said to be connected if there do not exist two nonempty disjoint open subsets **B** and **C** of **X** such that $\mathbf{X} = \mathbf{B} \cup \mathbf{C}$,
-
$$\chi_{\mathbf{A}}(x) = \begin{cases} 1 & \text{if } x \in \mathbf{A}, \\ 0 & \text{if } x \in (\mathbf{X} \setminus \mathbf{A}) \end{cases}$$

    is the characteristic function of **A**,
- if $\mathbf{A} \subset \mathbf{X}$ and if f is a function defined on **X**, then $f_{|\mathbf{A}}$ denotes the restriction of f to **A**.

If f and g are real functions of a real variable, then
- $f(a^+)$ and $f(a^-)$ denote the right-hand and the left-hand limit of f at a, respectively,
- if the quotient $f(x)/g(x)$ tends to zero (or remains bounded) as $x \to x_0$, then we write $f(x) = o(g(x))$ (or $f(x) = O(g(x))$),
- $C(\mathbf{A})$ - the set of all continuous functions on **A**,
- $C(a,b)$ - the set of all continuous functions on an open interval (a,b),
- $f^{(n)}$ - the nth derivative of f,
- $C^n(a,b)$ - the set of all functions n times continuously differentiable on (a,b),
- $f'_+(a)$, $f'_-(a)$ - the right- and the left-hand derivative of f at a, respectively,
- $C^1([a,b])$ denotes the set of all functions continuously differentiable on $[a,b]$, where at the endpoints the derivative is right- or left-hand, respectively. The set $C^n([a,b])$ of all functions n times continuously differentiable on $[a,b]$ is defined inductively,
- $C^\infty(a,b)$, $C^\infty([a,b])$ - the set of functions infinitely differentiable on (a,b) and $[a,b]$, respectively.

Problems

Chapter 1

Limits and Continuity

1.1. The Limit of a Function

We adopt the following definitions.

Definition 1. A real function f is said to be *increasing* (resp. *strictly increasing, decreasing, strictly decreasing*) on a nonempty set $\mathbf{A} \subset \mathbb{R}$ if $x_1 < x_2$, $x_1, x_2 \in \mathbf{A}$, implies $f(x_1) \leq f(x_2)$ (resp. $f(x_1) < f(x_2), f(x_1) \geq f(x_2), f(x_1) > f(x_2)$). A function which is either increasing or decreasing (resp. strictly increasing or strictly decreasing) is called *monotone* (resp. *strictly monotone*).

Definition 2. By a *deleted neighborhood* of a point $a \in \mathbb{R}$ we mean the set $(a - \varepsilon, a + \varepsilon) \setminus \{a\}$, where $\varepsilon > 0$.

1.1.1. Find the limits or state that they do not exist.

(a) $\lim\limits_{x \to 0} x \cos \dfrac{1}{x}$,

(b) $\lim\limits_{x \to 0} x \left[\dfrac{1}{x} \right]$,

(c) $\lim\limits_{x \to 0} \dfrac{x}{a} \left[\dfrac{b}{x} \right]$, $a, b > 0$,

(d) $\lim\limits_{x \to 0} \dfrac{[x]}{x}$,

(e) $\lim\limits_{x \to +\infty} x \left(\sqrt{x^2 + 1} - \sqrt[3]{x^3 + 1} \right)$,

(f) $\lim\limits_{x \to 0} \dfrac{\cos \left(\frac{\pi}{2} \cos x \right)}{\sin(\sin x)}$.

1.1.2. Assume that $f : (-a, a) \setminus \{0\} \to \mathbb{R}$. Show that
(a) $\lim_{x \to 0} f(x) = l$ if and only if $\lim_{x \to 0} f(\sin x) = l$,
(b) if $\lim_{x \to 0} f(x) = l$, then $\lim_{x \to 0} f(|x|) = l$. Does the other implication hold?

1.1.3. Suppose a function $f : (-a, a) \setminus \{0\} \to (0, +\infty)$ satisfies $\lim_{x \to 0} \left(f(x) + \frac{1}{f(x)} \right) = 2$. Show that $\lim_{x \to 0} f(x) = 1$.

1.1.4. Assume f is defined on a deleted neighborhood of a and $\lim_{x \to a} \left(f(x) + \frac{1}{|f(x)|} \right) = 0$. Determine $\lim_{x \to a} f(x)$.

1.1.5. Prove that if f is a bounded function on $[0, 1]$ satisfying $f(ax) = bf(x)$ for $0 \leq x \leq \frac{1}{a}$ and $a, b > 1$, then $\lim_{x \to 0^+} f(x) = f(0)$.

1.1.6. Calculate

(a) $\lim_{x \to 0} \left(x^2 \left(1 + 2 + 3 + \cdots + \left[\frac{1}{|x|} \right] \right) \right)$,

(b) $\lim_{x \to 0^+} \left(x \left(\left[\frac{1}{x} \right] + \left[\frac{2}{x} \right] + \cdots + \left[\frac{k}{x} \right] \right) \right)$, $k \in \mathbb{N}$.

1.1.7. Compute $\lim_{x \to \infty} \frac{[P(x)]}{P([x])}$, where $P(x)$ is a polynomial with positive coefficients.

1.1.8. Show by an example that the condition

(∗) $$\lim_{x \to 0} (f(x) + f(2x)) = 0$$

does not imply that f has a limit at 0. Prove that if there exists a function φ such that in a deleted neighborhood of zero the inequality $f(x) \geq \varphi(x)$ is satisfied and $\lim_{x \to 0} \varphi(x) = 0$, then (∗) implies $\lim_{x \to 0} f(x) = 0$.

1.1.9.

(a) Give an example of a function f satisfying the condition

$$\lim_{x \to 0} (f(x) f(2x)) = 0$$

1.1. The Limit of a Function

and such that $\lim_{x\to 0} f(x)$ does not exist.

(b) Show that if in a deleted neighborhood of zero the inequalities $f(x) \geq |x|^\alpha$, $\frac{1}{2} < \alpha < 1$, and $f(x)f(2x) \leq |x|$ hold, then $\lim_{x\to 0} f(x) = 0$.

1.1.10. Given a real α, assume that $\lim_{x\to\infty} \frac{f(ax)}{x^\alpha} = g(a)$ for each positive a. Show that there is c such that $g(a) = ca^\alpha$.

1.1.11. Suppose that $f : \mathbb{R} \to \mathbb{R}$ is a monotonic function such that $\lim_{x\to\infty} \frac{f(2x)}{f(x)} = 1$. Show that also $\lim_{x\to\infty} \frac{f(cx)}{f(x)} = 1$ for each $c > 0$.

1.1.12. Prove that if $a > 1$ and $\alpha \in \mathbb{R}$, then

(a) $\lim_{x\to\infty} \frac{a^x}{x} = +\infty$, (b) $\lim_{x\to\infty} \frac{a^x}{x^\alpha} = +\infty$.

1.1.13. Show that if $\alpha > 0$, then $\lim_{x\to\infty} \frac{\ln x}{x^\alpha} = 0$.

1.1.14. For $a > 0$, show that $\lim_{x\to 0} a^x = 1$. Use this equality to prove the continuity of the exponential function.

1.1.15. Show that

(a) $\lim_{x\to\infty} \left(1 + \frac{1}{x}\right)^x = e$, (b) $\lim_{x\to-\infty} \left(1 + \frac{1}{x}\right)^x = e$,

(c) $\lim_{x\to 0} (1+x)^{\frac{1}{x}} = e$.

1.1.16. Show that $\lim_{x\to 0} \ln(1+x) = 0$. Using this equality, deduce that the logarithmic function is continuous on $(0, \infty)$.

1.1.17. Determine the following limits:

(a) $\lim_{x\to 0} \frac{\ln(1+x)}{x}$, (b) $\lim_{x\to 0} \frac{a^x - 1}{x}$, $a > 0$,

(c) $\lim_{x\to 0} \frac{(1+x)^\alpha - 1}{x}$, $\alpha \in \mathbb{R}$.

1.1.18. Find

(a) $\lim\limits_{x\to\infty} (\ln x)^{\frac{1}{x}}$,

(b) $\lim\limits_{x\to 0^+} x^{\sin x}$,

(c) $\lim\limits_{x\to 0} (\cos x)^{\frac{1}{\sin^2 x}}$,

(d) $\lim\limits_{x\to\infty} (e^x - 1)^{\frac{1}{x}}$,

(e) $\lim\limits_{x\to 0^+} (\sin x)^{\frac{1}{\ln x}}$.

1.1.19. Find the following limits:

(a) $\lim\limits_{x\to 0} \dfrac{\sin 2x + 2\arctan 3x + 3x^2}{\ln(1+3x+\sin^2 x) + xe^x}$,

(b) $\lim\limits_{x\to 0} \dfrac{\ln \cos x}{\tan x^2}$,

(c) $\lim\limits_{x\to 0^+} \dfrac{\sqrt{1-e^{-x}} - \sqrt{1-\cos x}}{\sqrt{\sin x}}$,

(d) $\lim\limits_{x\to 0} (1+x^2)^{\cot x}$.

1.1.20. Calculate

(a) $\lim\limits_{x\to\infty} \left(\tan \dfrac{\pi x}{2x+1}\right)^{\frac{1}{x}}$,

(b) $\lim\limits_{x\to\infty} x \left(\ln\left(1 + \dfrac{x}{2}\right) - \ln \dfrac{x}{2}\right)$.

1.1.21. Suppose that $\lim\limits_{x\to 0^+} g(x) = 0$ and that there are $\alpha \in \mathbb{R}$ and positive m and M such that $m \leq \dfrac{f(x)}{x^\alpha} \leq M$ for positive x from a neighborhood of zero. Show that if $\alpha \lim\limits_{x\to 0^+} g(x) \ln x = \gamma$, then $\lim\limits_{x\to 0^+} f(x)^{g(x)} = e^\gamma$. In the case where $\gamma = \infty$ or $\gamma = -\infty$ we assume that $e^\infty = \infty$ and $e^{-\infty} = 0$.

1.1.22. Assume that $\lim\limits_{x\to 0} f(x) = 1$ and $\lim\limits_{x\to 0} g(x) = \infty$. Show that if $\lim\limits_{x\to 0} g(x)(f(x) - 1) = \gamma$, then $\lim\limits_{x\to 0} f(x)^{g(x)} = e^\gamma$.

1.1.23. Calculate

(a) $\lim\limits_{x\to 0^+} \left(2\sin\sqrt{x} + \sqrt{x}\sin\dfrac{1}{x}\right)^x$,

(b) $\lim\limits_{x\to 0} \left(1 + xe^{-\frac{1}{x^2}}\sin\dfrac{1}{x^4}\right)^{e^{\frac{1}{x^2}}}$,

(c) $\lim\limits_{x\to 0} \left(1 + e^{-\frac{1}{x^2}}\arctan\dfrac{1}{x^2} + xe^{-\frac{1}{x^2}}\sin\dfrac{1}{x^4}\right)^{e^{\frac{1}{x^2}}}$.

1.1. The Limit of a Function   7

1.1.24. Let $f : [0, +\infty) \to \mathbb{R}$ be a function such that each sequence $\{f(a+n)\}$, $a \geq 0$, converges to zero. Does the limit $\lim_{x \to \infty} f(x)$ exist?

1.1.25. Let $f : [0, +\infty) \to \mathbb{R}$ be a function such that, for any positive a, the sequence $\{f(an)\}$ converges to zero. Does the limit $\lim_{x \to \infty} f(x)$ exist?

1.1.26. Let $f : [0, +\infty) \to \mathbb{R}$ be a function such that, for each $a \geq 0$ and each $b > 0$, the sequence $\{f(a+bn)\}$ converges to zero. Does the limit $\lim_{x \to \infty} f(x)$ exist?

1.1.27. Prove that if $\lim_{x \to 0} f(x) = 0$ and $\lim_{x \to 0} \frac{f(2x)-f(x)}{x} = 0$, then $\lim_{x \to 0} \frac{f(x)}{x} = 0$.

1.1.28. Suppose that f defined on $(a, +\infty)$ is bounded on each finite interval (a, b), $a < b$. Prove that if $\lim_{x \to +\infty} (f(x+1) - f(x)) = l$, then also $\lim_{x \to +\infty} \frac{f(x)}{x} = l$.

1.1.29. Let f defined on $(a, +\infty)$ be bounded below on each finite interval (a, b), $a < b$. Show that if $\lim_{x \to +\infty} (f(x+1) - f(x)) = +\infty$, then also $\lim_{x \to +\infty} \frac{f(x)}{x} = +\infty$.

1.1.30. Let f defined on $(a, +\infty)$ be bounded on each finite interval (a, b), $a < b$. If for a nonnegative integer k, $\lim_{x \to +\infty} \frac{f(x+1)-f(x)}{x^k}$ exists, then
$$\lim_{x \to +\infty} \frac{f(x)}{x^{k+1}} = \frac{1}{k+1} \lim_{x \to +\infty} \frac{f(x+1) - f(x)}{x^k}.$$

1.1.31. Let f defined on $(a, +\infty)$ be bounded on each finite interval (a, b), $a < b$, and assume that $f(x) \geq c > 0$ for $x \in (a, +\infty)$. Show that if $\lim_{x \to +\infty} \frac{f(x+1)}{f(x)}$ exists, then $\lim_{x \to +\infty} (f(x))^{\frac{1}{x}}$ also exists and
$$\lim_{x \to +\infty} (f(x))^{\frac{1}{x}} = \lim_{x \to +\infty} \frac{f(x+1)}{f(x)}.$$

1.1.32. Assume that $\lim_{x \to 0} f\left(\left[\frac{1}{x}\right]^{-1}\right) = 0$. Does this imply that the limit $\lim_{x \to 0} f(x)$ exists?

1.1.33. Let $f : \mathbb{R} \to \mathbb{R}$ be such that, for any $a \in \mathbb{R}$, the sequence $\{f\left(\frac{a}{n}\right)\}$ converges to zero. Does f have a limit at zero?

1.1.34. Prove that if $\lim_{x \to 0} f\left(x\left(\frac{1}{x} - \left[\frac{1}{x}\right]\right)\right) = 0$, then $\lim_{x \to 0} f(x) = 0$.

1.1.35. Show that if f is monotonically increasing (decreasing) on (a, b), then for any $x_0 \in (a, b)$,

(a) $f(x_0^+) = \lim_{x \to x_0^+} f(x) = \inf_{x > x_0} f(x) \quad \left(f(x_0^+) = \sup_{x > x_0} f(x)\right)$,

(b) $f(x_0^-) = \lim_{x \to x_0^-} f(x) = \sup_{x < x_0} f(x) \quad \left(f(x_0^-) = \inf_{x < x_0} f(x)\right)$,

(c) $f(x_0^-) \leq f(x_0) \leq f(x_0^+) \quad (f(x_0^-) \geq f(x_0) \geq f(x_0^+))$.

1.1.36. Show that if f is monotonically increasing on (a, b), then for any $x_0 \in (a, b)$,

(a) $\quad\quad\quad\quad\quad\quad\quad\quad \lim_{x \to x_0^+} f(x^-) = f(x_0^+)$,

(b) $\quad\quad\quad\quad\quad\quad\quad\quad \lim_{x \to x_0^-} f(x^+) = f(x_0^-)$.

1.1.37. Prove the following *Cauchy theorem*. In order that f have a finite limit when x tends to a, a necessary and sufficient condition is that for every $\varepsilon > 0$ there exists $\delta > 0$ such that $|f(x) - f(x')| < \varepsilon$ whenever $0 < |x - a| < \delta$ and $0 < |x' - a| < \delta$. Formulate and prove an analogous necessary and sufficient condition in order that $\lim_{x \to \infty} f(x)$ exist.

1.1.38. Show that if $\lim_{x \to a} f(x) = A$ and $\lim_{y \to A} g(y) = B$, then $\lim_{x \to a} g(f(x)) = B$, provided $(g \circ f)(x) = g(f(x))$ is well defined and f does not attain A in a deleted neighborhood of a.

1.1.39. Find functions f and g such that $\lim_{x \to a} f(x) = A$ and $\lim_{y \to A} g(y) = B$, but $\lim_{x \to a} g(f(x)) \neq B$.

1.1.40. Suppose $f : \mathbb{R} \to \mathbb{R}$ is an increasing function and $x \mapsto f(x) - x$ has the period 1. Denote by f^n the nth *iterate of* f; that is, $f^1 = f$ and $f^n = f \circ f^{n-1}$ for $n \geq 2$. Prove that if $\lim_{n \to \infty} \frac{f^n(0)}{n}$ exists, then for every $x \in \mathbb{R}$, $\lim_{n \to \infty} \frac{f^n(x)}{n} = \lim_{n \to \infty} \frac{f^n(0)}{n}$.

1.1.41. Suppose $f : \mathbb{R} \to \mathbb{R}$ is an increasing function and $x \mapsto f(x) - x$ has the period 1. Moreover, suppose that $f(0) > 0$ and p is a fixed positive integer. Let f^n denote the nth iterate of f. Prove that if m_p is the least positive integer such that $f^{m_p}(0) > p$, then

$$\frac{p}{m_p} \leq \varliminf_{n \to \infty} \frac{f^n(0)}{n} \leq \varlimsup_{n \to \infty} \frac{f^n(0)}{n} \leq \frac{p}{m_p} + \frac{1 + f(0)}{m_p}.$$

1.1.42. Suppose $f : \mathbb{R} \to \mathbb{R}$ is an increasing function and $x \mapsto f(x) - x$ has the period 1. Show that $\lim_{n \to \infty} \frac{f^n(x)}{n}$ exists and its value is the same for each $x \in \mathbb{R}$, where f^n denotes the nth iterate of f.

1.2. Properties of Continuous Functions

1.2.1. Find all points of continuity of f defined by

$$f(x) = \begin{cases} 0 & \text{if } x \text{ irrational}, \\ \sin |x| & \text{if } x \text{ is rational}. \end{cases}$$

1.2.2. Determine the set of points of continuity of f given by

$$f(x) = \begin{cases} x^2 - 1 & \text{if } x \text{ is irrational}, \\ 0 & \text{if } x \text{ is rational}. \end{cases}$$

1.2.3. Study the continuity of the following functions:

(a) $\quad f(x) = \begin{cases} 0 & \text{if } x \text{ is irrational or } x = 0, \\ 1/q & \text{if } x = p/q, \ p \in \mathbb{Z}, \ q \in \mathbb{N}, \text{ and} \\ & p, q \text{ are co-prime}, \end{cases}$

(b) $$f(x) = \begin{cases} |x| & \text{if } x \text{ is irrational or } x = 0, \\ qx/(q+1) & \text{if } x = p/q, \ p \in \mathbb{Z}, \ q \in \mathbb{N}, \text{ and} \\ & p, q \text{ are co-prime.} \end{cases}$$

(The function defined in (a) is called *the Riemann function*.)

1.2.4. Prove that if $f \in C([a,b])$, then $|f| \in C([a,b])$. Show by an example that the converse is not true.

1.2.5. Determine all a_n and b_n for which the function defined by
$$f(x) = \begin{cases} a_n + \sin \pi x & \text{if } x \in [2n, 2n+1], \ n \in \mathbb{Z}, \\ b_n + \cos \pi x & \text{if } x \in (2n-1, 2n), \ n \in \mathbb{Z}, \end{cases}$$
is continuous on $\mathbb{R}$.

1.2.6. Let $f(x) = [x^2] \sin \pi x$ for $x \in \mathbb{R}$. Study the continuity of f.

1.2.7. Let
$$f(x) = [x] + (x - [x])^{[x]} \quad \text{for } x \geq \frac{1}{2}.$$
Show that f is continuous and that it is strictly increasing on $[1, \infty)$.

1.2.8. Study the continuity of the following functions and sketch their graphs:

(a) $\quad f(x) = \lim\limits_{n \to \infty} \dfrac{n^x - n^{-x}}{n^x + n^{-x}}, \quad x \in \mathbb{R}$,

(b) $\quad f(x) = \lim\limits_{n \to \infty} \dfrac{x^2 e^{nx} + x}{e^{nx} + 1}, \quad x \in \mathbb{R}$,

(c) $\quad f(x) = \lim\limits_{n \to \infty} \dfrac{\ln(e^n + x^n)}{n}, \quad x \geq 0$,

(d) $\quad f(x) = \lim\limits_{n \to \infty} \sqrt[n]{4^n + x^{2n} + \dfrac{1}{x^{2n}}}, \quad x \neq 0$,

(e) $\quad f(x) = \lim\limits_{n \to \infty} \sqrt[2n]{\cos^{2n} x + \sin^{2n} x}, \quad x \in \mathbb{R}$.

1.2.9. Show that if $f : \mathbb{R} \to \mathbb{R}$ is continuous and periodic, then it attains its supremum and infimum.

1.2. Properties of Continuous Functions

1.2.10. For $P(x) = x^{2n} + a_{2n-1}x^{2n-1} + \cdots + a_1 x + a_0$, show that there is $x_\star \in \mathbb{R}$ such that $P(x_\star) = \inf\{P(x) : x \in \mathbb{R}\}$. Show also that the absolute value of any polynomial P attains its infimum; that is, there is $x^\star \in \mathbb{R}$ such that $|P(x^\star)| = \inf\{|P(x)| : x \in \mathbb{R}\}$.

1.2.11.

(a) Give an example of a bounded function on $[0, 1]$ which achieves neither an infimum nor a supremum.

(b) Give an example of a bounded function on $[0, 1]$ which does not achieve its infimum on any $[a, b] \subset [0, 1]$, $a < b$.

1.2.12. For $f : \mathbb{R} \to \mathbb{R}$, $x_0 \in \mathbb{R}$ and $\delta > 0$, set
$$\omega_f(x_0, \delta) = \sup\{|f(x) - f(x_0)| : x \in \mathbb{R}, |x - x_0| < \delta\}$$
and $\omega_f(x_0) = \lim_{\delta \to 0^+} \omega_f(x_0, \delta)$. Show that f is continuous at x_0 if and only if $\omega_f(x_0) = 0$.

1.2.13.

(a) Let $f, g \in C([a, b])$ and for $x \in [a, b]$ let $h(x) = \min\{f(x), g(x)\}$ and $H(x) = \max\{f(x), g(x)\}$. Show that $h, H \in C([a, b])$.

(b) Let $f_1, f_2, f_3 \in C([a, b])$ and for $x \in [a, b]$ let $f(x)$ denote that one of the three values $f_1(x), f_2(x)$ and $f_3(x)$ that lies between the other two. Show that $f \in C([a, b])$.

1.2.14. Prove that if $f \in C([a, b])$, then the functions defined by setting
$$m(x) = \inf\{f(\zeta) : \zeta \in [a, x]\} \quad \text{and} \quad M(x) = \sup\{f(\zeta) : \zeta \in [a, x]\}$$
are also continuous on $[a, b]$.

1.2.15. Let f be a bounded function on $[a, b]$. Show that the functions defined by
$$m(x) = \inf\{f(\zeta) : \zeta \in [a, x)\} \quad \text{and} \quad M(x) = \sup\{f(\zeta) : \zeta \in [a, x)\}$$
are continuous from the left on (a, b).

1.2.16. Verify whether under the assumptions of the foregoing problem the functions

$$m^\star(x) = \inf\{f(\zeta) : \zeta \in [a,x]\} \quad \text{and} \quad M^\star(x) = \sup\{f(\zeta) : \zeta \in [a,x]\}$$

are continuous from the left on (a,b).

1.2.17. Suppose f is continuous on $[a,\infty)$ and $\lim\limits_{x\to\infty} f(x)$ is finite. Show that f is bounded on $[a,\infty)$.

1.2.18. Let f be continuous on $\mathbb{R}$ and let $\{x_n\}$ be a bounded sequence. Do the equalities

$$\varliminf_{n\to\infty} f(x_n) = f(\varliminf_{n\to\infty} x_n) \quad \text{and} \quad \varlimsup_{n\to\infty} f(x_n) = f(\varlimsup_{n\to\infty} x_n)$$

hold?

1.2.19. Let $f : \mathbb{R} \to \mathbb{R}$ be increasing and continuous and let $\{x_n\}$ be a bounded sequence. Show that

(a) $\quad \varliminf\limits_{n\to\infty} f(x_n) = f(\varliminf\limits_{n\to\infty} x_n),$

(b) $\quad \varlimsup\limits_{n\to\infty} f(x_n) = f(\varlimsup\limits_{n\to\infty} x_n).$

1.2.20. Let $f : \mathbb{R} \to \mathbb{R}$ be decreasing and continuous and let $\{x_n\}$ be a bounded sequence. Show that

(a) $\quad \varliminf\limits_{n\to\infty} f(x_n) = f(\varlimsup\limits_{n\to\infty} x_n),$

(b) $\quad \varlimsup\limits_{n\to\infty} f(x_n) = f(\varliminf\limits_{n\to\infty} x_n).$

1.2.21. Suppose that f is continuous on $\mathbb{R}$, $\lim\limits_{x\to-\infty} f(x) = -\infty$ and $\lim\limits_{x\to\infty} f(x) = +\infty$. Define g by setting

$$g(x) = \sup\{t : f(t) < x\} \quad \text{for} \quad x \in \mathbb{R}.$$

(a) Prove that g is continuous from the left.
(b) Is g continuous?

1.2. Properties of Continuous Functions

1.2.22. Let $f : \mathbb{R} \to \mathbb{R}$ be a continuous periodic function with two *incommensurate* periods T_1 and T_2; that is, $\frac{T_1}{T_2}$ is irrational. Prove that f is a constant function. Give an example of a nonconstant periodic function with two incommensurate periods.

1.2.23.

(a) Show that if $f : \mathbb{R} \to \mathbb{R}$ is nonconstant, periodic and continuous, then it has a smallest positive period, the so-called *fundamental period*.

(b) Give an example of a nonconstant periodic function without a fundamental period.

(c) Prove that if $f : \mathbb{R} \to \mathbb{R}$ is a periodic function without a fundamental period, then the set of all periods of f is dense in $\mathbb{R}$.

1.2.24.

(a) Prove that the theorem in part (a) of the preceding problem remains true when the continuity of f on $\mathbb{R}$ is replaced by the continuity at one point.

(b) Show that if $f : \mathbb{R} \to \mathbb{R}$ is a periodic function without a fundamental period and if it is continuous at least at one point, then it is constant.

1.2.25. Show that if $f, g : \mathbb{R} \to \mathbb{R}$ are continuous and periodic and $\lim\limits_{x \to \infty} (f(x) - g(x)) = 0$, then $f = g$.

1.2.26. Give an example of two periodic functions f and g such that any period of f is not commensurate with any period of g and such that $f + g$

(a) is not periodic,

(b) is periodic.

1.2.27. Let $f, g : \mathbb{R} \to \mathbb{R}$ be continuous and periodic with positive fundamental periods T_1 and T_2, respectively. Prove that if $\frac{T_1}{T_2} \notin \mathbb{Q}$, then $h = f + g$ is not a periodic function.

1.2.28. Let $f, g : \mathbb{R} \to \mathbb{R}$ be periodic and suppose that f is continuous and no period of g is commensurate with the fundamental period of f. Prove that $f + g$ is not a periodic function.

1.2.29. Prove that the set of points of discontinuity of a monotonic function $f : \mathbb{R} \to \mathbb{R}$ is at most countable.

1.2.30. Suppose f is continuous on $[0, 1]$. Prove that

$$\lim_{n \to \infty} \frac{1}{n} \sum_{k=1}^{n} (-1)^k f\left(\frac{k}{n}\right) = 0.$$

1.2.31. Let f be continuous on $[0, 1]$. Prove that

$$\lim_{n \to \infty} \frac{1}{2^n} \sum_{k=0}^{n} (-1)^k \binom{n}{k} f\left(\frac{k}{n}\right) = 0.$$

1.2.32. Suppose $f : (0, \infty) \to \mathbb{R}$ is a continuous function such that $f(x) \leq f(nx)$ for all positive x and natural n. Show that $\lim_{x \to \infty} f(x)$ exists (finite or infinite).

1.2.33. A function f defined on an interval $\mathbf{I} \subset \mathbb{R}$ is said to be *convex* on $\mathbf{I}$ if

$$f(\lambda x_1 + (1 - \lambda) x_2) \leq \lambda f(x_1) + (1 - \lambda) f(x_2)$$

whenever $x_1, x_2 \in \mathbf{I}$ and $\lambda \in (0, 1)$. Prove that if f is convex on an open interval, then it is continuous. Must a convex function on an arbitrary interval be continuous?

1.2.34. Prove that if a sequence $\{f_n\}$ of continuous functions on $\mathbf{A}$ converges uniformly to f on $\mathbf{A}$, then f is continuous on $\mathbf{A}$.

1.3. Intermediate Value Property

Recall the following:

Definition. A real function f has the *intermediate value property* on an interval $\mathbf{I}$ containing $[a, b]$ if $f(a) < v < f(b)$ or $f(b) < v < f(a)$;

1.3. Intermediate Value Property

that is, if v is between $f(a)$ and $f(b)$, there is between a and b a c such that $f(c) = v$.

1.3.1. Give examples of functions which have the intermediate value property on an interval $\mathbf{I}$ but are not continuous on this interval.

1.3.2. Prove that a strictly increasing function $f : [a,b] \to \mathbb{R}$ which has the intermediate value property is continuous on $[a,b]$.

1.3.3. Let $f : [0,1] \to [0,1]$ be continuous. Show that f has a *fixed point* in $[0,1]$; that is, there exists $x_0 \in [0,1]$ such that $f(x_0) = x_0$.

1.3.4. Assume that $f, g : [a,b] \to \mathbb{R}$ are continuous and such that $f(a) < g(a)$ and $f(b) > g(b)$. Prove that there exists $x_0 \in (a,b)$ for which $f(x_0) = g(x_0)$.

1.3.5. Let $f : \mathbb{R} \to \mathbb{R}$ be continuous and periodic with period $T > 0$. Prove that there is x_0 such that

$$f\left(x_0 + \frac{T}{2}\right) = f(x_0).$$

1.3.6. A function $f : (a,b) \to \mathbb{R}$ is continuous. Prove that, given $x_1, x_2, \ldots, x_n$ in (a,b), there exists $x_0 \in (a,b)$ such that

$$f(x_0) = \frac{1}{n}\left(f(x_1) + f(x_2) + \cdots + f(x_n)\right).$$

1.3.7.

(a) Prove that the equation $(1-x)\cos x = \sin x$ has at least one solution in $(0,1)$.

(b) For a nonzero polynomial P, show that the equation $|P(x)| = e^x$ has at least one solution.

1.3.8. For $a_0 < b_0 < a_1 < b_1 < \cdots < a_n < b_n$, show that all roots of the polynomial

$$P(x) = \prod_{k=0}^{n}(x + a_k) + 2\prod_{k=0}^{n}(x + b_k), \quad x \in \mathbb{R},$$

are real.

1.3.9. Suppose that f and g have the intermediate value property on $[a,b]$. Must $f+g$ possess the intermediate value property on that interval?

1.3.10. Assume that $f \in C([0,2])$ and $f(0) = f(2)$. Prove that there exist x_1 and x_2 in $[0,2]$ such that
$$x_2 - x_1 = 1 \quad \text{and} \quad f(x_2) = f(x_1).$$
Give a geometric interpretation of this fact.

1.3.11. Let $f \in C([0,2])$. Show that there are x_1 and x_2 in $[0,2]$ such that
$$x_2 - x_1 = 1 \quad \text{and} \quad f(x_2) - f(x_1) = \frac{1}{2}(f(2) - f(0)).$$

1.3.12. For $n \in \mathbb{N}$, let $f \in C([0,n])$ be such that $f(0) = f(n)$. Prove that there are x_1 and x_2 in $[0,n]$ satisfying
$$x_2 - x_1 = 1 \quad \text{and} \quad f(x_2) = f(x_1).$$

1.3.13. A continuous function f on $[0,n]$, $n \in \mathbb{N}$, satisfies $f(0) = f(n)$. Show that for every $k \in \{1,2,\ldots,n-1\}$ there are x_k and x'_k such that $f(x_k) = f(x'_k)$, where $x_k - x'_k = k$ or $x_k - x'_k = n - k$. Is it true that for every $k \in \{1,2,\ldots,n-1\}$ there are x_k and x'_k such that $f(x_k) = f(x'_k)$, where $x_k - x'_k = k$?

1.3.14. For $n \in \mathbb{N}$, let $f \in C([0,n])$ be such that $f(0) = f(n)$. Prove that the equation $f(x) = f(y)$ has at least n solutions with $x - y \in \mathbb{N}$.

1.3.15. Suppose that real continuous functions f and g defined on $\mathbb{R}$ commute; that is, $f(g(x)) = g(f(x))$ for $x \in \mathbb{R}$. Prove that if the equation $f^2(x) = g^2(x)$ has a solution, then the equation $f(x) = g(x)$ also has (here $f^2(x) = f(f(x))$ and $g^2(x) = g(g(x))$).

Show by example that the assumption of continuity of f and g in the foregoing problem cannot be omitted.

1.3.16. Prove that a continuous injection $f : \mathbb{R} \to \mathbb{R}$ is either strictly decreasing or strictly increasing.

1.3. Intermediate Value Property

1.3.17. Assume that $f : \mathbb{R} \to \mathbb{R}$ is a continuous injection. Prove that if there exists n such that the nth iteration of f is an identity, that is, $f^n(x) = x$ for all $x \in \mathbb{R}$, then
(a) $f(x) = x$, $x \in \mathbb{R}$, if f is strictly increasing,
(b) $f^2(x) = x$, $x \in \mathbb{R}$, if f is strictly decreasing.

1.3.18. Assume $f : \mathbb{R} \to \mathbb{R}$ satisfies the condition $f(f(x)) = f^2(x) = -x$, $x \in \mathbb{R}$. Show that f cannot be continuous.

1.3.19. Find all functions $f : \mathbb{R} \to \mathbb{R}$ which have the intermediate value property and such that there is $n \in \mathbb{N}$ for which $f^n(x) = -x$, $x \in \mathbb{R}$, where f^n denotes the nth iteration of f.

1.3.20. Prove that if $f : \mathbb{R} \to \mathbb{R}$ has the intermediate value property and $f^{-1}(\{q\})$ is closed for every rational q, then f is continuous.

1.3.21. Assume that $f : (a, \infty) \to \mathbb{R}$ is continuous and bounded. Prove that, given T, there exists a sequence $\{x_n\}$ such that

$$\lim_{n \to \infty} x_n = +\infty \quad \text{and} \quad \lim_{n \to \infty} (f(x_n + T) - f(x_n)) = 0.$$

1.3.22. Give an example of a continuous function $f : \mathbb{R} \to \mathbb{R}$ which attains each of its values exactly three times. Does there exist a continuous function $f : \mathbb{R} \to \mathbb{R}$ which attains each of its values exactly two times?

1.3.23. Let $f : [0, 1] \to \mathbb{R}$ be continuous and *piecewise strictly monotone*. (A function f is said to be piecewise strictly monotone on $[0, 1]$, if there exists a partition of $[0, 1]$ into finitely many subintervals $[t_{i-1}, t_i]$, where $i = 1, 2, \ldots, n$ and $0 = t_0 < t_1 < \cdots < t_n = 1$, such that f is strictly monotone on each of these subintervals.) Prove that f attains at least one of its values an odd number of times.

1.3.24. A continuous function $f : [0, 1] \to \mathbb{R}$ attains each of its values finitely many times and $f(0) \neq f(1)$. Show that f attains at least one of its values an odd number of times.

1.3.25. Assume that $f : \mathbf{K} \to \mathbf{K}$ is continuous on a compact set $\mathbf{K} \subset \mathbb{R}$. Moreover, assume that an $x_0 \in \mathbf{K}$ is such that each limit point of the sequence of iterates $\{f^n(x_0)\}$ is a fixed point of f. Prove that $\{f^n(x_0)\}$ is convergent.

1.3.26. A function $f : \mathbb{R} \to \mathbb{R}$ is increasing, continuous, and such that F defined by $F(x) = f(x) - x$ is periodic with period 1. Prove that if $\alpha(f) = \lim\limits_{n\to\infty} \frac{f^n(0)}{n}$, then there is $x_0 \in [0,1]$ such that $F(x_0) = \alpha(f)$. Prove also that f has a fixed point in $[0,1]$ if and only if $\alpha(f) = 0$. (See Problems 1.1.40 - 1.1.42.)

1.3.27. A function $f : [0,1] \to \mathbb{R}$ satisfies $f(0) < 0$ and $f(1) > 0$, and there exists a function g continuous on $[0,1]$ and such that $f + g$ is decreasing. Prove that the equation $f(x) = 0$ has a solution in the open interval $(0,1)$.

1.3.28. Show that every bijection $f : \mathbb{R} \to [0, \infty)$ has infinitely many points of discontinuity.

1.3.29. Recall that each $x \in (0,1)$ can be represented by a binary fraction $.a_1 a_2 a_3 \dots$, where $a_i \in \{0,1\}$, $i = 1, 2, \dots$. In the case where x has two distinct binary expansions we choose the one with infinitely many digits equal to 1. Next let a function $f : (0,1) \to [0,1]$ be defined by
$$f(x) = \varlimsup_{n\to\infty} \frac{1}{n} \sum_{i=1}^{n} a_i.$$
Prove that f is discontinuous at each $x \in (0,1)$ but nevertheless it has the intermediate value property.

1.4. Semicontinuous Functions

Definition 1. The *extended real number system* $\overline{\mathbb{R}}$ consists of the real number system to which two symbols, $+\infty$ and $-\infty$, have been adjoined, with the following properties:
  (i) If x is real, then $-\infty < x < +\infty$, and $x + \infty = +\infty$, $x - \infty = -\infty$ and $\frac{x}{+\infty} = \frac{x}{-\infty} = 0$.
  (ii) If $x > 0$, then $x \cdot (+\infty) = +\infty$, $x \cdot (-\infty) = -\infty$.

1.4. Semicontinuous Functions

(iii) If $x < 0$, then $x \cdot (+\infty) = -\infty$, $x \cdot (-\infty) = +\infty$.

Definition 2. If $\mathbf{A} \subset \overline{\mathbb{R}}$ is a nonempty set, then sup $\mathbf{A}$ (resp. inf $\mathbf{A}$) is the smallest (resp. greatest) extended real number which is greater (resp. smaller) than or equal to each element of $\mathbf{A}$.

Let f be a real-valued function defined on a nonempty set $\mathbf{A} \subset \mathbb{R}$.

Definition 3. If x_0 is a limit point of $\mathbf{A}$, then the *limit inferior* (resp. the *limit superior*) of $f(x)$ as $x \to x_0$ is defined as the infimum (resp. the supremum) of the set of all $y \in \overline{\mathbb{R}}$ such that there is a sequence $\{x_n\}$ of points in $\mathbf{A}$ which is convergent to x_0, whose terms are all different from x_0 and $y = \lim_{n \to \infty} f(x_n)$. The limit inferior and the limit superior of $f(x)$ as $x \to x_0$ are denoted by $\underline{\lim}_{x \to x_0} f(x)$ and $\overline{\lim}_{x \to x_0} f(x)$, respectively.

Definition 4. A real-valued function is said to be *lower* (resp. *upper*) *semicontinuous* at an $x_0 \in \mathbf{A}$ which is a limit point of $\mathbf{A}$ if $\underline{\lim}_{x \to x_0} f(x) \geq f(x_0)$ (resp. $\overline{\lim}_{x \to x_0} f(x) \leq f(x_0)$). If x_0 is an isolated point of $\mathbf{A}$, then we assume that f is lower and upper semicontinuous at that point.

1.4.1. Show that if x_0 is a limit point of $\mathbf{A}$ and $f : \mathbf{A} \to \mathbb{R}$, then

(a) $\quad \underline{\lim}_{x \to x_0} f(x) = \sup_{\delta > 0} \inf\{f(x) : x \in \mathbf{A},\ 0 < |x - x_0| < \delta\}$,

(b) $\quad \overline{\lim}_{x \to x_0} f(x) = \inf_{\delta > 0} \sup\{f(x) : x \in \mathbf{A},\ 0 < |x - x_0| < \delta\}$.

1.4.2. Show that if x_0 is a limit point of $\mathbf{A}$ and $f : \mathbf{A} \to \mathbb{R}$, then

(a) $\quad \underline{\lim}_{x \to x_0} f(x) = \lim_{\delta \to 0^+} \inf\{f(x) : x \in \mathbf{A},\ 0 < |x - x_0| < \delta\}$,

(b) $\quad \overline{\lim}_{x \to x_0} f(x) = \lim_{\delta \to 0^+} \sup\{f(x) : x \in \mathbf{A},\ 0 < |x - x_0| < \delta\}$.

1.4.3. Prove that $y_0 \in \mathbb{R}$ is the limit inferior of $f : \mathbf{A} \to \mathbb{R}$ at a limit point x_0 of $\mathbf{A}$ if and only if for every $\varepsilon > 0$ the following two conditions are satisfied:
  (i) there is $\delta > 0$ such that $f(x) > y_0 - \varepsilon$ for all $x \in \mathbf{A}$ with $0 < |x - x_0| < \delta$,
  (ii) for every $\delta > 0$ there is $x' \in \mathbf{A}$ such that $0 < |x' - x_0| < \delta$ and $f(x') < y_0 + \varepsilon$.

Establish an analogous statement for the limit superior of f at x_0.

1.4.4. Let $f : \mathbf{A} \to \mathbb{R}$ and let x_0 be a limit point of $\mathbf{A}$. Prove that
  (a) $\varliminf\limits_{x \to x_0} f(x) = -\infty$ if and only if for any real y and for any $\delta > 0$ there exists $x' \in \mathbf{A}$ such that $0 < |x' - x_0| < \delta$ and $f(x') < y$.
  (b) $\varlimsup\limits_{x \to x_0} f(x) = +\infty$ if and only if for any real y and for any $\delta > 0$ there exists $x' \in \mathbf{A}$ such that $0 < |x' - x_0| < \delta$ and $f(x') > y$.

1.4.5. Suppose $f : \mathbf{A} \to \mathbb{R}$ and x_0 is a limit point of $\mathbf{A}$. Show that if $l = \varliminf\limits_{x \to x_0} f(x)$ (resp. $L = \varlimsup\limits_{x \to x_0} f(x)$), then there is a sequence $\{x_n\}$, $x_n \in \mathbf{A}$, $x_n \neq x_0$, converging to x_0 such that $l = \lim\limits_{n \to \infty} f(x_n)$ (resp. $L = \lim\limits_{n \to \infty} f(x_n)$).

1.4.6. Let $f : \mathbf{A} \to \mathbb{R}$ and let x_0 be a limit point of $\mathbf{A}$. Prove that

$$\varliminf_{x \to x_0} (-f(x)) = - \varlimsup_{x \to x_0} f(x) \quad \text{and} \quad \varlimsup_{x \to x_0} (-f(x)) = - \varliminf_{x \to x_0} f(x).$$

1.4.7. Let $f : \mathbf{A} \to (0, \infty)$ and let x_0 be a limit point of $\mathbf{A}$. Show that

$$\varliminf_{x \to x_0} \frac{1}{f(x)} = \frac{1}{\varlimsup\limits_{x \to x_0} f(x)} \quad \text{and} \quad \varlimsup_{x \to x_0} \frac{1}{f(x)} = \frac{1}{\varliminf\limits_{x \to x_0} f(x)}.$$

(We assume that $\frac{1}{+\infty} = 0$ and $\frac{1}{0^+} = +\infty$.)

1.4.8. Assume that $f, g : \mathbf{A} \to \mathbb{R}$ and that x_0 is a limit point of $\mathbf{A}$. Prove that (excluding the indeterminate forms of the type $+\infty - \infty$

1.4. Semicontinuous Functions

and $-\infty + \infty$) the following inequalities hold:

$$\varliminf_{x \to x_0} f(x) + \varliminf_{x \to x_0} g(x) \le \varliminf_{x \to x_0} (f(x) + g(x)) \le \varliminf_{x \to x_0} f(x) + \varlimsup_{x \to x_0} g(x)$$
$$\le \varlimsup_{x \to x_0} (f(x) + g(x)) \le \varlimsup_{x \to x_0} f(x) + \varlimsup_{x \to x_0} g(x).$$

Give examples of functions for which "$\le$" in the above inequalities is replaced by "$<$".

1.4.9. Assume that $f, g : \mathbf{A} \to [0, \infty)$ and that x_0 is a limit point of $\mathbf{A}$. Prove that (excluding the indeterminate forms of the type $0 \cdot (+\infty)$ and $(+\infty) \cdot 0$) the following inequalities hold:

$$\varliminf_{x \to x_0} f(x) \cdot \varliminf_{x \to x_0} g(x) \le \varliminf_{x \to x_0} (f(x) \cdot g(x)) \le \varliminf_{x \to x_0} f(x) \cdot \varlimsup_{x \to x_0} g(x)$$
$$\le \varlimsup_{x \to x_0} (f(x) \cdot g(x)) \le \varlimsup_{x \to x_0} f(x) \cdot \varlimsup_{x \to x_0} g(x).$$

Give examples of functions for which "$\le$" in the above inequalities is replaced by "$<$".

1.4.10. Prove that if $\lim_{x \to x_0} f(x)$ exists, then (excluding the indeterminate forms of the type $+\infty - \infty$ and $-\infty + \infty$)

$$\varliminf_{x \to x_0} (f(x) + g(x)) = \lim_{x \to x_0} f(x) + \varliminf_{x \to x_0} g(x),$$
$$\varlimsup_{x \to x_0} (f(x) + g(x)) = \lim_{x \to x_0} f(x) + \varlimsup_{x \to x_0} g(x).$$

Moreover, if f and g are nonnegative, then (excluding the indeterminate forms of the type $0 \cdot (+\infty)$ and $(+\infty) \cdot 0$)

$$\varliminf_{x \to x_0} (f(x) \cdot g(x)) = \lim_{x \to x_0} f(x) \cdot \varliminf_{x \to x_0} g(x),$$
$$\varlimsup_{x \to x_0} (f(x) \cdot g(x)) = \lim_{x \to x_0} f(x) \cdot \varlimsup_{x \to x_0} g(x).$$

1.4.11. Prove that if f is continuous on (a, b), $l = \varliminf_{x \to a} f(x)$ and $L = \varlimsup_{x \to a} f(x)$, then for every $\lambda \in [l, L]$ there is a sequence $\{x_n\}$ of points in (a, b) converging to a and such that $\lim_{n \to \infty} f(x_n) = \lambda$.

1.4.12. Find the points at which $f : \mathbb{R} \to \mathbb{R}$ defined by
$$f(x) = \begin{cases} 0 & \text{if } x \text{ is irrational,} \\ \sin x & \text{if } x \text{ is rational} \end{cases}$$
is semicontinuous.

1.4.13. Determine points at which the function f defined by
$$f(x) = \begin{cases} x^2 - 1 & \text{if } x \text{ is irrational,} \\ 0 & \text{if } x \text{ is rational} \end{cases}$$
is semicontinuous.

1.4.14. Show that the function given by setting
$$f(x) = \begin{cases} 0 & \text{if } x \text{ is irrational or } x = 0, \\ \frac{1}{q} & \text{if } x = \frac{p}{q}, \ p \in \mathbb{Z}, q \in \mathbb{N}, \\ & \text{and } p, q \text{ are co-prime} \end{cases}$$
is upper semicontinuous.

1.4.15. Find the points at which the function defined by

(a) $$f(x) = \begin{cases} |x| & \text{if } x \text{ is irrational or } x = 0, \\ \frac{qx}{q+1} & \text{if } x = \frac{p}{q}, \ p \in \mathbb{Z}, q \in \mathbb{N}, \\ & \text{and } p, q \text{ are co-prime,} \end{cases}$$

(b) $$f(x) = \begin{cases} \frac{(-1)^q p}{q+1} & \text{if } x \in \mathbb{Q} \cap (0, 1] \text{ and } x = \frac{p}{q}, \ p, q \in \mathbb{N}, \\ & \text{and } p, q \text{ are co-prime,} \\ 0 & \text{if } x \in (0, 1) \text{ is irrational} \end{cases}$$

is neither upper nor lower semicontinuous.

1.4.16. Let $f, g : \mathbf{A} \to \mathbb{R}$ be lower (resp. upper) semicontinuous at $x_0 \in \mathbf{A}$. Show that
(a) if $a > 0$, then af is lower (resp. upper) semicontinuous at x_0. If $a < 0$, then af is upper (resp. lower) semicontinuous at x_0.
(b) $f + g$ is lower (resp. upper) semicontinuous at x_0.

1.4. Semicontinuous Functions

1.4.17. Assume that $f_n : \mathbf{A} \to \mathbb{R}$, $n \in \mathbb{N}$, are lower (resp. upper) semicontinuous at $x_0 \in \mathbf{A}$. Show that $\sup_{n \in \mathbb{N}} f_n$ (resp. $\inf_{n \in \mathbb{N}} f_n$) is lower (resp. upper) semicontinuous at x_0.

1.4.18. Prove that a pointwise limit of an increasing (resp. decreasing) sequence of lower (resp. upper) semicontinuous functions is lower (resp. upper) semicontinuous.

1.4.19. For $f : \mathbf{A} \to \mathbb{R}$ and x a limit point of $\mathbf{A}$ define the *oscillation of f at x* by

$$o_f(x) = \lim_{\delta \to 0^+} \sup\{|f(z) - f(u)| : z, u \in \mathbf{A},\ |z - x| < \delta,\ |u - x| < \delta\}$$

Show that $o_f(x) = f_1(x) - f_2(x)$, where

$$f_1(x) = \max\{f(x), \overline{\lim_{z \to x}} f(z)\} \quad \text{and} \quad f_2(x) = \min\{f(x), \underline{\lim_{z \to x}} f(z)\}.$$

1.4.20. Let f_1, f_2, and o_f be as in the foregoing problem. Show that f_1 and o_f are upper semicontinuous, and f_2 is lower semicontinuous.

1.4.21. Prove that in order that $f : \mathbf{A} \to \mathbb{R}$ be lower (resp. upper) semicontinuous at $x_0 \in \mathbf{A}$, a necessary and sufficient condition is that for every $a < f(x_0)$ (resp. $a > f(x_0)$) there is $\delta > 0$ such that $f(x) > a$ (resp. $f(x) < a$) whenever $|x - x_0| < \delta$, $x \in \mathbf{A}$.

1.4.22. Prove that in order that $f : \mathbf{A} \to \mathbb{R}$ be lower (resp. upper) semicontinuous on $\mathbf{A}$, a necessary and sufficient condition is that for every $a \in \mathbb{R}$ the set $\{x \in \mathbf{A} : f(x) > a\}$ (resp. $\{x \in \mathbf{A} : f(x) < a\}$) be open in $\mathbf{A}$.

1.4.23. Prove that $f : \mathbb{R} \to \mathbb{R}$ is lower semicontinuous if and only if the set $\{(x, y) \in \mathbb{R}^2 : y \geq f(x)\}$ is closed in $\mathbb{R}^2$.

Formulate and prove an analogous necessary and sufficient condition for upper semicontinuity of f on $\mathbb{R}$.

1.4.24. Prove the following *theorem of Baire*. Every lower (resp. upper) semicontinuous $f : \mathbf{A} \to \mathbb{R}$ is the pointwise limit of an increasing (resp. decreasing) sequence of continuous functions on $\mathbf{A}$.

1.4.25. Prove that if $f : \mathbf{A} \to \mathbb{R}$ is upper semicontinuous, $g : \mathbf{A} \to \mathbb{R}$ is lower semicontinuous and $f(x) \leq g(x)$ everywhere on $\mathbf{A}$, then there is a continuous function h on $\mathbf{A}$ such that
$$f(x) \leq h(x) \leq g(x), \quad x \in \mathbf{A}.$$

1.5. Uniform Continuity

Definition. A real function f defined on $\mathbf{A} \subset \mathbb{R}$ is said to be *uniformly continuous* on $\mathbf{A}$ if, given $\varepsilon > 0$, there exists $\delta > 0$ such that for all x and y in $\mathbf{A}$ with $|x - y| < \delta$ we have $|f(x) - f(y)| < \varepsilon$.

1.5.1. Verify whether the following functions are uniformly continuous on $(0, 1)$:

(a) $f(x) = e^x$,
(b) $f(x) = \sin \dfrac{1}{x}$,
(c) $f(x) = x \sin \dfrac{1}{x}$,
(d) $f(x) = e^{\frac{1}{x}}$,
(e) $f(x) = e^{-\frac{1}{x}}$,
(f) $f(x) = e^x \cos \dfrac{1}{x}$,
(g) $f(x) = \ln x$,
(h) $f(x) = \cos x \cdot \cos \dfrac{\pi}{x}$,
(i) $f(x) = \cot x$.

1.5.2. Which of the following functions are uniformly continuous on $[0, \infty)$?

(a) $f(x) = \sqrt{x}$,
(b) $f(x) = x \sin x$,
(c) $f(x) = \sin^2 x$,
(d) $f(x) = \sin(x^2)$,
(e) $f(x) = e^x$,
(f) $f(x) = e^{\sin(x^2)}$,
(g) $f(x) = \sin(\sin x)$,
(h) $f(x) = \sin(x \sin x)$,
(i) $f(x) = \sin \sqrt{x}$.

1.5.3. Show that if f is uniformly continuous on (a, b), $a, b \in \mathbb{R}$, then $\lim\limits_{x \to a^+} f(x)$ and $\lim\limits_{x \to b^-} f(x)$ exist as finite limits.

1.5. Uniform Continuity

1.5.4. Suppose f and g are uniformly continuous on (a,b) ($[a,\infty)$). Does this imply the uniform continuity on (a,b) ($[a,\infty)$) of the functions

(a) $f+g$,   (b) fg,   (c) $x \mapsto f(x)\sin x$?

1.5.5.

(a) Show that if f is uniformly continuous on $(a,b]$ and on $[b,c)$, then it is also uniformly continuous on (a,c).

(b) Suppose $\mathbf{A}$ and $\mathbf{B}$ are closed sets in $\mathbb{R}$ and let $f : \mathbf{A} \cup \mathbf{B} \to \mathbb{R}$ be uniformly continuous on $\mathbf{A}$ and on $\mathbf{B}$. Must f be uniformly continuous on $\mathbf{A} \cup \mathbf{B}$?

1.5.6. Prove that any function continuous and periodic on $\mathbb{R}$ must be uniformly continuous on $\mathbb{R}$.

1.5.7.

(a) Show that if $f : \mathbb{R} \to \mathbb{R}$ is continuous and such that $\lim\limits_{x \to -\infty} f(x)$ and $\lim\limits_{x \to \infty} f(x)$ are finite, then f is uniformly continuous on $\mathbb{R}$.

(b) Show that if $f : [a,\infty) \to \mathbb{R}$ is continuous and $\lim\limits_{x \to \infty} f(x)$ is finite, then f is uniformly continuous on $[a,\infty)$.

1.5.8. Examine the uniform continuity of

(a) $f(x) = \arctan x$ on $(-\infty,\infty)$,

(b) $f(x) = x\sin\frac{1}{x}$ on $(0,\infty)$,

(c) $f(x) = e^{-\frac{1}{x}}$ on $(0,\infty)$.

1.5.9. Assume that f is uniformly continuous on $(0,\infty)$. Must the limits $\lim\limits_{x \to 0^+} f(x)$ and $\lim\limits_{x \to \infty} f(x)$ exist?

1.5.10. Prove that any function which is bounded, monotonic and continuous on an interval $\mathbf{I} \subset \mathbb{R}$ is uniformly continuous on $\mathbf{I}$.

1.5.11. Assume f is uniformly continuous and unbounded on $[0,\infty)$. Is it true that either $\lim\limits_{x \to \infty} f(x) = +\infty$ or $\lim\limits_{x \to \infty} f(x) = -\infty$?

1.5.12. A function $f : [0,\infty) \to \mathbb{R}$ is uniformly continuous and for any $x \geq 0$ the sequence $\{f(x+n)\}$ converges to zero. Prove that $\lim_{x \to \infty} f(x) = 0$.

1.5.13. Suppose that $f : [1,\infty) \to \mathbb{R}$ is uniformly continuous. Prove that there is a positive M such that $\frac{|f(x)|}{x} \leq M$ for $x \geq 1$.

1.5.14. Let $f : [0,\infty) \to \mathbb{R}$ be uniformly continuous. Prove that there is a positive M with the following property:

$$\sup_{u>0}\{|f(x+u) - f(u)|\} \leq M(x+1) \quad \text{for every} \quad x \geq 0.$$

1.5.15. Let $f : \mathbf{A} \to \mathbb{R}$, $\mathbf{A} \subset \mathbb{R}$, be uniformly continuous. Prove that if $\{x_n\}$ is any Cauchy sequence of elements in $\mathbf{A}$, then $\{f(x_n)\}$ is also a Cauchy sequence.

1.5.16. Suppose $\mathbf{A} \subset \mathbb{R}$ is bounded. Prove that if $f : \mathbf{A} \to \mathbb{R}$ transforms Cauchy sequences of elements of $\mathbf{A}$ into Cauchy sequences, then f is uniformly continuous on $\mathbf{A}$. Is the boundedness of $\mathbf{A}$ an essential assumption?

1.5.17. Prove that f is uniformly continuous on $\mathbf{A} \subset \mathbb{R}$ if and only if for any sequences $\{x_n\}$ and $\{y_n\}$ of elements of $\mathbf{A}$,

$$\lim_{n \to \infty}(x_n - y_n) = 0 \quad \text{implies} \quad \lim_{n \to \infty}(f(x_n) - f(y_n)) = 0.$$

1.5.18. Suppose that $f : (0,\infty) \to (0,\infty)$ is uniformly continuous. Does this imply that

$$\lim_{x \to \infty} \frac{f\left(x + \frac{1}{x}\right)}{f(x)} = 1?$$

1.5.19. A function $f : \mathbb{R} \to \mathbb{R}$ is continuous at zero and satisfies the following conditions

$$f(0) = 0 \quad \text{and} \quad f(x_1 + x_2) \leq f(x_1) + f(x_2) \quad \text{for any } x_1, x_2 \in \mathbb{R}.$$

Prove that f is uniformly continuous on $\mathbb{R}$.

1.5.20. For $f : \mathbf{A} \to \mathbb{R}$, $\mathbf{A} \subset \mathbb{R}$, we define
$$\omega_f(\delta) = \sup\{|f(x_1) - f(x_2)| : \; x_1, x_2 \in \mathbf{A}, \; |x_1 - x_2| < \delta\}$$
and call ω_f the *modulus of continuity of f*. Show that f is uniformly continuous on $\mathbf{A}$ if and only if $\lim\limits_{\delta \to 0^+} \omega_f(\delta) = 0$.

1.5.21. Let $f : \mathbb{R} \to \mathbb{R}$ be uniformly continuous. Prove that the following statements are equivalent.
 (a) For any uniformly continuous function $g : \mathbb{R} \to \mathbb{R}$, $f \cdot g$ is uniformly continuous on $\mathbb{R}$.
 (b) The function $x \mapsto |x|f(x)$ is uniformly continuous on $\mathbb{R}$.

1.5.22. Prove that the following condition is necessary and sufficient for f to be uniformly continuous on an interval $\mathbf{I}$. Given $\varepsilon > 0$, there is $N > 0$ such that for every $x_1, x_2 \in \mathbf{I}$, $x_1 \neq x_2$,
$$\left|\frac{f(x_1) - f(x_2)}{x_1 - x_2}\right| > N \quad \text{implies} \quad |f(x_1) - f(x_2)| < \varepsilon.$$

1.6. Functional Equations

1.6.1. Prove that the only functions continuous on $\mathbb{R}$ and satisfying the *Cauchy functional equation*
$$f(x + y) = f(x) + f(y)$$
are the linear functions of the form $f(x) = ax$.

1.6.2. Prove that if $f : \mathbb{R} \to \mathbb{R}$ satisfies the Cauchy functional equation
$$f(x + y) = f(x) + f(y)$$
and one of the conditions
 (a) f is continuous at an $x_0 \in \mathbb{R}$,
 (b) f is bounded above on some interval (a, b),
 (c) f is monotonic on $\mathbb{R}$,
then $f(x) = ax$.

1.6.3. Determine all continuous functions $f : \mathbb{R} \to \mathbb{R}$ such that $f(1) > 0$ and
$$f(x+y) = f(x)f(y).$$

1.6.4. Show that the only solutions of the functional equation
$$f(xy) = f(x) + f(y)$$
which are not identically zero and are continuous on $(0, \infty)$ are the logarithmic functions.

1.6.5. Show that the only solutions of the functional equation
$$f(xy) = f(x)f(y)$$
which are not identically zero and are continuous on $(0, \infty)$ are the power functions of the form $f(x) = x^a$.

1.6.6. Find all continuous functions $f : \mathbb{R} \to \mathbb{R}$ such that $f(x) - f(y)$ is rational for rational $x - y$.

1.6.7. For $|q| < 1$, find all functions $f : \mathbb{R} \to \mathbb{R}$ continuous at zero and satisfying the functional equation
$$f(x) + f(qx) = 0.$$

1.6.8. Find all functions $f : \mathbb{R} \to \mathbb{R}$ continuous at zero and satisfying the equation
$$f(x) + f\left(\frac{2}{3}x\right) = x.$$

1.6.9. Determine all solutions $f : \mathbb{R} \to \mathbb{R}$ of the functional equation
$$2f(2x) = f(x) + x$$
which are continuous at zero.

1.6.10. Find all continuous functions $f : \mathbb{R} \to \mathbb{R}$ satisfying the *Jensen equation*
$$f\left(\frac{x+y}{2}\right) = \frac{f(x) + f(y)}{2}.$$

1.6. Functional Equations

1.6.11. Find all functions continuous on (a, b), $a, b \in \mathbb{R}$, satisfying the Jensen equation
$$f\left(\frac{x+y}{2}\right) = \frac{f(x) + f(y)}{2}.$$

1.6.12. Determine all solutions $f : \mathbb{R} \to \mathbb{R}$ of the functional equation
$$f(2x+1) = f(x)$$
which are continuous at -1.

1.6.13. For a real a, show that if $f : \mathbb{R} \to \mathbb{R}$ is a continuous solution of the equation
$$f(x+y) = f(x) + f(y) + axy,$$
then $f(x) = \frac{a}{2}x^2 + bx$, where $b = f(1) - \frac{a}{2}$.

1.6.14. Determine all continuous at zero solutions of the functional equation
$$f(x) = f\left(\frac{x}{1-x}\right), \quad x \neq 1.$$

1.6.15. Let $f : [0,1] \to [0,1]$ be continuous, monotonically decreasing and such that $f(f(x)) = x$ for $x \in [0,1]$. Is $f(x) = 1 - x$ the only such function?

1.6.16. Suppose that f and g satisfy the equation
$$f(x+y) + f(x-y) = 2f(x)g(y), \quad x, y \in \mathbb{R}.$$
Show that if f is not identically zero and $|f(x)| \leq 1$ for $x \in \mathbb{R}$, then also $|g(x)| \leq 1$ for $x \in \mathbb{R}$.

1.6.17. Find all continuous functions $f : \mathbb{R} \to \mathbb{R}$ satisfying the functional equation
$$f(x+y) = f(x)e^y + f(y)e^x.$$

1.6.18. Determine all continuous at zero solutions $f : \mathbb{R} \to \mathbb{R}$ of
$$f(x+y) - f(x-y) = f(x)f(y).$$

1.6.19. Solve the functional equation
$$f(x) + f\left(\frac{x-1}{x}\right) = 1 + x \quad \text{for} \quad x \neq 0, 1.$$

1.6.20. A sequence $\{x_n\}$ *converges in the Cesàro sense* if
$$C\text{-}\lim_{n\to\infty} x_n = \lim_{n\to\infty} \frac{x_1 + x_2 + x_3 + \cdots + x_n}{n}$$
exists and is finite. Find all functions which are *Cesàro continuous*, that is,
$$f(C\text{-}\lim_{n\to\infty} x_n) = C\text{-}\lim_{n\to\infty} f(x_n)$$
for every Cesàro convergent sequence $\{x_n\}$.

1.6.21. Let $f : [0,1] \to [0,1]$ be an injection such that $f(2x - f(x)) = x$ for $x \in [0,1]$. Prove that $f(x) = x$, $x \in [0,1]$.

1.6.22. For m different from zero, prove that if a continuous function $f : \mathbb{R} \to \mathbb{R}$ satisfies the equation
$$f\left(2x - \frac{f(x)}{m}\right) = mx,$$
then $f(x) = m(x - c)$.

1.6.23. Show that the only solutions of the functional equation
$$f(x+y) + f(y-x) = 2f(x)f(y)$$
continuous on $\mathbb{R}$ and not identically zero are $f(x) = \cos(ax)$ and $f(x) = \cosh(ax)$ with a real.

1.6.24. Determine all continuous on $(-1,1)$ solutions of
$$f\left(\frac{x+y}{1+xy}\right) = f(x) + f(y).$$

1.6.25. Find all polynomials P such that
$$P(2x - x^2) = (P(x))^2.$$

1.6. Functional Equations

1.6.26. Let $m, n \geq 2$ be integers. Find all functions $f : [0, \infty) \to \mathbb{R}$ continuous at at least one point in $[0, \infty)$ and such that

$$f\left(\frac{1}{n}\sum_{i=1}^{n} x_i^m\right) = \frac{1}{n}\sum_{i=1}^{n} (f(x_i))^m \quad \text{for} \quad x_i \geq 0, \ i = 1, 2, \ldots, n.$$

1.6.27. Find all not identically zero functions $f : \mathbb{R} \to \mathbb{R}$ satisfying the equations

$$f(xy) = f(x)f(y) \quad \text{and} \quad f(x+z) = f(x) + f(z)$$

with some $z \neq 0$.

1.6.28. Find all functions $f : \mathbb{R} \setminus \{0\} \to \mathbb{R}$ such that

$$f(x) = -f\left(\frac{1}{x}\right), \quad x \neq 0.$$

1.6.29. Find all solutions $f : \mathbb{R} \setminus \{0\} \to \mathbb{R}$ of the functional equation

$$f(x) + f(x^2) = f\left(\frac{1}{x}\right) + f\left(\frac{1}{x^2}\right), \quad x \neq 0.$$

1.6.30. Prove that the functions $f, g, \phi : \mathbb{R} \to \mathbb{R}$ satisfy the equation

$$\frac{f(x) - g(y)}{x - y} = \phi\left(\frac{x+y}{2}\right), \quad y \neq x,$$

if and only if there exist a, b and c such that

$$f(x) = g(x) = ax^2 + bx + c, \quad \phi(x) = 2ax + b.$$

1.6.31. Prove that there is a function $f : \mathbb{R} \to \mathbb{Q}$ satisfying the following three conditions:

(a) $f(x+y) = f(x) + f(y)$ for $x, y \in \mathbb{R}$,
(b) $f(x) = x$ for $x \in \mathbb{Q}$,
(c) f is not continuous on $\mathbb{R}$.

1.7. Continuous Functions in Metric Spaces

In this section **X** and **Y** will stand for metric spaces $(\mathbf{X}, d_1)$ and $(\mathbf{Y}, d_2)$, respectively. To shorten notation we say that **X** is a metric space instead of saying that $(\mathbf{X}, d_1)$ is a metric space. If not stated otherwise, $\mathbb{R}$ and $\mathbb{R}^n$ are always assumed to be equipped with the Euclidean metric.

1.7.1. Let $(\mathbf{X}, d_1)$ and $(\mathbf{Y}, d_2)$ be metric spaces and let $f : \mathbf{X} \to \mathbf{Y}$. Prove that the following conditions are equivalent.

(a) The function f is continuous.
(b) For each closed set $\mathbf{F} \subset \mathbf{Y}$ the set $f^{-1}(\mathbf{F})$ is closed in **X**.
(c) For each open set $\mathbf{G} \subset \mathbf{Y}$ the set $f^{-1}(\mathbf{G})$ is open in **X**.
(d) For each subset **A** of **X**, $f(\overline{\mathbf{A}}) \subset \overline{f(\mathbf{A})}$.
(e) For each subset **B** of **Y**, $\overline{f^{-1}(\mathbf{B})} \subset f^{-1}(\overline{\mathbf{B}})$.

1.7.2. Let $(\mathbf{X}, d_1)$ and $(\mathbf{Y}, d_2)$ be metric spaces and let $f : \mathbf{X} \to \mathbf{Y}$ be continuous. Prove that the inverse image $f^{-1}(\mathbf{B})$ of a Borel set **B** in $(\mathbf{Y}, d_2)$ is a Borel set in $(\mathbf{X}, d_1)$.

1.7.3. Give an example of a continuous function $f : \mathbf{X} \to \mathbf{Y}$ such that the image $f(\mathbf{F})$ (resp. $f(\mathbf{G})$) is not closed (resp. open) in **Y** for a closed **F** (resp. open **G**) in **X**.

1.7.4. Let $(\mathbf{X}, d_1)$ and $(\mathbf{Y}, d_2)$ be metric spaces and let $f : \mathbf{X} \to \mathbf{Y}$ be continuous. Prove that the image of each compact set **F** in **X** is compact in **Y**.

1.7.5. Let f be defined on the union of closed sets $\mathbf{F}_1, \mathbf{F}_2, \ldots, \mathbf{F}_m$. Prove that if the restriction of f to each $\mathbf{F}_i$, $i = 1, 2, \ldots, m$, is continuous, then f is continuous on $\mathbf{F}_1 \cup \mathbf{F}_2 \cup \cdots \cup \mathbf{F}_m$.

Show by example that the statement does not hold in the case of infinitely many sets $\mathbf{F}_i$.

1.7.6. Let f be defined on the union of open sets $\mathbf{G}_t$, $t \in \mathbf{T}$. Prove that if for each $t \in \mathbf{T}$ the restriction $f_{|\mathbf{G}_t}$ is continuous, then f is continuous on $\bigcup_{t \in \mathbf{T}} \mathbf{G}_t$.

1.7. Continuous Functions in Metric Spaces

1.7.7. Let $(\mathbf{X}, d_1)$ and $(\mathbf{Y}, d_2)$ be metric spaces. Prove that $f : \mathbf{X} \to \mathbf{Y}$ is continuous if and only if for each compact $\mathbf{A}$ in $\mathbf{X}$ the function $f_{|\mathbf{A}}$ is continuous.

1.7.8. Assume that f is a continuous bijection of a compact metric space $\mathbf{X}$ onto a metric space $\mathbf{Y}$. Prove that the inverse function f^{-1} is continuous on $\mathbf{Y}$. Prove also that compactness cannot be omitted from the hypotheses.

1.7.9. Let f be a continuous mapping of a compact metric space $\mathbf{X}$ into a metric space $\mathbf{Y}$. Show that f is uniformly continuous on $\mathbf{X}$.

1.7.10. Let $(\mathbf{X}, d)$ be a metric space and let $\mathbf{A}$ be a nonempty subset of $\mathbf{X}$. Prove that the function $f : \mathbf{X} \to [0, \infty)$ defined by

$$f(x) = \text{dist}(x, \mathbf{A}) = \inf\{d(x, y) : y \in \mathbf{A}\}$$

is uniformly continuous on $\mathbf{X}$.

1.7.11. Assume that f is a continuous mapping of a connected metric space $\mathbf{X}$ into a metric space $\mathbf{Y}$. Show that $f(\mathbf{X})$ is connected in $\mathbf{Y}$.

1.7.12. Let $f : \mathbf{A} \to \mathbf{Y}$, $\emptyset \neq \mathbf{A} \subset \mathbf{X}$. For $x \in \overline{\mathbf{A}}$ define

$$o_f(x, \delta) = \text{diam}(f(\mathbf{A} \cap \mathbf{B}(x, \delta))).$$

The *oscillation of f at x* is defined as

$$o_f(x) = \lim_{\delta \to 0^+} o_f(x, \delta).$$

Prove that f is continuous at $x_0 \in \mathbf{A}$ if and only if $o_f(x_0) = 0$ (compare with 1.4.19 and 1.4.20).

1.7.13. Let $f : \mathbf{A} \to \mathbf{Y}$, $\emptyset \neq \mathbf{A} \subset \mathbf{X}$ and for $x \in \overline{\mathbf{A}}$ let $o_f(x)$ be the oscillation of f at x defined in the foregoing problem. Prove that for each $\varepsilon > 0$ the set $\{x \in \overline{\mathbf{A}} : o_f(x) \geq \varepsilon\}$ is closed in $\mathbf{X}$.

1.7.14. Show that the set of points of continuity of $f : \mathbf{X} \to \mathbf{Y}$ is a countable intersection of open sets, that is, a $\mathcal{G}_\delta$ in $(\mathbf{X}, d_1)$. Show also that the set of points of discontinuity of f is a countable union of closed sets, that is, an $\mathcal{F}_\sigma$ in $(\mathbf{X}, d_1)$.

1.7.15. Give an example of a function $f : \mathbb{R} \to \mathbb{R}$ whose set of points of discontinuity is $\mathbb{Q}$.

1.7.16. Prove that every $\mathcal{F}_\sigma$ subset of $\mathbb{R}$ is the set of points of discontinuity for some $f : \mathbb{R} \to \mathbb{R}$.

1.7.17. Let $\mathbf{A}$ be an $\mathcal{F}_\sigma$ subset of a metric space $\mathbf{X}$. Must there exist a function $f : \mathbf{X} \to \mathbb{R}$ whose set of points of discontinuity is $\mathbf{A}$?

1.7.18. Let $\chi_\mathbf{A}$ be the characteristic function of $\mathbf{A} \subset \mathbf{X}$. Show that $\{x \in \mathbf{X} : o_{\chi_\mathbf{A}}(x) > 0\} = \partial \mathbf{A}$, where $o_f(x)$ is the oscillation of f at x defined in 1.7.12. Conclude that $\chi_\mathbf{A}$ is continuous on $\mathbf{X}$ if and only if $\mathbf{A}$ is both open and closed in $\mathbf{X}$.

1.7.19. Assume that g_1 and g_2 are continuous mappings of a metric space $(\mathbf{X}, d_1)$ into a metric space $(\mathbf{Y}, d_2)$, and that a set $\mathbf{A}$ with a void interior is dense in $\mathbf{X}$. Prove that if

$$f(x) = \begin{cases} g_1(x) & \text{for } x \in \mathbf{A}, \\ g_2(x) & \text{for } x \in \mathbf{X} \setminus \mathbf{A}, \end{cases}$$

then

$$o_f(x) = d_2(g_1(x), g_2(x)), \quad x \in \mathbf{X},$$

where $o_f(x)$ is the oscillation of f at x defined in 1.7.12.

1.7.20. We say that a real function f defined on a metric space $\mathbf{X}$ is in the *first Baire class* if f is a pointwise limit of a sequence of continuous functions on $\mathbf{X}$. Prove that if f is in the first Baire class, then the set of points of discontinuity of f is a set of the first category; that is, it is the union of countably many nowhere dense sets.

1.7.21. Prove that if $\mathbf{X}$ is a complete metric space and f is in the first Baire class on $\mathbf{X}$, then the set of points of continuity of f is dense in $\mathbf{X}$.

1.7.22. Let $f : (0, \infty) \to \mathbb{R}$ be continuous and such that, for each positive x, the sequence $\{f\left(\frac{x}{n}\right)\}$ converges to zero. Does this imply that $\lim\limits_{x \to 0^+} f(x) = 0$? (Compare with 1.1.33.)

1.7. Continuous Functions in Metric Spaces

1.7.23. Let $\mathcal{F}$ denote a family of real functions continuous on a complete metric space $\mathbf{X}$ such that for every $x \in \mathbf{X}$ there is M_x such that
$$|f(x)| \leq M_x \quad \text{for all } f \in \mathcal{F}.$$
Prove that there exist a positive constant M and a nonempty open set $\mathbf{G} \subset \mathbf{X}$ such that
$$|f(x)| \leq M \quad \text{for every } f \in \mathcal{F} \text{ and every } x \in \mathbf{G}.$$

1.7.24. Let $\mathbf{F}_1 \supset \mathbf{F}_2 \supset \mathbf{F}_3 \supset \ldots$ be a nested collection of nonempty closed subsets of a complete metric space $\mathbf{X}$ such that $\lim\limits_{n \to \infty} \text{diam } \mathbf{F}_n = 0$. Prove that if f is continuous on $\mathbf{X}$, then
$$f\left(\bigcap_{n=1}^{\infty} \mathbf{F}_n\right) = \bigcap_{n=1}^{\infty} f(\mathbf{F}_n).$$

1.7.25. Let $(\mathbf{X}, d_1)$ be a metric space and p a fixed point in $\mathbf{X}$. For $u \in \mathbf{X}$ define the function f_u by $f_u(x) = d_1(u, x) - d_1(p, x)$, $x \in \mathbf{X}$. Prove that $u \mapsto f_u$ is a distance preserving mapping, that is, an isometry of $(\mathbf{X}, d_1)$ into the space $C(\mathbf{X}, \mathbb{R})$ of real functions continuous on $\mathbf{X}$ endowed with the metric $d(f, g) = \sup\{|f(x) - g(x)| : x \in \mathbf{X}\}$.

1.7.26. Prove that a metric space $\mathbf{X}$ is compact if and only if every continuous function $f : \mathbf{X} \to \mathbb{R}$ is bounded.

1.7.27. Let $(\mathbf{X}, d_1)$ be a metric space and for $x \in \mathbf{X}$ define $\rho(x) = \text{dist}(x, \mathbf{X} \setminus \{x\})$. Prove that the following two conditions are equivalent.

(a) Each continuous function $f : \mathbf{X} \to \mathbb{R}$ is uniformly continuous.

(b) Every sequence $\{x_n\}$ of elements in $\mathbf{X}$ such that
$$\lim_{n \to \infty} \rho(x_n) = 0$$
contains a convergent subsequence.

1.7.28. Show that a metric space $\mathbf{X}$ is compact if and only if every real function continuous on $\mathbf{X}$ is uniformly continuous and for every $\varepsilon > 0$ the set $\{x \in \mathbf{X} : \rho(x) > \varepsilon\}$, where ρ is defined in 1.7.27, is finite.

1.7.29. Give an example of a noncompact metric space $\mathbf{X}$ such that every continuous $f : \mathbf{X} \to \mathbb{R}$ is uniformly continuous on $\mathbf{X}$.

Chapter 2

Differentiation

2.1. The Derivative of a Real Function

2.1.1. Find the derivatives (if they exist) of the following functions:

(a) $f(x) = x|x|$, $x \in \mathbb{R}$,
(b) $f(x) = \sqrt{|x|}$, $x \in \mathbb{R}$,
(c) $f(x) = [x]\sin^2(\pi x)$, $x \in \mathbb{R}$,
(d) $f(x) = (x - [x])\sin^2(\pi x)$, $x \in \mathbb{R}$,
(e) $f(x) = \ln|x|$, $x \in \mathbb{R} \setminus \{0\}$,
(f) $f(x) = \arccos\dfrac{1}{|x|}$, $|x| > 1$.

2.1.2. Find the derivatives of the following functions:

(a) $f(x) = \log_x 2$, $x > 0$, $x \neq 1$,
(b) $f(x) = \log_x \cos x$, $x \in \left(0, \dfrac{\pi}{2}\right) \setminus \{1\}$.

2.1.3. Study differentiability of the following functions:

(a) $f(x) = \begin{cases} \arctan x & \text{if } |x| \leq 1, \\ \frac{\pi}{4}\operatorname{sgn} x + \frac{x-1}{2} & \text{if } |x| > 1, \end{cases}$

37

(b) $$f(x) = \begin{cases} x^2 e^{-x^2} & \text{if } |x| \le 1, \\ \frac{1}{e} & \text{if } |x| > 1, \end{cases}$$

(c) $$f(x) = \begin{cases} \arctan \frac{1}{|x|} & \text{if } x \ne 0, \\ \frac{\pi}{2} & \text{if } x = 0. \end{cases}$$

2.1.4. Show that the function given by
$$f(x) = \begin{cases} x^2 \left|\cos \frac{\pi}{x}\right| & \text{if } x \ne 0, \\ 0 & \text{if } x = 0 \end{cases}$$
is not differentiable at $x_n = \frac{2}{2n+1}$, $n \in \mathbb{Z}$, but is differentiable at zero, which is a limit point of $\{x_n, n \in \mathbb{Z}\}$.

2.1.5. Determine the constants a, b, c and d so that f is differentiable on $\mathbb{R}$:

(a) $$f(x) = \begin{cases} 4x & \text{if } x \le 0, \\ ax^2 + bx + c & \text{if } 0 < x < 1, \\ 3 - 2x & \text{if } x \ge 1, \end{cases}$$

(b) $$f(x) = \begin{cases} ax + b & \text{if } x \le 0, \\ cx^2 + dx & \text{if } 0 < x \le 1, \\ 1 - \frac{1}{x} & \text{if } x > 1, \end{cases}$$

(c) $$f(x) = \begin{cases} ax + b & \text{if } x \le 1, \\ ax^2 + c & \text{if } 1 < x \le 2, \\ \frac{dx^2+1}{x} & \text{if } x > 2. \end{cases}$$

2.1.6. Find the following sums:

(a) $$\sum_{k=0}^{n} k e^{kx}, \quad x \in \mathbb{R},$$

(b) $$\sum_{k=0}^{2n} (-1)^k \binom{2n}{k} k^n, \quad n \ge 1,$$

(c) $$\sum_{k=1}^{n} k \cos(kx), \quad x \in \mathbb{R}.$$

2.1. The Derivative of a Real Function

2.1.7. Prove that if $|a_1 \sin x + a_2 \sin 2x + \cdots + a_n \sin nx| \leq |\sin x|$ for $x \in \mathbb{R}$, then $|a_1 + 2a_2 + \cdots + na_n| \leq 1$.

2.1.8. Assume that f and g are differentiable at a. Find

(a) $\lim\limits_{x \to a} \dfrac{xf(a) - af(x)}{x - a}$,

(b) $\lim\limits_{x \to a} \dfrac{f(x)g(a) - f(a)g(x)}{x - a}$.

2.1.9. Suppose that $f(a) > 0$ and that f is differentiable at a. Determine the limits

(a) $\lim\limits_{n \to \infty} \left(\dfrac{f\left(a + \frac{1}{n}\right)}{f(a)} \right)^{\frac{1}{n}}$,

b) $\lim\limits_{x \to a} \left(\dfrac{f(x)}{f(a)} \right)^{\frac{1}{\ln x - \ln a}}$, $a > 0$.

2.1.10. Let f be differentiable at a. Find the following limits:

(a) $\lim\limits_{x \to a} \dfrac{a^n f(x) - x^n f(a)}{x - a}$, $n \in \mathbb{N}$,

(b) $\lim\limits_{x \to a} \dfrac{f(x)e^x - f(a)}{f(x)\cos x - f(a)}$, $a = 0$, $f'(0) \neq 0$,

(c) $\lim\limits_{n \to \infty} n \left(f\left(a + \dfrac{1}{n}\right) + f\left(a + \dfrac{2}{n}\right) + \cdots + f\left(a + \dfrac{k}{n}\right) - kf(a) \right)$, $k \in \mathbb{N}$,

(d) $\lim\limits_{n \to \infty} \left(f\left(a + \dfrac{1}{n^2}\right) + f\left(a + \dfrac{2}{n^2}\right) + \cdots + f\left(a + \dfrac{n}{n^2}\right) - nf(a) \right)$.

2.1.11. For $a > 0$ and $m, k \in \mathbb{N}$, calculate

(a) $\lim\limits_{n \to \infty} \left(\dfrac{(n+1)^m + (n+2)^m + \cdots + (n+k)^m}{n^{m-1}} - kn \right)$,

(b) $\lim\limits_{n \to \infty} \dfrac{\left(a + \frac{1}{n}\right)^n \left(a + \frac{2}{n}\right)^n \cdots \left(a + \frac{k}{n}\right)^n}{a^{nk}}$,

(c) $\lim\limits_{n \to \infty} \left(\left(1 + \dfrac{a}{n^2}\right)\left(1 + \dfrac{2a}{n^2}\right) \cdots \left(1 + \dfrac{na}{n^2}\right) \right)$.

2.1.12. Assume that $f(0) = 0$ and that f is differentiable at zero. For a positive integer k find

$$\lim_{x \to 0} \frac{1}{x}\left(f(x) + f\left(\frac{x}{2}\right) + f\left(\frac{x}{3}\right) + \cdots + f\left(\frac{x}{k}\right)\right).$$

2.1.13. Let f be differentiable at a and let $\{x_n\}$ and $\{z_n\}$ be two sequences converging to a and such that $x_n \neq a$, $z_n \neq a$, $x_n \neq z_n$ for $n \in \mathbb{N}$. Give an example of f for which

$$\lim_{n \to \infty} \frac{f(x_n) - f(z_n)}{x_n - z_n}$$

(a) is equal to $f'(a)$,
(b) does not exist or exists but is different from $f'(a)$.

2.1.14. Let f be differentiable at a and let $\{x_n\}$ and $\{z_n\}$ be two sequences converging to a and such that $x_n < a < z_n$ for $n \in \mathbb{N}$. Prove that

$$\lim_{n \to \infty} \frac{f(x_n) - f(z_n)}{x_n - z_n} = f'(a).$$

2.1.15.
(a) Show that f defined on $(0, 2)$ by setting

$$f(x) = \begin{cases} x^2 & \text{for rational } x \in (0, 2), \\ 2x - 1 & \text{for irrational } x \in (0, 2) \end{cases}$$

is differentiable only at $x = 1$ and that $f'(1) \neq 0$. Is the inverse function differentiable at $1 = y = f(1)$?

(b) Let

$$\mathbf{A} = \{y \in (0, 3) : y \in \mathbb{Q}, \sqrt{y} \notin \mathbb{Q}\},$$
$$\mathbf{B} = \{x : x = \frac{1}{2}(y + 4),\ y \in \mathbf{A}\}.$$

Define f by setting

$$f(x) = \begin{cases} x^2 & \text{for rational } x \in (0, 2), \\ 2x - 1 & \text{for irrational } x \in (0, 2), \\ 2x - 4 & \text{for } x \in \mathbf{B}. \end{cases}$$

2.1. The Derivative of a Real Function

Show that the interval $(0,3)$ is contained in the range of f, and that the inverse function is not differentiable at 1.

2.1.16. Consider the function f defined on $\mathbb{R}$ as follows:

$$f(x) = \begin{cases} 0 & \text{if } x \text{ is irrational or } x = 0, \\ a_q & \text{if } x = \frac{p}{q},\ p \in \mathbb{Z},\ q \in \mathbb{N},\ \text{and } p,q \text{ are co-prime,} \end{cases}$$

where the sequence $\{a_q\}$ is such that $\lim_{n \to \infty} n^k a_n = 0$ for some integer $k \geq 2$. Prove that f is differentiable at each irrational which is algebraic of degree at most k (that is, at each algebraic surd of degree at most k).

2.1.17. Let P be a polynomial of degree n with n different real roots $x_1, x_2, \ldots, x_n$ and let Q be a polynomial of degree at most $n-1$. Show that

$$\frac{Q(x)}{P(x)} = \sum_{k=1}^{n} \frac{Q(x_k)}{P'(x_k)(x - x_k)}$$

for $x \in \mathbb{R} \setminus \{x_1, x_2, \ldots, x_n\}$ and find the sum $\sum_{k=1}^{n} \frac{1}{P'(x_k)}$, $n \geq 2$.

2.1.18. Using the result of the foregoing problem, establish the following equalities:

(a)
$$\sum_{k=0}^{n} \binom{n}{k} \frac{(-1)^k}{x+k} = \frac{n!}{x(x+1)(x+2)\cdots(x+n)}$$
$$\text{for} \quad x \in \mathbb{R} \setminus \{-n, -(n-1), \ldots, -1, 0\},$$

(b)
$$\sum_{k=0}^{n} \binom{n}{k} \frac{(-1)^k}{x+2k} = \frac{n! 2^n}{x(x+2)(x+4)\cdots(x+2n)}$$
$$\text{for} \quad x \in \mathbb{R} \setminus \{-2n, -2(n-1), \ldots, -2, 0\}.$$

2.1.19. Let f be differentiable on $\mathbb{R}$. Describe the points of differentiability of $|f|$.

2.1.20. Assume that $f_1, f_2, \ldots, f_n$ are defined in some neighborhood of x, are different from zero at x, and are differentiable at x. Prove that

$$\frac{\left(\prod_{k=1}^{n} f_k\right)'}{\prod_{k=1}^{n} f_k}(x) = \sum_{k=1}^{n} \frac{f_k'(x)}{f_k(x)}.$$

2.1.21. Assume that functions $f_1, f_2, \ldots, f_n$; $g_1, g_2, \ldots, g_n$ are defined in some neighborhood of x, are different from zero at x, and are differentiable at x. Prove that

$$\left(\prod_{k=1}^{n} \frac{f_k}{g_k}\right)'(x) = \prod_{k=1}^{n} \frac{f_k}{g_k}(x) \sum_{k=1}^{n} \left(\frac{f_k'(x)}{f_k(x)} - \frac{g_k'(x)}{g_k(x)}\right).$$

2.1.22. Study the differentiability of f and $|f|$ when

(a) $$f(x) = \begin{cases} x & \text{if } x \in \mathbb{Q}, \\ \sin x & \text{if } x \in \mathbb{R} \setminus \mathbb{Q}. \end{cases}$$

(b) $$f(x) = \begin{cases} x - \frac{3}{2^k} & \text{if } x \in \mathbb{Q} \cap \left[\frac{1}{2^{k-1}}, \frac{1}{2^{k-2}}\right), \ k \geq 2, \\ \sin\left(x - \frac{3}{2^k}\right) & \text{if } x \in (\mathbb{R} \setminus \mathbb{Q}) \cap \left[\frac{1}{2^{k-1}}, \frac{1}{2^{k-2}}\right), \ k \geq 2. \end{cases}$$

2.1.23. Show that if the one-sided derivatives $f_-'(x_0)$ and $f_+'(x_0)$ exist, then f is continuous at x_0.

2.1.24. Prove that if $f : (a, b) \to \mathbb{R}$ assumes its largest value at $c \in (a, b)$, that is, $f(c) = \max\{f(x) : x \in (a, b)\}$, and there exist one-sided derivatives $f_-'(c)$ and $f_+'(c)$, then $f_-'(c) \geq 0$ and $f_+'(c) \leq 0$. Establish the analogous necessary condition that f assumes its smallest value.

2.1.25. Prove that if $f \in C([a,b])$, $f(a) = f(b)$, and f_-' exists on (a, b), then

$$\inf\{f_-'(x) : x \in (a,b)\} \leq 0 \leq \sup\{f_-'(x) : x \in (a,b)\}.$$

2.1. The Derivative of a Real Function

2.1.26. Prove that if $f \in C([a,b])$ and f'_- exists on (a,b), then

$$\inf\{f'_-(x) : x \in (a,b)\} \leq \frac{f(b) - f(a)}{b - a} \leq \sup\{f'_-(x) : x \in (a,b)\}.$$

2.1.27. Prove that if f'_- exists and is continuous on (a,b), then f is differentiable on (a,b) and $f'(x) = f'_-(x)$ for $x \in (a,b)$.

2.1.28. Does there exist a function $f : (1,2) \to \mathbb{R}$ such that $f'_-(x) = x$ and $f'_+(x) = 2x$ for $x \in (1,2)$?

2.1.29. Let f be differentiable on $[a,b]$ and such that

(i) $\qquad\qquad f(a) = f(b) = 0$,
(ii) $\qquad\qquad f'(a) = f'_+(a) > 0, \ f'(b) = f'_-(b) > 0.$

Prove that there is $c \in (a,b)$ such that $f(c) = 0$ and $f'(c) \leq 0$.

2.1.30. Show that $f(x) = \arctan x$ satisfies the equation

$$(1 + x^2)f^{(n)}(x) + 2(n-1)xf^{(n-1)}(x) + (n-2)(n-1)f^{(n-2)}(x) = 0$$

for $x \in \mathbb{R}$ and $n \geq 2$. Show also that for $m \geq 0$,

$$f^{(2m)}(0) = 0, \quad f^{(2m+1)}(0) = (-1)^m (2m)!.$$

2.1.31. Show that

(a) $(e^x \sin x)^{(n)} = 2^{\frac{n}{2}} e^x \sin\left(x + n\frac{\pi}{4}\right)$, $x \in \mathbb{R}$, $n \geq 1$,

(b) $(x^n \ln x)^{(n)} = n!\left(\ln x + 1 + \frac{1}{2} + \cdots + \frac{1}{n}\right)$, $x > 0$, $n \geq 1$,

(c) $\left(\frac{\ln x}{x}\right)^{(n)} = (-1)^n n! x^{-n-1}\left(\ln x - 1 - \frac{1}{2} - \cdots - \frac{1}{n}\right)$, $x > 0$, $n \geq 1$,

(d) $\left(x^{n-1} e^{\frac{1}{x}}\right)^{(n)} = (-1)^n \frac{e^{\frac{1}{x}}}{x^{n+1}}$, $x \neq 0$, $n \geq 1$.

2.1.32. Prove the following identities:

(a) $\sum_{k=0}^{n} \binom{n}{k} \sin\left(x + k\frac{\pi}{2}\right) = 2^{\frac{n}{2}} \sin\left(x + n\frac{\pi}{4}\right)$, $x \in \mathbb{R}$, $n \geq 1$,

(b) $\sum_{k=1}^{n} (-1)^{k+1} \frac{1}{k} \binom{n}{k} = 1 + \frac{1}{2} + \cdots + \frac{1}{n}$, $n \geq 1$.

2.1.33. Let $f(x) = \sqrt{x^2 - 1}$ for $x > 1$. Show that $f^{(n)}(x) > 0$ if n is odd, and $f^{(n)}(x) < 0$ if n is positive and even.

2.1.34. For $f_{2n}(x) = \ln(1 + x^{2n})$, $n \in \mathbb{N}$, show that $f_{2n}^{(2n)}(-1) = 0$.

2.1.35. For a polynomial P of degree n, prove that

$$\sum_{k=0}^{n} \frac{P^{(k)}(0)}{(k+1)!} x^{k+1} = \sum_{k=0}^{n} (-1)^k \frac{P^{(k)}(x)}{(k+1)!} x^{k+1}.$$

2.1.36. Let $\lambda_1, \lambda_2, \ldots, \lambda_n$ be such that $\lambda_1^k + \lambda_2^k + \cdots + \lambda_n^k > 0$ for any $k \in \mathbb{N}$. Then f given by

$$f(x) = \frac{1}{(1 - \lambda_1 x)(1 - \lambda_2 x) \cdots (1 - \lambda_n x)}$$

is well defined in some neighborhood of zero. Show that for $k \in \mathbb{N}$, $f^{(k)}(0) > 0$

2.1.37. Let f be n times differentiable on $(0, \infty)$. Prove that for positive x,

$$\frac{1}{x^{n+1}} f^{(n)}\left(\frac{1}{x}\right) = (-1)^n \left(x^{n-1} f\left(\frac{1}{x}\right)\right)^{(n)}.$$

2.1.38. Let $\mathbf{I}, \mathbf{J}$ be open intervals and let $f : \mathbf{J} \to \mathbb{R}$, $g : \mathbf{I} \to \mathbf{J}$ be infinitely differentiable on $\mathbf{J}$ and $\mathbf{I}$, respectively. Prove the *Faà di Bruno formula* for the nth derivative of $h = f \circ g$:

$$h^{(n)}(t) = \sum \frac{n!}{k_1! k_2! \cdots k_n!} f^{(k)}(g(t)) \left(\frac{g^{(1)}(t)}{1!}\right)^{k_1} \left(\frac{g^{(2)}(t)}{2!}\right)^{k_2} \cdots \left(\frac{g^{(n)}(t)}{n!}\right)^{k_n},$$

where $k = k_1 + k_2 + \cdots + k_n$ and summation is over all $k_1, k_2, \ldots, k_n$ such that $k_1 + 2k_2 + \cdots + nk_n = n$.

2.2. Mean Value Theorems

2.1.39. Show that the functions

(a) $$f(x) = \begin{cases} e^{-\frac{1}{x^2}} & \text{if } x \neq 0, \\ 0 & \text{if } x = 0, \end{cases}$$

(b) $$g(x) = \begin{cases} e^{-\frac{1}{x}} & \text{if } x > 0, \\ 0 & \text{if } x \leq 0, \end{cases}$$

(c) $$h(x) = \begin{cases} e^{-\frac{1}{x-a} + \frac{1}{x-b}} & \text{if } x \in (a,b), \\ 0 & \text{if } x \notin (a,b), \end{cases}$$

are in $C^\infty(\mathbb{R})$.

2.1.40. Let f be differentiable on (a,b) and such that for $x \in (a,b)$ we have $f'(x) = g(f(x))$, where $g \in C^\infty(\mathbb{R})$. Prove that $f \in C^\infty(a,b)$.

2.1.41. Assume that f is twice differentiable on (a,b) and that for some real α, β, γ such that $\alpha^2 + \beta^2 > 0$,
$$\alpha f''(x) + \beta f'(x) + \gamma f(x) = 0, \quad x \in (a,b).$$
Prove that $f \in C^\infty((a,b))$.

2.2. Mean Value Theorems

2.2.1. Prove that if f is continuous on a closed interval $[a,b]$, differentiable on the open interval (a,b), and if $f(a) = f(b) = 0$, then for a real α there is an $x \in (a,b)$ such that
$$\alpha f(x) + f'(x) = 0.$$

2.2.2. Let f and g be functions continuous on $[a,b]$, differentiable on the open interval (a,b), and let $f(a) = f(b) = 0$. Show that there is a point $x \in (a,b)$ such that $g'(x)f(x) + f'(x) = 0$.

2.2.3. Assume that f is continuous on $[a,b]$, $a > 0$, and differentiable on the open interval (a,b). Show that if
$$\frac{f(a)}{a} = \frac{f(b)}{b},$$
then there is $x_0 \in (a,b)$ such that $x_0 f'(x_0) = f(x_0)$.

2.2.4. Suppose f is continuous on $[a,b]$ and differentiable on the open interval (a,b). Prove that if $f^2(b) - f^2(a) = b^2 - a^2$, then the equation $f'(x)f(x) = x$ has at least one root in (a,b).

2.2.5. Assume that f and g are continuous and never vanishing on $[a,b]$, and differentiable on (a,b). Prove that if $f(a)g(b) = f(b)g(a)$, then there is $x_0 \in (a,b)$ such that
$$\frac{f'(x_0)}{f(x_0)} = \frac{g'(x_0)}{g(x_0)}.$$

2.2.6. Assume that $a_0, a_1, \ldots, a_n$ are real numbers such that
$$\frac{a_0}{n+1} + \frac{a_1}{n} + \cdots + \frac{a_{n-1}}{2} + a_n = 0.$$
Prove that the polynomial $P(x) = a_0 x^n + a_1 x^{n-1} + \cdots + a_n$ has at least one root in $(0,1)$.

2.2.7. For real constants $a_0, a_1, \ldots, a_n$ such that
$$\frac{a_0}{1} + \frac{2a_1}{2} + \frac{2^2 a_2}{3} + \cdots + \frac{2^{n-1} a_{n-1}}{n} + \frac{2^n a_n}{n+1} = 0,$$
show that the function
$$f(x) = a_n \ln^n x + \cdots + a_2 \ln^2 x + a_1 \ln x + a_0$$
has at least one root in $(1, e^2)$.

2.2.8. Prove that if all roots of a polynomial P of degree $n \geq 2$ are real, then all roots of P' are also real.

2.2.9. Let f be continuously differentiable on $[a,b]$ and twice differentiable on (a,b), and suppose that $f(a) = f'(a) = f(b) = 0$. Prove that there is $x_1 \in (a,b)$ such that $f''(x_1) = 0$.

2.2. Mean Value Theorems

2.2.10. Let f be continuously differentiable on $[a,b]$ and twice differentiable on (a,b), and suppose that $f(a) = f(b)$ and $f'(a) = f'(b) = 0$. Show that there are $x_1, x_2 \in (a,b)$, $x_1 \neq x_2$, such that
$$f''(x_1) = f''(x_2).$$

2.2.11. Show that each of the equations

(a) $\qquad x^{13} + 7x^3 - 5 = 0,$

(b) $\qquad 3^x + 4^x = 5^x$

has exactly one real root.

2.2.12. For nonzero $a_1, a_2, \ldots, a_n$ and for $\alpha_1, \alpha_2, \ldots, \alpha_n$ such that $\alpha_i \neq \alpha_j$ for $i \neq j$, prove that the equation
$$a_1 x^{\alpha_1} + a_2 x^{\alpha_2} + \cdots + a_n x^{\alpha_n} = 0, \quad x \in (0, \infty),$$
has at most $n-1$ roots in $(0, \infty)$.

2.2.13. Prove that under the assumptions of the foregoing problem the equation
$$a_1 e^{\alpha_1 x} + a_2 e^{\alpha_2 x} + \cdots + a_n e^{\alpha_n x} = 0$$
has at most $n-1$ real roots.

2.2.14. For functions f, g and h continuous on $[a,b]$ and differentiable on (a,b), define
$$F(x) = \det \begin{vmatrix} f(x) & g(x) & h(x) \\ f(a) & g(a) & h(a) \\ f(b) & g(b) & h(b) \end{vmatrix}, \quad x \in [a,b].$$
Show that there is $x_0 \in (a,b)$ such that $F'(x_0) = 0$. Use this to derive the mean value theorem and the generalized mean value theorem.

2.2.15. Let f be continuous on $[0,2]$ and twice differentiable on $(0,2)$. Show that if $f(0) = 0$, $f(1) = 1$ and $f(2) = 2$, then there is $x_0 \in (0,2)$ such that $f''(x_0) = 0$.

2.2.16. Suppose that f is continuous on $[a,b]$ and differentiable on (a,b). Prove that if f is not a linear function, then there are x_1 and x_2 in (a,b) such that

$$f'(x_1) < \frac{f(b)-f(a)}{b-a} < f'(x_2).$$

2.2.17. Let f be continuous on $[0,1]$ and differentiable on $(0,1)$. Suppose that $f(0) = f(1) = 0$ and that there is $x_0 \in (0,1)$ such that $f(x_0) = 1$. Prove that $|f'(c)| > 2$ for some $c \in (0,1)$.

2.2.18. Let f be continuous on $[a,b]$, $a > 0$, and differentiable on (a,b). Show that there is $x_1 \in (a,b)$ such that

$$\frac{bf(a)-af(b)}{b-a} = f(x_1) - x_1 f'(x_1).$$

2.2.19. Show that the functions $x \mapsto \ln(1+x)$, $x \mapsto \ln(1+x^2)$ and $x \mapsto \arctan x$ are uniformly continuous on $[0,\infty)$.

2.2.20. Assume that f is twice differentiable on (a,b), and that there is $M \geq 0$ such that $|f''(x)| \leq M$ for all $x \in (a,b)$. Prove that f is uniformly continuous on (a,b).

2.2.21. Suppose that $f : [a,b] \to \mathbb{R}$, $b - a \geq 4$, is differentiable on the open interval (a,b). Prove that there is $x_0 \in (a,b)$ such that

$$f'(x_0) < 1 + f^2(x_0).$$

2.2.22. Prove that if f is differentiable on (a,b), and if

(i) $\quad \lim\limits_{x \to a^+} f(x) = +\infty, \quad \lim\limits_{x \to b^-} f(x) = -\infty,$

(ii) $\quad f'(x) + f^2(x) + 1 \geq 0 \quad$ for $\quad x \in (a,b)$,

then $b - a \geq \pi$.

2.2.23. Let f be continuous on $[a,b]$ and differentiable on (a,b). Show that if $\lim\limits_{x \to b^-} f'(x) = A$, then $f'_-(b) = A$.

2.2. Mean Value Theorems

2.2.24. Suppose f is differentiable on $(0,\infty)$ and $f'(x) = O(x)$ as $x \to \infty$. Prove that $f(x) = O(x^2)$ as $x \to \infty$.

2.2.25. Let $f_1, f_2, \ldots, f_n$ and $g_1, g_2, \ldots, g_n$ be continuous on $[a,b]$ and differentiable on (a,b). Suppose, further, that $g_k(a) \neq g_k(b)$ for $k = 1, 2, \ldots, n$. Prove that there is $c \in (a,b)$ for which

$$\sum_{k=1}^n f'_k(c) = \sum_{k=1}^n g'_k(c) \frac{f_k(b) - f_k(a)}{g_k(b) - g_k(a)}.$$

2.2.26. Assume that f is differentiable on an open interval **I** and that $[a,b] \subset \mathbf{I}$. We say that *f is uniformly differentiable on* $[a,b]$, if for any $\varepsilon > 0$ there is $\delta > 0$ such that

$$\left| \frac{f(x+h) - f(x)}{h} - f'(x) \right| < \varepsilon$$

for all $x \in [a,b]$ and $|h| < \delta$, $x+h \in \mathbf{I}$. Prove that f is uniformly differentiable on $[a,b]$ if and only if f' is continuous on $[a,b]$.

2.2.27. Let f be continuous on $[a,b]$, g differentiable on $[a,b]$, and $g(a) = 0$. Prove that if there is a $\lambda \neq 0$ such that

$$|g(x)f(x) + \lambda g'(x)| \leq |g(x)| \quad \text{for} \quad x \in [a,b],$$

then $g(x) \equiv 0$ on $[a,b]$.

2.2.28. Let f be differentiable on $(0,\infty)$. Show that if $\lim_{x \to +\infty} \frac{f(x)}{x} = 0$, then $\lim_{x \to +\infty} |f'(x)| = 0$.

2.2.29. Show that the only functions $f : \mathbb{R} \to \mathbb{R}$ satisfying the equation

$$\frac{f(x+h) - f(x)}{h} = f'\left(x + \frac{1}{2}h\right) \quad \text{for} \quad x, h \in \mathbb{R},\ h \neq 0,$$

are polynomials of second degree.

2.2.30. For positive p and q such that $p+q=1$, find all functions $f: \mathbb{R} \to \mathbb{R}$ satisfying the equation
$$\frac{f(x)-f(y)}{x-y} = f'(px+qy) \quad \text{for} \quad x,y \in \mathbb{R},\ x \neq y.$$

2.2.31. Prove that if f is differentiable on an interval $\mathbf{I}$, then f' enjoys the intermediate value property on $\mathbf{I}$.

2.2.32. Let f be differentiable on $(0, \infty)$. Show that
(a) if $\lim\limits_{x \to +\infty} (f(x) + f'(x)) = 0$, then $\lim\limits_{x \to +\infty} f(x) = 0$,
(b) if $\lim\limits_{x \to +\infty} (f(x) + 2\sqrt{x}f'(x)) = 0$, then $\lim\limits_{x \to +\infty} f(x) = 0$.

2.2.33. Prove that if $f \in C^2([a,b])$ has at least three distinct zeros in $[a,b]$, then the equation $f(x) + f''(x) = 2f'(x)$ has at least one root in $[a,b]$.

2.2.34. Prove that if a polynomial P has n distinct zeros greater than 1, then the polynomial
$$Q(x) = (x^2+1)P(x)P'(x) + x\left((P(x))^2 + (P'(x))^2\right)$$
has at least $2n-1$ distinct real zeros.

2.2.35. Let a polynomial $P(x) = a_m x^m + a_{m-1}x^{m-1} + \cdots + a_1 x + a_0$ with $a_m > 0$ have m distinct real zeros. Show that the polynomial $Q(x) = (P(x))^2 - P'(x)$ has
(1) exactly $m+1$ distinct real zeros if m is odd,
(2) exactly m distinct real zeros if m is even.

2.2.36. Assume that all zeros of a polynomial P of degree $n \geq 3$ are real and write
$$P(x) = (x-a_1)(x-a_2)\cdots(x-a_n),$$
where $a_i \leq a_{i+1},\ i = 1, \ldots, n-1$, and
$$P'(x) = n(x-c_1)(x-c_2)\cdots(x-c_{n-1}),$$

2.2. Mean Value Theorems

where $a_i \leq c_i \leq a_{i+1}$, $i = 1, \ldots, n-1$. Show that if
$$Q(x) = (x-a_1)(x-a_2)\cdots(x-a_{n-1}),$$
$$Q'(x) = (n-1)(x-d_1)(x-d_2)\cdots(x-d_{n-2}),$$
then $d_i \geq c_i$ for $i = 1, \ldots, n-2$. Moreover, show that if
$$R(x) = (x-a_2)(x-a_3)\cdots(x-a_n),$$
$$R'(x) = (n-1)(x-e_1)(x-e_2)\cdots(x-e_{n-2}),$$
then $e_i \leq c_{i+1}$ for $i = 1, 2, \ldots, n-2$.

2.2.37. Under the assumptions of the foregoing problem, show that
(1) if $S(x) = (x-a_1-\varepsilon)(x-a_2)\cdots(x-a_n)$, where $\varepsilon \geq 0$ is such that $a_1 + \varepsilon \leq a_{n-1}$, and if $S'(x) = n(x-f_1)(x-f_2)\cdots(x-f_{n-1})$, then $f_{n-1} \geq c_{n-1}$,
(2) if $T(x) = (x-a_1)(x-a_2)\cdots(x-a_n+\varepsilon)$, where $\varepsilon \geq 0$ is such that $a_n - \varepsilon \geq a_2$, and if $T'(x) = n(x-g_1)(x-g_2)\cdots(x-g_{n-1})$, then $g_1 \leq c_1$.

2.2.38. Show that under the assumptions of 2.2.36,
$$a_i + \frac{a_{i+1} - a_i}{n-i+1} \leq c_i \leq a_{i+1} - \frac{a_{i+1} - a_i}{i+1}, \quad i = 1, 2, \ldots, n-1.$$

2.2.39. Prove that if f is differentiable on $[0, 1]$, and if
(i) $f(0) = 0$,
(ii) there is $K > 0$ such that $|f'(x)| \leq K|f(x)|$ for $x \in [0, 1]$,
then $f(x) \equiv 0$.

2.2.40. Let f be in C^∞ on the interval $(-1, 1)$, and let $\mathbf{J} \subset (-1, 1)$ be an interval whose length is λ. Suppose that $\mathbf{J}$ is decomposed into three consecutive intervals $\mathbf{J_1}, \mathbf{J_2}$ and $\mathbf{J_3}$ whose lengths are λ_1, λ_2 and λ_3, respectively. (So we have $\mathbf{J_1} \cup \mathbf{J_2} \cup \mathbf{J_3} = \mathbf{J}$ and $\lambda_1 + \lambda_2 + \lambda_3 = \lambda$.) Prove that if
$$m_k(\mathbf{J}) = \inf\{|f^{(k)}(x)| : x \in \mathbf{J}\}, \quad k \in \mathbb{N},$$
then
$$m_k(\mathbf{J}) \leq \frac{1}{\lambda_2}(m_{k-1}(\mathbf{J_1}) + m_{k-1}(\mathbf{J_3})).$$

2.2.41. Prove that under the assumptions of the foregoing problem, if $|f(x)| \leq 1$ for $x \in (-1, 1)$, then

$$m_k(\mathbf{J}) \leq \frac{2^{\frac{k(k+1)}{2}} k^k}{\lambda^k}, \quad k \in \mathbb{N}.$$

2.2.42. Assume that a polynomial $P(x) = a_n x^n + a_{n-1} x^{n-1} + \cdots + a_1 x + a_0$ has n distinct real zeros. Prove that if there is p, $1 \leq p \leq n-1$, such that $a_p = 0$ and $a_i \neq 0$ for all $i \neq p$, then $a_{p-1} a_{p+1} < 0$.

2.3. Taylor's Formula and L'Hospital's Rule

2.3.1. Suppose that $f : [a,b] \to \mathbb{R}$ is $n-1$ times differentiable on $[a,b]$. If $f^{(n)}(x_0)$ exists, then for every $x \in [a,b]$,

$$f(x) = f(x_0) + \frac{f'(x_0)}{1!}(x - x_0) + \frac{f''(x_0)}{2!}(x - x_0)^2 + \cdots + \frac{f^{(n)}(x_0)}{n!}(x - x_0)^n + o((x - x_0)^n).$$

(This formula is called *Taylor's formula with the Peano form for the remainder*.)

2.3.2. Suppose that $f : [a,b] \to \mathbb{R}$ is n times continuously differentiable on $[a,b]$, and $f^{(n+1)}$ exists in the open interval (a,b). Prove that for any $x, x_0 \in [a,b]$ and any $p > 0$ there exists $\theta \in (0,1)$ such that

$$f(x) = f(x_0) + \frac{f'(x_0)}{1!}(x - x_0) + \frac{f''(x_0)}{2!}(x - x_0)^2 + \cdots + \frac{f^{(n)}(x_0)}{n!}(x - x_0)^n + r_n(x),$$

where

$$r_n(x) = \frac{f^{(n+1)}(x_0 + \theta(x - x_0))}{n!p}(1 - \theta)^{n+1-p}(x - x_0)^{n+1}$$

is the *Schlömilch-Roche form for the remainder*.

2.3. Taylor's Formula and L'Hospital's Rule

2.3.3. Using the above result, derive the following forms for the remainder:

(a) $\quad r_n(x) = \dfrac{f^{(n+1)}(x_0 + \theta(x - x_0))}{(n+1)!}(x - x_0)^{n+1}$

(the *Lagrange form*),

(b) $\quad r_n(x) = \dfrac{f^{(n+1)}(x_0 + \theta(x - x_0))}{n!}(1 - \theta)^n (x - x_0)^{n+1}$

(the *Cauchy form*).

2.3.4. Let $f : [a, b] \to \mathbb{R}$ be $n + 1$ times differentiable on $[a, b]$. For $x, x_0 \in [a, b]$ prove the following *Taylor formula with integral remainder*:

$$f(x) = f(x_0) + \frac{f'(x_0)}{1!}(x - x_0) + \frac{f''(x_0)}{2!}(x - x_0)^2$$
$$+ \cdots + \frac{f^{(n)}(x_0)}{n!}(x - x_0)^n + \frac{1}{n!}\int_{x_0}^{x} f^{(n+1)}(t)(x - t)^n dt.$$

2.3.5. Let $f : [a, b] \to \mathbb{R}$ be $n + 1$ times differentiable on $[a, b]$. For $x, x_0 \in [a, b]$ prove the following Taylor formula:

$$f(x) = f(x_0) + \frac{f'(x_0)}{1!}(x - x_0) + \frac{f''(x_0)}{2!}(x - x_0)^2$$
$$+ \cdots + \frac{f^{(n)}(x_0)}{n!}(x - x_0)^n + R_{n+1}(x),$$

where

$$R_{n+1}(x) = \int_{x_0}^{x} \int_{x_0}^{t_{n+1}} \int_{x_0}^{t_n} \cdots \int_{x_0}^{t_2} f^{(n+1)}(t_1) dt_1 \cdots dt_n dt_{n+1}.$$

2.3.6. Show that the approximation formula

$$\sqrt{1 + x} \approx 1 + \frac{1}{2}x - \frac{1}{8}x^2$$

gives $\sqrt{1 + x}$ with the error not greater than $\frac{1}{2}|x|^3$, if $|x| < \frac{1}{2}$.

2.3.7. For $x > -1$, $x \neq 0$, show that

(a) $\quad (1+x)^\alpha > 1 + \alpha x \quad$ if $\quad \alpha > 1 \quad$ or $\quad \alpha < 0$,

(b) $\quad (1+x)^\alpha < 1 + \alpha x \quad$ if $\quad 0 < \alpha < 1$.

2.3.8. Suppose that $f, g \in C^2([0,1])$ and $g'(x) \neq 0$ for $x \in (0,1)$, and $f'(0)g''(0) \neq f''(0)g'(0)$. For $x \in (0,1)$, let $\theta(x)$ be one of the numbers for which the assertion of the generalized mean value theorem holds, that is,
$$\frac{f(x) - f(0)}{g(x) - g(0)} = \frac{f'(\theta(x))}{g'(\theta(x))}.$$
Compute
$$\lim_{x \to 0^+} \frac{\theta(x)}{x}.$$

2.3.9. Let $f : \mathbb{R} \to \mathbb{R}$ be $n+1$ times differentiable on $\mathbb{R}$. Prove that for every $x \in \mathbb{R}$ there is $\theta \in (0,1)$ such that

(a)
$$f(x) = f(0) + xf'(x) - \frac{x^2}{2}f''(x) + \cdots + (-1)^{n+1}\frac{x^n}{n!}f^{(n)}(x)$$
$$+ (-1)^{n+2}\frac{x^{n+1}}{(n+1)!}f^{(n+1)}(\theta x),$$

(b)
$$f\left(\frac{x}{1+x}\right) = f(x) - \frac{x^2}{1+x}f'(x) + \cdots + (-1)^n \frac{x^{2n}}{(1+x)^n} \frac{f^{(n)}(x)}{n!}$$
$$+ (-1)^{n+1}\frac{x^{2n+2}}{(1+x)^{n+1}} \frac{f^{(n+1)}\left(\frac{x+\theta x^2}{1+x}\right)}{(n+1)!}, \quad x \neq -1.$$

2.3.10. Let $f : \mathbb{R} \to \mathbb{R}$ be $2n+1$ times differentiable on $\mathbb{R}$. Prove that for every $x \in \mathbb{R}$ there is $\theta \in (0,1)$ such that
$$f(x) = f(0) + \frac{2}{1!}f'\left(\frac{x}{2}\right)\left(\frac{x}{2}\right) + \frac{2}{3!}f^{(3)}\left(\frac{x}{2}\right)\left(\frac{x}{2}\right)^3$$
$$+ \cdots + \frac{2}{(2n-1)!}f^{(2n-1)}\left(\frac{x}{2}\right)\left(\frac{x}{2}\right)^{2n-1}$$
$$+ \frac{2}{(2n+1)!}f^{(2n+1)}(\theta x)\left(\frac{x}{2}\right)^{2n+1}.$$

2.3. Taylor's Formula and L'Hospital's Rule

2.3.11. Using the result in the foregoing problem, prove that

$$\ln(1+x) > 2\sum_{k=0}^{n} \frac{1}{2k+1}\left(\frac{x}{2+x}\right)^{2k+1}$$

for $n = 0, 1, \ldots$ and $x > 0$

2.3.12. Show that if $f''(x)$ exists, then

(a) $\quad \lim\limits_{h \to 0} \dfrac{f(x+h) - 2f(x) + f(x-h)}{h^2} = f''(x),$

(b) $\quad \lim\limits_{h \to 0} \dfrac{f(x+2h) - 2f(x+h) + f(x)}{h^2} = f''(x).$

2.3.13. Show that if $f'''(x)$ exists, then

$$\lim_{h \to 0} \frac{f(x+3h) - 3f(x+2h) + 3f(x+h) - f(x)}{h^3} = f'''(x).$$

2.3.14. For $x > 0$, establish the following inequalities:

(a) $\quad e^x > \sum\limits_{k=0}^{n} \dfrac{x^k}{k!},$

(b) $\quad x - \dfrac{x^2}{2} + \dfrac{x^3}{3} - \dfrac{x^4}{4} < \ln(1+x) < x - \dfrac{x^2}{2} + \dfrac{x^3}{3},$

(c) $\quad 1 + \dfrac{1}{2}x - \dfrac{1}{8}x^2 < \sqrt{1+x} < 1 + \dfrac{1}{2}x - \dfrac{1}{8}x^2 + \dfrac{1}{16}x^3.$

2.3.15. Prove that if $f^{(n+1)}(x)$ exists and is different from zero, and $\theta(h)$ is a number defined by the Taylor formula

$$f(x+h) = f(x) + hf'(x) + \cdots + \frac{h^{n-1}}{(n-1)!}f^{(n-1)}(x) + \frac{h^n}{n!}f^{(n)}(x+\theta(h)h),$$

then

$$\lim_{h \to 0} \theta(h) = \frac{1}{n+1}.$$

2.3.16. Suppose that f is differentiable on $[0,1]$ and that $f(0) = f(1) = 0$. Suppose, further, that f'' exists on $(0,1)$ and is bounded (say $|f''(x)| \leq A$ for $x \in (0,1)$). Prove that

$$|f'(x)| \leq \frac{A}{2} \quad \text{for} \quad x \in [0,1].$$

2.3.17. Suppose $f : [-c, c] \to \mathbb{R}$ is twice differentiable on $[-c, c]$, and set $M_k = \sup\{|f^{(k)}(x)| : x \in [-c, c]\}$ for $k = 0, 1, 2$. Prove that

(a) $\quad |f'(x)| \leq \dfrac{M_0}{c} + (x^2 + c^2)\dfrac{M_2}{2c} \quad \text{for} \quad x \in [-c, c],$

(b) $\quad M_1 \leq 2\sqrt{M_0 M_2} \quad \text{for} \quad c \geq \sqrt{\dfrac{M_0}{M_2}}.$

2.3.18. Let f be twice differentiable on (a, ∞), $a \in \mathbb{R}$, and let

$$M_k = \sup\{|f^{(k)}(x)| : x \in (a, \infty)\} < \infty, \quad k = 0, 1, 2.$$

Prove that $M_1 \leq 2\sqrt{M_0 M_2}$. Give an example of a function for which the equality $M_1 = 2\sqrt{M_0 M_2}$ holds.

2.3.19. Let f be twice differentiable on $\mathbb{R}$, and let

$$M_k = \sup\{|f^{(k)}(x)| : x \in \mathbb{R}\} < \infty, \quad k = 0, 1, 2.$$

Prove that $M_1 \leq \sqrt{2 M_0 M_2}$.

2.3.20. Let f be p times differentiable on $\mathbb{R}$, and let

$$M_k = \sup\{|f^{(k)}(x)| : x \in \mathbb{R}\} < \infty, \quad k = 0, 1, \ldots, p, \ p \geq 2.$$

Prove that

$$M_k \leq 2^{\frac{k(p-k)}{2}} M_0^{1-\frac{k}{p}} M_p^{\frac{k}{p}} \quad \text{for} \quad k = 1, 2, \ldots, p-1.$$

2.3.21. Assume that f'' exists and is bounded on $(0, \infty)$. Prove that if $\lim\limits_{x \to \infty} f(x) = 0$, then $\lim\limits_{x \to \infty} f'(x) = 0$.

2.3. Taylor's Formula and L'Hospital's Rule

2.3.22. Assume that f is twice continuously differentiable on $(0, \infty)$,
$$\lim_{x \to +\infty} xf(x) = 0 \quad \text{and} \quad \lim_{x \to +\infty} xf''(x) = 0.$$
Prove that $\lim\limits_{x \to +\infty} xf'(x) = 0$.

2.3.23. Assume f is twice continuously differentiable on $(0,1)$ and such that
  (i) $\lim\limits_{x \to 1^-} f(x) = 0$,
  (ii) there is $M > 0$ such that $(1-x)^2 |f''(x)| \leq M$ for $x \in (0,1)$.
Prove that $\lim\limits_{x \to 1^-} (1-x)f'(x) = 0$.

2.3.24. Let f be differentiable on $[a,b]$ and let $f'(a) = f'(b) = 0$. Prove that if f'' exists in (a,b), then there is $c \in (a,b)$ such that
$$|f''(c)| \geq \frac{4}{(b-a)^2}|f(b) - f(a)|.$$

2.3.25. Let $f : [-1,1] \to \mathbb{R}$ be three times differentiable and let $f(-1) = f(0) = 0$, $f(1) = 1$ and $f'(0) = 0$. Show that there exists $c \in (-1,1)$ such that $f'''(c) \geq 3$.

2.3.26. Let f be n times continuously differentiable on $[a,b]$ and let
$$Q(t) = \frac{f(x) - f(t)}{x - t}, \quad x, t \in [a,b], \ t \neq x.$$
Prove the following version of Taylor's formula:
$$f(x) = f(x_0) + \frac{f'(x_0)}{1!}(x - x_0) + \cdots + \frac{f^{(n)}(x_0)}{n!}(x - x_0)^n + r_n(x),$$
where $r_n(x) = \frac{Q^{(n)}(x_0)}{n!}(x - x_0)^{n+1}$.

2.3.27. Assume that $f : (-1,1) \to \mathbb{R}$ is differentiable at zero. For $-1 < x_n < y_n < 1$, $n \in \mathbb{N}$, such that $\lim\limits_{n \to \infty} x_n = \lim\limits_{n \to \infty} y_n = 0$, form the quotient
$$D_n = \frac{f(y_n) - f(x_n)}{y_n - x_n}.$$
Prove that
  (a) if $x_n < 0 < y_n$, then $\lim\limits_{n \to \infty} D_n = f'(0)$,

(b) if $0 < x_n < y_n$ and the sequence $\{\frac{y_n}{y_n - x_n}\}$ is bounded, then $\lim_{n \to \infty} D_n = f'(0)$,

(c) if f' exists on $(-1, 1)$ and is continuous at 0, then $\lim_{n \to \infty} D_n = f'(0)$.

(Compare with 2.1.13 and 2.1.14.)

2.3.28. For $m \in \mathbb{N}$, define the polynomial P by setting

$$P(x) = \sum_{k=0}^{m+1} \binom{m+1}{k} (-1)^k (x-k)^m, \quad x \in \mathbb{R}.$$

Show that $P(x) \equiv 0$.

2.3.29. Suppose that $f^{(n+2)}$ is continuous on $[0, x]$. Prove that there is $\theta \in (0, 1)$ such that

$$f(x) = f(0) + \frac{f'(0)}{1!} x + \cdots + \frac{f^{(n-1)}(0)}{(n-1)!} x^{n-1} + \frac{f^{(n)}\left(\frac{x}{n+1}\right)}{n!} x^n$$
$$+ \frac{n}{2(n+1)} f^{(n+2)}(\theta x) \frac{x^{n+2}}{(n+2)!}.$$

2.3.30. Suppose that $f^{(n+p)}$ exists in $[a, b]$ and is continuous at an $x_0 \in [a, b]$. Prove that if $f^{(n+j)}(x_0) = 0$ for $j = 1, 2, \ldots, p-1$, and $f^{(n+p)}(x_0) \neq 0$, and

$$f(x) = f(x_0) + \frac{f'(x_0)}{1!}(x - x_0) + \cdots + \frac{f^{(n-1)}(x_0)}{(n-1)!}(x - x_0)^{n-1}$$
$$+ \frac{f^{(n)}(x_0 + \theta(x)(x - x_0))}{n!}(x - x_0)^n,$$

then

$$\lim_{x \to x_0} \theta(x) = \binom{n+p}{n}^{-\frac{1}{p}}.$$

2.3.31. Suppose that f is twice continuously differentiable on $(-1, 1)$ and that $f(0) = 0$. Find

$$\lim_{x \to 0^+} \sum_{k=1}^{\left[\frac{1}{\sqrt{x}}\right]} f(kx).$$

2.3. Taylor's Formula and L'Hospital's Rule

2.3.32. Let f be infinitely differentiable on (a,b). Prove that if f vanishes at infinitely many points in the closed interval $[c,d] \subset (a,b)$, and $\sup\{|f^{(n)}(x)| : x \in (a,b)\} = O(n!)$ as $n \to \infty$, then f vanishes on an open subinterval of (a,b).

2.3.33. Suppose
  (i) f is infinitely differentiable on $\mathbb{R}$,
  (ii) there is $L > 0$ such that $|f^{(n)}(x)| \leq L$ for all $x \in \mathbb{R}$ and all $n \in \mathbb{N}$,
  (iii) $f\left(\frac{1}{n}\right) = 0$ for $n \in \mathbb{N}$.
Prove that $f(x) \equiv 0$ on $\mathbb{R}$.

2.3.34. Use l'Hospital's rule to evaluate the following limits:

(a) $\displaystyle\lim_{x \to 1} \frac{\arctan \frac{x^2-1}{x^2+1}}{x-1}$, (b) $\displaystyle\lim_{x \to +\infty} x\left(\left(1 + \frac{1}{x}\right)^x - e\right)$,

(c) $\displaystyle\lim_{x \to 5} (6-x)^{\frac{1}{x-5}}$, (d) $\displaystyle\lim_{x \to 0^+} \left(\frac{\sin x}{x}\right)^{\frac{1}{x}}$,

(e) $\displaystyle\lim_{x \to 0^+} \left(\frac{\sin x}{x}\right)^{\frac{1}{x^2}}$.

2.3.35. Prove that if f is twice continuously differentiable on $\mathbb{R}$ such that $f(0) = 1$, $f'(0) = 0$, and $f''(0) = -1$, then for $a \in \mathbb{R}$,

$$\lim_{x \to +\infty} \left(f\left(\frac{a}{\sqrt{x}}\right)\right)^x = e^{-\frac{a^2}{2}}.$$

2.3.36. For $a > 0$, $a \neq 1$, evaluate

$$\lim_{x \to +\infty} \left(\frac{a^x - 1}{x(a-1)}\right)^{\frac{1}{x}}.$$

2.3.37. Can l'Hospital's rule be applied to evaluate the following limits?

(a) $\quad \lim\limits_{x\to\infty} \dfrac{x - \sin x}{2x + \sin x}$,

(b) $\quad \lim\limits_{x\to\infty} \dfrac{2x + \sin 2x + 1}{(2x + \sin 2x)(\sin x + 3)^2}$,

(c) $\quad \lim\limits_{x\to 0^+} \left(2\sin\sqrt{x} + \sqrt{x}\sin\dfrac{1}{x}\right)^x$,

(d) $\quad \lim\limits_{x\to 0} \left(1 + xe^{-\frac{1}{x^2}}\sin\dfrac{1}{x^4}\right)^{e^{\frac{1}{x^2}}}$.

2.3.38. Is the function given by

$$f(x) = \begin{cases} \dfrac{1}{x\ln 2} - \dfrac{1}{2^x - 1} & \text{if } x \neq 0, \\ \dfrac{1}{2} & \text{if } x = 0 \end{cases}$$

differentiable at zero?

2.3.39. Suppose f is n times continuously differentiable on $\mathbb{R}$. For $a \in \mathbb{R}$, establish the equality

$$f^{(n)}(a) = \lim_{h\to 0} \dfrac{1}{h^n} \sum_{k=0}^{n} \left((-1)^{n-k}\binom{n}{k}f(a + kh)\right).$$

2.3.40. Prove the following version of l'Hospital's rule. Suppose $f, g : (a, b) \to \mathbb{R}$, $-\infty \leq a < b \leq +\infty$, are differentiable on (a, b). Suppose, further, that
 (i) $g'(x) \neq 0$ for $x \in (a, b)$,
 (ii) $\lim\limits_{x\to a^+} g(x) = +\infty\ (-\infty)$,
 (iii) $\lim\limits_{x\to a^+} \dfrac{f'(x)}{g'(x)} = L$, $-\infty \leq L \leq +\infty$.
Then
$$\lim_{x\to a^+} \dfrac{f(x)}{g(x)} = L.$$

2.3.41. Use the above version of l'Hospital's rule to prove the following generalizations of the results given in 2.2.32. Let f be differentiable on $(0, \infty)$, and let $a > 0$.

(a) If $\lim_{x \to +\infty} (af(x) + f'(x)) = L$, then $\lim_{x \to +\infty} f(x) = \frac{L}{a}$.

(b) If $\lim_{x \to +\infty} (af(x) + 2\sqrt{x} f'(x)) = L$, then $\lim_{x \to +\infty} f(x) = \frac{L}{a}$.

Are the above statements true for negative a?

2.3.42. Assume that f is three times differentiable on $(0, \infty)$ and such that $f(x) > 0$, $f'(x) > 0$, $f''(x) > 0$ for $x > 0$. Prove that if

$$\lim_{x \to \infty} \frac{f'(x) f'''(x)}{(f''(x))^2} = c, \quad c \neq 1,$$

then

$$\lim_{x \to \infty} \frac{f(x) f''(x)}{(f'(x))^2} = \frac{1}{2 - c}.$$

2.3.43. Assume that f is in C^∞ on $(-1, 1)$, and that $f(0) = 0$. Prove that if g is defined on $(-1, 1) \setminus \{0\}$ by $g(x) = \frac{f(x)}{x}$, then there exists an extension of g which is in C^∞ on $(-1, 1)$.

2.4. Convex Functions

A function f is said to be *convex* on an interval $\mathbf{I} \subset \mathbb{R}$ if

(1) $\qquad f(\lambda x_1 + (1 - \lambda) x_2) \leq \lambda f(x_1) + (1 - \lambda) f(x_2)$

whenever $x_1, x_2 \in \mathbf{I}$ and $\lambda \in (0, 1)$. A convex function f is said to be *strictly convex* on $\mathbf{I}$ if strict inequality holds in (1) for $x_1 \neq x_2$. f is *concave* on $\mathbf{I}$ if $-f$ is convex.

2.4.1. Prove that a function f differentiable on an open interval $\mathbf{I}$ is convex if and only if f' is increasing on $\mathbf{I}$.

2.4.2. Prove that a function f twice differentiable on an open interval $\mathbf{I}$ is convex if and only if $f''(x) \geq 0$ for $x \in \mathbf{I}$.

2.4.3. Prove that if a function f is convex on an interval **I**, then the following *Jensen inequality*

$$f(\lambda_1 x_1 + \lambda_2 x_2 + \cdots + \lambda_n x_n) \leq \lambda_1 f(x_1) + \lambda_2 f(x_2) + \cdots + \lambda_n f(x_n)$$

holds for any points $x_1, \ldots, x_n$ in **I** and any nonnegative numbers $\lambda_1, \ldots, \lambda_n$ such that $\lambda_1 + \lambda_2 + \cdots + \lambda_n = 1$.

2.4.4. For $x, y > 0$ and $p, q > 0$ such that $\frac{1}{p} + \frac{1}{q} = 1$, establish the inequality

$$xy \leq \frac{x^p}{p} + \frac{y^q}{q}.$$

2.4.5. Prove that

$$\frac{1}{n}\sum_{k=1}^{n} x_k \geq \sqrt[n]{\prod_{k=1}^{n} x_k} \quad \text{for} \quad x_1, x_2, \ldots, x_n > 0.$$

2.4.6. Show that if $a \neq b$, then

$$\frac{e^b - e^a}{b - a} < \frac{e^a + e^b}{2}.$$

2.4.7. For positive x and y, establish the inequality

$$x \ln x + y \ln y \geq (x + y) \ln \frac{x + y}{2}.$$

2.4.8. For $\alpha > 1$ and for positive $x_1, x_2, \ldots, x_n$, prove that

$$\left(\frac{1}{n}\sum_{k=1}^{n} x_k\right)^{\alpha} \leq \frac{1}{n}\sum_{k=1}^{n} x_k^{\alpha}.$$

2.4. Convex Functions

2.4.9. Let $x_1, \ldots, x_n \in (0,1)$ and let $p_1, \ldots, p_n$ be positive and such that $\sum_{k=1}^{n} p_k = 1$. Prove that

(a) $$1 + \left(\sum_{k=1}^{n} p_k x_k\right)^{-1} \leq \prod_{k=1}^{n} \left(\frac{1 + x_k}{x_k}\right)^{p_k},$$

(b) $$\frac{1 + \sum_{k=1}^{n} p_k x_k}{1 - \sum_{k=1}^{n} p_k x_k} \leq \prod_{k=1}^{n} \left(\frac{1 + x_k}{1 - x_k}\right)^{p_k}.$$

2.4.10. Let $x = \frac{1}{n}\sum_{k=1}^{n} x_k$ with $x_1, \ldots, x_n \in (0, \pi)$. Show that

(a) $$\prod_{k=1}^{n} \sin x_k \leq (\sin x)^n,$$

(b) $$\prod_{k=1}^{n} \frac{\sin x_k}{x_k} \leq \left(\frac{\sin x}{x}\right)^n.$$

2.4.11. Prove that if $a > 1$, and $x_1, \ldots, x_n \in (0,1)$ are such that $x_1 + \cdots + x_n = 1$, then

$$\sum_{k=1}^{n} \left(x_k + \frac{1}{x_k}\right)^a \geq \frac{(n^2 + 1)^a}{n^{a-1}}.$$

2.4.12. For $n \geq 2$, verify the following claim:

$$\prod_{k=2}^{n} \frac{2^k - 1}{2^{k-1}} \leq \left(2 - \frac{2}{n} + \frac{1}{n \cdot 2^{n-1}}\right)^n.$$

2.4.13. Establish the following inequalities:

(a)
$$\frac{n^2}{x_1+\cdots+x_n} \leq \frac{1}{x_1}+\cdots+\frac{1}{x_n}, \quad x_1,\ldots,x_n > 0,$$

(b)
$$\frac{1}{\frac{\alpha_1}{x_1}+\cdots+\frac{\alpha_n}{x_n}} \leq x_1^{\alpha_1}\cdots x_n^{\alpha_n} \leq \alpha_1 x_1 + \cdots + \alpha_n x_n$$

for $\alpha_k, x_k > 0$, $k = 1, 2, \ldots, n$, such that $\sum_{k=1}^{n} \alpha_k = 1$,

(c)
$$x_1^{\alpha_1}\cdots x_n^{\alpha_n} + y_1^{\alpha_1}\cdots y_n^{\alpha_n} \leq (x_1+y_1)^{\alpha_1}\cdots(x_n+y_n)^{\alpha_n}$$

for $x_k, y_k \geq 0$, $\alpha_k > 0$, $k = 1, 2, \ldots, n$, such that $\sum_{k=1}^{n} \alpha_k = 1$,

(d)
$$\sum_{j=1}^{m}\prod_{i=1}^{n} x_{i,j}^{\alpha_i} \leq \prod_{i=1}^{n}\left(\sum_{j=1}^{m} x_{i,j}\right)^{\alpha_i}$$

for $x_{i,j} \geq 0$, $\alpha_i > 0$, $i = 1, 2, \ldots, n$, $j = 1, 2, \ldots, m$, such that $\sum_{k=1}^{n} \alpha_k = 1$.

2.4.14. Show that if $f : \mathbb{R} \to \mathbb{R}$ is convex and bounded above, then f is constant on $\mathbb{R}$.

2.4.15. Must a convex and bounded function on (a, ∞) or on $(-\infty, a)$ be constant?

2.4.16. Suppose that $f : (a,b) \to \mathbb{R}$ is convex on (a,b) (the cases $a = -\infty$ or $b = \infty$ are admitted). Prove that either f is monotonic on (a,b), or there is $c \in (a,b)$ such that
$$f(c) = \min\{f(x) : x \in (a,b)\}$$
and f is decreasing on $(a,c]$ and increasing on $[c,b)$.

2.4. Convex Functions

2.4.17. Let $f : (a,b) \to \mathbb{R}$ be convex on (a,b) (the cases $a = -\infty$ or $b = \infty$ are admitted). Show that finite or infinite limits
$$\lim_{x \to a^+} f(x) \quad \text{and} \quad \lim_{x \to b^-} f(x)$$
exist.

2.4.18. Suppose that $f : (a,b) \to \mathbb{R}$ is convex and bounded on (a,b) (the cases $a = -\infty$ or $b = \infty$ are admitted). Prove that f is uniformly continuous on (a,b). (Compare with 2.4.14).

2.4.19. Let $f : (a,b) \to \mathbb{R}$ be convex on (a,b) (the cases $a = -\infty$ or $b = \infty$ are admitted). Prove that one-sided derivatives of f exist on (a,b) and are monotonic. Prove moreover that the right- and left-hand derivatives are equal to each other except on a countable set.

2.4.20. Assume that f is twice differentiable on $\mathbb{R}$ and f, f' and f'' are strictly increasing on $\mathbb{R}$. For fixed a, b, $a \leq b$, let $x \to \xi(x)$, $x > 0$, be defined by the mean value theorem, that is,
$$\frac{f(b+x) - f(a-x)}{b - a + 2x} = f'(\xi).$$
Prove that the function ξ is increasing on $(0, \infty)$.

2.4.21. Using the result in 2.4.4 prove *Hölder's inequality*: If $p, q > 1$ and $\frac{1}{p} + \frac{1}{q} = 1$, then
$$\sum_{i=1}^n |x_i y_i| \leq \left(\sum_{i=1}^n |x_i|^p\right)^{\frac{1}{p}} \left(\sum_{i=1}^n |y_i|^q\right)^{\frac{1}{q}}.$$

2.4.22. Using Hölder's inequality, prove the *Minkowski inequality*: If $p \geq 1$, then
$$\left(\sum_{i=1}^n |x_i + y_i|^p\right)^{\frac{1}{p}} \leq \left(\sum_{i=1}^n |x_i|^p\right)^{\frac{1}{p}} + \left(\sum_{i=1}^n |y_i|^p\right)^{\frac{1}{p}}.$$

2.4.23. Prove that, if a series $\sum\limits_{n=1}^{\infty} a_n^4$ converges, then $\sum\limits_{n=1}^{\infty} \frac{a_n}{n^{\frac{4}{5}}}$ also converges.

2.4.24. For $x_i, y_i \geq 0$, $i = 1, 2, \ldots, n$, and $p > 1$, establish the inequality

$$((x_1 + \cdots + x_n)^p + (y_1 + \cdots + y_n)^p)^{\frac{1}{p}} \leq (x_1^p + y_1^p)^{\frac{1}{p}} + \cdots + (x_n^p + y_n^p)^{\frac{1}{p}}.$$

2.4.25. Prove the following *generalized Minkowski inequality*: For $x_{i,j} \geq 0$, $i = 1, 2, \ldots, n$, $j = 1, 2, \ldots, m$, and for $p > 1$,

$$\left(\sum_{i=1}^{n}\left(\sum_{j=1}^{m} x_{i,j}\right)^p\right)^{\frac{1}{p}} \leq \sum_{j=1}^{m}\left(\sum_{i=1}^{n} x_{i,j}^p\right)^{\frac{1}{p}}.$$

2.4.26. Assume that f, continuous on an interval $\mathbf{I}$, is *midpoint-convex* on that interval, that is,

$$f\left(\frac{x+y}{2}\right) \leq \frac{f(x) + f(y)}{2} \quad \text{for} \quad x, y \in \mathbf{I}.$$

Prove that f is convex on $\mathbf{I}$.

2.4.27. Show that the continuity is an essential hypothesis in 2.4.26.

2.4.28. Let f be continuous on an interval $\mathbf{I}$ and such that

$$f\left(\frac{x+y}{2}\right) < \frac{f(x) + f(y)}{2}$$

for $x, y \in \mathbf{I}$, $x \neq y$. Show that f is strictly convex on $\mathbf{I}$.

2.4.29. Assume that f is a convex function on an open interval $\mathbf{I}$. Prove that f satisfies the Lipschitz condition locally on $\mathbf{I}$.

2.4.30. Let $f : (0, \infty) \to \mathbb{R}$ be convex and let

$$\lim_{x \to 0^+} f(x) = 0.$$

Prove that the function $x \mapsto \frac{f(x)}{x}$ is increasing on $(0, \infty)$.

2.4. Convex Functions

2.4.31. We say that f is *subadditive* on $(0,\infty)$ if for $x_1, x_2 \in (0,\infty)$,
$$f(x_1 + x_2) \leq f(x_1) + f(x_2).$$
Prove that
(a) if $x \mapsto \frac{f(x)}{x}$ is decreasing on $(0,\infty)$, then f is subadditive,
(b) if f is convex and subadditive on $(0,\infty)$, then $x \mapsto \frac{f(x)}{x}$ is a decreasing function on that interval.

2.4.32. Suppose f is differentiable on (a,b) and for $x, y \in (a,b)$, $x \neq y$, there is exactly one ζ such that
$$\frac{f(y) - f(x)}{y - x} = f'(\zeta).$$
Prove that f is strictly convex or strictly concave on (a,b).

2.4.33. Let $f : \mathbb{R} \to \mathbb{R}$ be continuous and such that for each $d \in \mathbb{R}$ the function $g_d(x) = f(x+d) - f(x)$ is in C^∞ on $\mathbb{R}$. Prove that f is in C^∞ on $\mathbb{R}$.

2.4.34. Assume that $a_n \leq \cdots \leq a_2 \leq a_1$, and f is convex on the interval $[a_n, a_1]$. Prove that
$$\sum_{k=1}^{n} f(a_{k+1}) a_k \leq \sum_{k=1}^{n} f(a_k) a_{k+1},$$
where $a_{n+1} = a_1$.

2.4.35. Suppose that f is concave and strictly increasing on an interval (a,b) (the cases $a = -\infty$ or $b = \infty$ are admitted). Prove that if $a < f(x) < x$ for $x \in (a,b)$ and
$$\lim_{x \to a^+} f'_+(x) = 1,$$
then for $x, y \in (a,b)$,
$$\lim_{n \to \infty} \frac{f^{n+1}(x) - f^n(x)}{f^{n+1}(y) - f^n(y)} = 1,$$
where f^n denotes the nth iterate of f (see, e.g., 1.1.40).

2.5. Applications of Derivatives

2.5.1. Using the generalized mean value theorem, show that

(a) $\quad 1 - \dfrac{x^2}{2!} < \cos x \quad \text{for} \quad x \neq 0,$

(b) $\quad x - \dfrac{x^3}{3!} < \sin x \quad \text{for} \quad x > 0,$

(c) $\quad \cos x < 1 - \dfrac{x^2}{2!} + \dfrac{x^4}{4!} \quad \text{for} \quad x \neq 0,$

(d) $\quad \sin x < x - \dfrac{x^3}{3!} + \dfrac{x^5}{5!} \quad \text{for} \quad x > 0,$

2.5.2. For $n \in \mathbb{N}$ and $x > 0$, verify the following claims:

(a) $\quad x - \dfrac{x^3}{3!} + \dfrac{x^5}{5!} - \cdots + \dfrac{x^{4n-3}}{(4n-3)!} - \dfrac{x^{4n-1}}{(4n-1)!} < \sin x$

$\quad < x - \dfrac{x^3}{3!} + \cdots + \dfrac{x^{4n-3}}{(4n-3)!} - \dfrac{x^{4n-1}}{(4n-1)!} + \dfrac{x^{4n+1}}{(4n+1)!},$

(b) $\quad 1 - \dfrac{x^2}{2!} + \dfrac{x^4}{4!} - \cdots + \dfrac{x^{4n-4}}{(4n-4)!} - \dfrac{x^{4n-2}}{(4n-2)!} < \cos x$

$\quad < 1 - \dfrac{x^2}{2!} + \cdots + \dfrac{x^{4n-4}}{(4n-4)!} - \dfrac{x^{4n-2}}{(4n-2)!} + \dfrac{x^{4n}}{(4n)!}.$

2.5.3. Let f be continuous on $[a, b]$ and differentiable on the open interval (a, b). Show that if $a \geq 0$, then there are $x_1, x_2, x_3 \in (a, b)$ such that

$$f'(x_1) = (b + a)\dfrac{f'(x_2)}{2x_2} = (b^2 + ba + a^2)\dfrac{f'(x_3)}{3x_3^2}.$$

2.5.4. Prove the following generalization of the result in 2.2.32. Let f be a complex valued function on $(0, \infty)$, and let α be a complex number with a positive real part. Prove that if f is differentiable and $\lim\limits_{x \to +\infty} (\alpha f(x) + f'(x)) = 0$, then $\lim\limits_{x \to +\infty} f(x) = 0$.

2.5. Applications of Derivatives

2.5.5. Let f be twice differentiable on the interval $(0, \infty)$. Prove that if $\lim_{x \to +\infty} (f(x) + f'(x) + f''(x)) = L$, then $\lim_{x \to +\infty} f(x) = L$.

2.5.6. Let f be three times differentiable on $(0, \infty)$. Does the existence of the limit
$$\lim_{x \to +\infty} (f(x) + f'(x) + f''(x) + f'''(x))$$
imply the existence of $\lim_{x \to +\infty} f(x)$?

2.5.7.

(a) Let f be continuously differentiable on $(0, \infty)$ and let $f(0) = 1$. Show that if $|f(x)| \leq e^{-x}$ for $x \geq 0$, then there is $x_0 > 0$ such that $f'(x_0) = -e^{-x_0}$.

(b) Let f be continuously differentiable on $(1, \infty)$ and let $f(1) = 1$. Show that if $|f(x)| \leq \frac{1}{x}$ for $x \geq 1$, then there is $x_0 > 0$ such that $f'(x_0) = -\frac{1}{x_0^2}$.

2.5.8. Assume that f and g are differentiable on $[0, a]$ and such that $f(0) = g(0) = 0$ and $g(x) > 0$, $g'(x) > 0$ for $x \in (0, a]$. Prove that if $\frac{f'}{g'}$ increases on $(0, a]$, then also $\frac{f}{g}$ increases on that interval.

2.5.9. Show that each of the equations
$$\sin(\cos x) = x \quad \text{and} \quad \cos(\sin x) = x$$
has exactly one root in $[0, \pi/2]$. Moreover, show that if x_1 and x_2 are the roots of the former and the latter equation, respectively, then $x_1 < x_2$.

2.5.10. Prove that if f is differentiable on $[a, b]$, $f(a) = 0$, and there is a constant $C \geq 0$ such that $|f'(x)| \leq C|f(x)|$ for $x \in [a, b]$, then $f(x) \equiv 0$.

2.5.11. Using the mean value theorem, prove that if $0 < p < q$, then
$$\left(1 + \frac{x}{p}\right)^p < \left(1 + \frac{x}{q}\right)^q$$
for $x > 0$.

2.5.12. Show first that $e^x \geq 1 + x$ for $x \in \mathbb{R}$, and then, using this result, prove the arithmetic-geometric mean inequality.

2.5.13. Show that
$$xy \leq e^x + y(\ln y - 1)$$
for $x \in \mathbb{R}$ and positive y. Show also that the equality holds if and only if $y = e^x$.

2.5.14. Suppose that $f : \mathbb{R} \to [-1, 1]$ is in C^2 on $\mathbb{R}$, and that $(f(0))^2 + (f'(0))^2 = 4$. Prove that there exists $x_0 \in \mathbb{R}$ such that $f(x_0) + f''(x_0) = 0$.

2.5.15. Establish the following inequalities:

(a) $\left(x + \dfrac{1}{x}\right) \arctan x > 1$ for $x > 0$,

(b) $2\tan x - \sinh x > 0$ for $0 < x < \dfrac{\pi}{2}$,

(c) $\ln x < \dfrac{x}{e}$ for $x > 0, x \neq e$,

(d) $\dfrac{x \ln x}{x^2 - 1} < \dfrac{1}{2}$ for $x > 0, x \neq 1$.

2.5.16. Decide which of the two numbers is greater:

(a) e^π or π^e,

(b) $2^{\sqrt{2}}$ or e,

(c) $\ln 8$ or 2.

2.5.17. Verify the following claims:

(a) $\ln\left(1 + \dfrac{x}{a}\right) \ln\left(1 + \dfrac{b}{x}\right) < \dfrac{b}{a}$, $a, b, x > 0$,

(b) $\left(1 + \dfrac{x}{m}\right)^m \left(1 - \dfrac{x}{n}\right)^n < 1$, $x \in \mathbb{R} \setminus \{0\}$, $m, n \in \mathbb{N}$, $m, n \geq |x|$,

(c) $\ln(1 + \sqrt{1 + x^2}) < \dfrac{1}{x} + \ln x$, $x > 0$.

2.5. Applications of Derivatives

2.5.18. For $x > 0$, establish the following inequalities:

(a) $\ln(1+x) < \dfrac{x}{\sqrt{1+x}}$,

(b) $(x-1)^2 \geq x\ln^2 x$.

2.5.19. Show that

(a) $x + \dfrac{x^2}{2} - \dfrac{x^3}{6} < (x+1)\ln(1+x) < x + \dfrac{x^2}{2}$ for $x > 0$,

(b) $\ln(1+\cos x) < \ln 2 - \dfrac{x^2}{4}$ for $x \in (0, \pi)$.

2.5.20. For $x > 0$, verify the following claims:

(a) $e^x < 1 + xe^x$,   (b) $e^x - 1 - x < x^2 e^x$,

(c) $xe^{\frac{x}{2}} < e^x - 1$,   (d) $e^x < (1+x)^{1+x}$,

(e) $\left(\dfrac{x+1}{2}\right)^{x+1} \leq x^x$.

2.5.21. Show that $(e+x)^{e-x} > (e-x)^{e+x}$ for $x \in (0, e)$.

2.5.22. Show that if $x > 1$, then $e^{x-1} + \ln x - 2x + 1 > 0$.

2.5.23. Establish the following inequalities:

(a) $\dfrac{1}{3}\tan x + \dfrac{2}{3}\sin x > x$ for $0 < x < \dfrac{\pi}{2}$,

(b) $x(2+\cos x) > 3\sin x$ for $x > 0$,

(c) $\cos x < \dfrac{\sin^2 x}{x^2}$ for $0 < x < \dfrac{\pi}{2}$.

2.5.24. Show that if $\alpha > 1$, then for $0 \leq x \leq 1$,
$$\dfrac{1}{2^{\alpha-1}} \leq x^\alpha + (1-x)^\alpha \leq 1.$$

2.5.25. Show that if $0 < \alpha < 1$, then for $x, y > 0$,
$$(x+y)^\alpha < x^\alpha + y^\alpha.$$

2.5.26. For $\alpha \in (0,1)$ and $x \in [-1,1]$, show that
$$(1+x)^\alpha \leq 1 + \alpha x - \frac{\alpha(\alpha-1)}{8}x^2.$$

2.5.27. Prove the following generalizations of the result in the foregoing problem. For $B \geq 0$ and $x \in [-1, B]$,

(a) $\quad (1+x)^\alpha \leq 1 + \alpha x - \dfrac{\alpha(1-\alpha)}{2(1+B)^2}x^2 \quad$ if $\quad 0 < \alpha < 1$,

(b) $\quad (1+x)^\alpha \geq 1 + \alpha x - \dfrac{\alpha(1-\alpha)}{2(1+B)^2}x^2 \quad$ if $\quad 1 < \alpha < 2$.

2.5.28. Prove that

(a) $\quad \sin x \geq \dfrac{2}{\pi}x \quad$ for $\quad x \in [0, \frac{\pi}{2}]$,

(b) $\quad \sin x \geq \dfrac{2}{\pi}x + \dfrac{x}{\pi^3}(\pi^2 - 4x^2) \quad$ for $\quad x \in [0, \frac{\pi}{2}]$.

2.5.29. Prove that for $x \in (0,1)$,
$$\pi x(1-x) < \sin \pi x \leq 4x(1-x).$$

2.5.30. Prove that for a positive x and a positive integer n,
$$e^x - \sum_{k=0}^{n} \frac{x^k}{k!} < \frac{x}{n}(e^x - 1).$$

2.5.31. For a positive integer n, find all local extrema of the function
$$f(x) = \left(1 + x + \frac{x^2}{2!} + \cdots + \frac{x^n}{n!}\right)e^{-x}.$$

2.5. Applications of Derivatives

2.5.32. For positive integers m and n, find all local extrema of the function
$$f(x) = x^m(1-x)^n.$$

2.5.33. For positive integers m and n, find the maximum value of the function
$$f(x) = \sin^{2m} x \cdot \cos^{2n} x.$$

2.5.34. Determine all local extremum points of $f(x) = x^{\frac{1}{3}}(1-x)^{\frac{2}{3}}$.

2.5.35. Find the minimum and maximum values of the function $f(x) = x\arcsin x + \sqrt{1-x^2}$ on $[-1,1]$.

2.5.36. Find the maximum value of f on $\mathbb{R}$, where f is given by
$$f(x) = \frac{1}{1+|x|} + \frac{1}{1+|x-1|}.$$

2.5.37. Show that for nonnegative $a_1, a_2, \ldots, a_n$ the following inequalities hold:

(a) $\quad \dfrac{1}{n}\sum_{k=1}^{n} a_k e^{-a_k} \leq \dfrac{1}{e},$

(b) $\quad \dfrac{1}{n}\sum_{k=1}^{n} a_k^2 e^{-a_k} \leq \dfrac{4}{e^2},$

(c) $\quad \prod_{k=1}^{n} a_k \leq \left(\dfrac{3}{e}\right)^n \exp\left(\dfrac{1}{3}\sum_{k=1}^{n} a_k\right).$

2.5.38. Determine all local extremum points of the function
$$f(x) = \begin{cases} e^{-\frac{1}{|x|}}\left(\sqrt{2} + \sin\frac{1}{x}\right) & \text{if } x \neq 0, \\ 0 & \text{if } x = 0. \end{cases}$$

2.5.39. Let
$$f(x) = \begin{cases} x^4 \left(2 + \sin \frac{1}{x}\right) & \text{if } x \neq 0, \\ 0 & \text{if } x = 0. \end{cases}$$
Prove that f is differentiable on $\mathbb{R}$ and that at zero f attains its proper absolute minimum, but in no interval $(-\varepsilon, 0)$ or $(0, \varepsilon)$ is f monotone.

2.5.40. For $x > 0$, establish the inequalities
$$\frac{\sinh x}{\sqrt{\sinh^2 x + \cosh^2 x}} < \tanh x < x < \sinh x < \frac{1}{2} \sinh 2x.$$

2.5.41. Using the result in the preceding problem, prove that, if a and b are positive and $a \neq b$, then
$$\frac{2}{\frac{1}{a} + \frac{1}{b}} < \sqrt{ab} < \frac{b - a}{\ln b - \ln a} < \frac{a + b}{2} < \sqrt{\frac{a^2 + b^2}{2}}.$$
The number $L(a,b) = \frac{b-a}{\ln b - \ln a}$ is called the *logarithmic mean* of positive numbers a and b, $a \neq b$. (It is also convenient to adopt the convention that $L(a,a) = a$.)

2.5.42. The *power mean* of positive numbers x and y is defined by
$$M_p(x, y) = \left(\frac{x^p + y^p}{2}\right)^{\frac{1}{p}} \quad \text{if } p \neq 0.$$

(a) Show that
$$\lim_{p \to 0} M_p(x, y) = \sqrt{xy}.$$
(So it is natural to adopt the convention that $M_0(x,y) = \sqrt{xy}$.)

(b) Show that if $x \neq y$ and $p < q$, then $M_p(x,y) < M_q(x,y)$.

2.5.43. For $\lambda \geq 1$, and for positive x, y, and for an integer $n \geq 2$, prove that
$$\sqrt{xy} \leq \sqrt[n]{\frac{x^n + y^n + \lambda\left((x+y)^n - x^n - y^n\right)}{2 + \lambda(2^n - 2)}} \leq \frac{x+y}{2}.$$

2.5. Applications of Derivatives

2.5.44. Prove that

(a) $\sin(\tan x) \geq x$ for $x \in \left[0, \frac{\pi}{4}\right]$,

(b) $\tan(\sin x) \geq x$ for $x \in \left[0, \frac{\pi}{3}\right]$.

2.5.45. Prove that if $x \in (0, \pi/2]$, then
$$\frac{1}{\sin^2 x} \leq \frac{1}{x^2} + 1 - \frac{4}{\pi^2}.$$

2.5.46. For $x > 0$, show that
$$\arctan x > \frac{3x}{1 + 2\sqrt{1 + x^2}}.$$

2.5.47. Let a_k, b_k, $k = 1, 2, \ldots, n$, be positive. Prove that the inequality
$$\prod_{k=1}^{n} (xa_k + (1-x)b_k) \leq \max\left\{\prod_{k=1}^{n} a_k, \prod_{k=1}^{n} b_k\right\}$$
holds for $x \in [0, 1]$ if and only if
$$\left(\sum_{k=1}^{n} \frac{a_k - b_k}{a_k}\right)\left(\sum_{k=1}^{n} \frac{a_k - b_k}{b_k}\right) \geq 0.$$

2.5.48. Using the result in 2.5.1, show that
$$\cos x + \cos y \leq 1 + \cos(xy) \quad \text{for} \quad x^2 + y^2 \leq \pi.$$

2.5.49. For positive x and y, establish the inequality
$$x^y + y^x > 1.$$

2.5.50. For an integer $n \geq 2$, prove that if $0 < x < \frac{n}{n+1}$, then
$$(1 - 2x^n + x^{n+1})^n < (1 - x^n)^{n+1}.$$

2.5.51. Let f be defined by setting
$$f(x) = x - \frac{x^3}{6} + \frac{x^4}{24}\sin\frac{1}{x} \quad \text{for} \quad x > 0.$$
Prove that if y and z are positive and such that $y + z < 1$, then $f(y + z) < f(y) + f(z)$.

2.5.52. Prove the inequality
$$\sum_{k=0}^{n}(k - nx)^2 \binom{n}{k} x^k (1-x)^{n-k} \leq \frac{n}{4}.$$

2.5.53. Assume that f is in $C^2([a,b])$, $f(a)f(b) < 0$ and f' and f'' do not change their signs on $[a,b]$. Prove that the recursive sequence defined by
$$x_{n+1} = x_n - \frac{f(x_n)}{f'(x_n)}, \quad n = 0, 1, 2, \ldots,$$
where we put $x_0 = b$ if f' and f'' have the same sign, and $x_0 = a$ in the other case, converges to the unique root of the equation $f(x) = 0$ in (a, b). (This is the so-called *Newton's method* of approximating a root of the equation $f(x) = 0$.)

2.5.54. Under the assumptions of the foregoing problem, prove that if $M = \max\{|f''(x)| : x \in [a,b]\}$ and $m = \min\{|f'(x)| : x \in [a,b]\}$, then
$$|x_{n+1} - \xi| \leq \frac{M}{2m}(x_n - \xi)^2, \quad n = 0, 1, 2, \ldots,$$
where ξ is the unique root of $f(x) = 0$.

2.5.55. Find $\sup\{2^{-x} + 2^{-\frac{1}{x}} : x > 0\}$.

2.5.56. Let f be infinitely differentiable on $[0,1]$, and suppose that for each $x \in [0,1]$ there is an integer $n(x)$ such that $f^{(n(x))}(x) = 0$. Prove that f coincides on $[0,1]$ with some polynomial.

2.5.57. Show by example that in the foregoing problem the assumption that f is infinitely differentiable on $[0,1]$ is essential. Show also that the conclusion of 2.5.56 is not true if $\lim_{n\to\infty} f^{(n)}(x) = 0$ for each $x \in [0,1]$.

2.6. Strong Differentiability and Schwarz Differentiability

Definition 1. A real function defined on an open set $\mathbf{A} \subset \mathbb{R}$ is said to be *strongly differentiable* at $a \in \mathbf{A}$ if

$$\lim_{\substack{(x_1, x_2) \to (a,a) \\ x_1 \neq x_2}} \frac{f(x_1) - f(x_2)}{x_1 - x_2} = f^*(a)$$

exists and is finite. $f^*(a)$ is called the *strong derivative* of f at a.

Definition 2. A real function defined on an open set $\mathbf{A} \subset \mathbb{R}$ is said to be *Schwarz differentiable* at $a \in \mathbf{A}$ if

$$\lim_{h \to 0} \frac{f(a+h) - f(a-h)}{2h} = f^s(a)$$

exists as a finite limit. $f^s(a)$ is called the *Schwarz derivative* or the *symmetric derivative* of f at a.

The *upper* (resp. *lower*) *strong derivative of f at a* is defined by replacing lim by $\overline{\lim}$ (resp. $\underline{\lim}$) in Definition 1 and is denoted by $D^* f(a)$ (resp. $D_* f(a)$). *The upper and the lower Schwarz derivatives are defined analogously.* We denote them by $D^s f(a)$ and $D_s f(a)$, respectively.

2.6.1. Show that if f is strongly differentiable at a, then it is differentiable at a and $f^*(a) = f'(a)$. Show by example that the converse is not true.

2.6.2. Let $f : \mathbf{A} \to \mathbb{R}$ and let $\mathbf{A}^1$, $\mathbf{A}^*$ denote the sets of points at which f is differentiable and strongly differentiable, respectively. Prove that if $a \in \mathbf{A}^*$ is a limit point of $\mathbf{A}^*$, then

$$\lim_{\substack{x \to a \\ x \in \mathbf{A}^*}} f^*(x) = \lim_{\substack{x \to a \\ x \in \mathbf{A}^1}} f'(x) = f^*(a) = f'(a).$$

2.6.3. Prove that each function continuously differentiable at a is strongly differentiable at a.

2.6.4. Does the strong differentiability of f at a imply the continuity of f' at this point?

2.6.5. Suppose $\mathbf{G} \subset \mathbf{A}$ is open. Prove that f is strongly differentiable on $\mathbf{G}$ if and only if the derivative f' is continuous on $\mathbf{G}$.

2.6.6. Prove that if f is differentiable on $\mathbb{R}$, then it is strongly differentiable on a *residual set,* that is, on a set $\mathbb{R} \setminus \mathbf{B}$, where $\mathbf{B}$ is of the first category in $\mathbb{R}$.

2.6.7. Suppose f is continuous on $[a,b]$ and the Schwarz derivative f^s exists on the open interval (a,b). Show that if $f(b) > f(a)$, then there is $c \in (a,b)$ such that $f^s(c) \geq 0$.

2.6.8. Let f be continuous on $[a,b]$ and let $f(a) = f(b) = 0$. Show that if f is Schwarz differentiable on the open interval (a,b), then there are $x_1, x_2 \in (a,b)$ such that $f^s(x_1) \geq 0$ and $f^s(x_2) \leq 0$.

2.6.9. Let f be continuous on $[a,b]$ and Schwarz differentiable on the open interval (a,b). Show that there are $x_1, x_2 \in (a,b)$ such that

$$f^s(x_2) \leq \frac{f(b)-f(a)}{b-a} \leq f^s(x_1).$$

2.6.10. Assume that f is continuous and Schwarz differentiable on (a,b). Show that if the Schwarz derivative f^s is bounded on (a,b), then f satisfies a Lipschitz condition on this interval.

2.6.11. Suppose that f and f^s are continuous on (a,b). Show that f is differentiable and $f'(x) = f^s(x)$ for all $x \in (a,b)$.

2.6.12. Assume that f is continuous and Schwarz differentiable on an open interval $\mathbf{I}$. Prove that if $f^s(x) \geq 0$ for $x \in \mathbf{I}$, then f increases on $\mathbf{I}$.

2.6.13. Assume that f is continuous and Schwarz differentiable on an open interval $\mathbf{I}$. Prove that if $f^s(x) = 0$ for $x \in \mathbf{I}$, then f is constant on $\mathbf{I}$.

2.6.14. Let f be Schwarz differentiable on (a,b) and let $x_0 \in (a,b)$ be a local extremum of f. Must the Schwarz derivative vanish at x_0?

2.6. Strong and Schwarz Differentiability

2.6.15. A function $f : \mathbb{R} \to \mathbb{R}$ is said to have *Baire's property* if there exists a residual set $\mathbf{S} \subset \mathbb{R}$ on which f is continuous. Prove that if f has Baire's property, then there exists a residual set $\mathbf{B}$ such that for $x \in \mathbf{B}$,

$$D_s f(x) = D_* f(x) \quad \text{and} \quad D^s f(x) = D^* f(x).$$

2.6.16. Prove that if f has Baire's property and is Schwarz differentiable on $\mathbb{R}$, then f is strongly differentiable on a residual set.

2.6.17. Let f be Schwarz differentiable on an open interval $\mathbf{I}$ and let $[a,b] \subset \mathbf{I}$. We say that f *is uniformly Schwarz differentiable on* $[a,b]$, if for any $\varepsilon > 0$ there is $\delta > 0$ such that if $|h| < \delta$,

$$\left| \frac{f(x+h) - f(x-h)}{2h} - f^s(x) \right| < \varepsilon$$

whenever $x \in [a,b]$ and $x+h, x-h \in \mathbf{I}$. Assume that f is Schwarz differentiable on $\mathbf{I}$ and $[a,b] \subset \mathbf{I}$. Prove that if there is $x_0 \in (a,b)$ such that $\lim_{h \to 0} |f(x_0 + h)| = +\infty$, and there is x_1 such that f is locally bounded in $[x_1, x_0)$, then f is not uniformly Schwarz differentiable on the interval $[a,b]$.

2.6.18. Assume that f is continuous on an open interval $\mathbf{I}$ containing $[a,b]$. Show that f is uniformly Schwarz differentiable on $[a,b]$ if and only if f^s is continuous on $[a,b]$.

2.6.19. Show by example that in the foregoing problem the assumption of continuity of f is essential.

2.6.20. Prove that a function f locally bounded on an open interval $\mathbf{I}$ is uniformly Schwarz differentiable on every $[a,b] \subset \mathbf{I}$ if and only if f' is continuous on $\mathbf{I}$.

Chapter 3

Sequences and Series of Functions

3.1. Sequences of Functions, Uniform Convergence

We adopt the following definition.

Definition. We say that a sequence of functions $\{f_n\}$ *converges uniformly* on $\mathbf{A}$ to a function f if for every $\varepsilon > 0$ there is an $n_0 \in \mathbb{N}$ such that $n \geq n_0$ implies $|f_n(x) - f(x)| < \varepsilon$ for all $x \in \mathbf{A}$. We denote this symbolically by writing $f_n \underset{\mathbf{A}}{\rightrightarrows} f$.

3.1.1. Prove that a sequence of functions $\{f_n\}$ defined on $\mathbf{A}$ is uniformly convergent on $\mathbf{B} \subset \mathbf{A}$ to $f : \mathbf{B} \to \mathbb{R}$ if and only if the sequence of numbers $\{d_n\}$, where

$$d_n = \sup\{|f_n(x) - f(x)| : x \in \mathbf{B}\}, \quad n \in \mathbb{N},$$

converges to zero.

3.1.2. Assume that $f_n \underset{\mathbf{A}}{\rightrightarrows} f$ and $g_n \underset{\mathbf{A}}{\rightrightarrows} g$. Show that $f_n + g_n \underset{\mathbf{A}}{\rightrightarrows} f + g$. Is it true that $f_n \cdot g_n \underset{\mathbf{A}}{\rightrightarrows} f \cdot g$?

3.1.3. Assume that $f_n \underset{\mathbf{A}}{\rightrightarrows} f$, $g_n \underset{\mathbf{A}}{\rightrightarrows} g$, and there is $M > 0$ such that $|f(x)| < M$ and $|g(x)| < M$ for all $x \in \mathbf{A}$. Show that $f_n \cdot g_n \underset{\mathbf{A}}{\rightrightarrows} f \cdot g$.

3.1.4. Let $\{a_n\}$ be a convergent sequence of real numbers, and let $\{f_n\}$ be a sequence of functions satisfying

$$\sup\{|f_n(x) - f_m(x)| : x \in \mathbf{A}\} \leq |a_n - a_m|, \quad n, m \in \mathbb{N}.$$

Prove that $\{f_n\}$ converges uniformly on $\mathbf{A}$.

3.1.5. Show that the limit function of a uniformly convergent on $\mathbf{A}$ sequence of bounded functions is bounded. Does the assertion hold in the case of pointwise convergence?

3.1.6. Show that the sequence of functions $\{f_n\}$, where

$$f_n(x) = \begin{cases} \frac{x}{n} & \text{if } n \text{ is even,} \\ \frac{1}{n} & \text{if } n \text{ is odd,} \end{cases}$$

is pointwise convergent but not uniformly convergent on $\mathbb{R}$. Find a uniformly convergent subsequence.

3.1.7. Prove the following *Cauchy criterion for uniform convergence.* The sequence of functions $\{f_n\}$, defined on $\mathbf{A}$, converges uniformly on $\mathbf{A}$ if and only if for every $\varepsilon > 0$ there exists an $n_0 \in \mathbb{N}$ such that $m > n_0$ implies $|f_{n+m}(x) - f_m(x)| < \varepsilon$ for all $n \in \mathbb{N}$ and all $x \in \mathbf{A}$.

3.1.8. Study the uniform convergence on $[0, 1]$ of the sequence of functions $\{f_n\}$ defined by setting

(a) $\quad f_n(x) = \dfrac{1}{1 + (nx - 1)^2},$

(b) $\quad f_n(x) = \dfrac{x^2}{x^2 + (nx - 1)^2},$

(c) $\quad f_n(x) = x^n(1 - x),$

3.1. Sequences of Functions, Uniform Convergence

(d) $\quad f_n(x) = nx^n(1-x),$

(e) $\quad f_n(x) = n^3 x^n (1-x)^4,$

(f) $\quad f_n(x) = \dfrac{nx^2}{1+nx},$

(g) $\quad f_n(x) = \dfrac{1}{1+x^n}.$

3.1.9. Study the uniform convergence of $\{f_n\}$ on **A** and **B** when

(a) $\quad f_n(x) = \cos^n x(1 - \cos^n x), \quad \mathbf{A} = [0, \pi/2], \ \mathbf{B} = [\pi/4, \pi/2],$

(b) $\quad f_n(x) = \cos^n x \sin^{2n} x, \quad \mathbf{A} = \mathbb{R}, \ \mathbf{B} = [0, \pi/4].$

3.1.10. Determine whether the sequence $\{f_n\}$ converges uniformly on **A** when

(a) $\quad f_n(x) = \arctan \dfrac{2x}{x^2 + n^3}, \quad \mathbf{A} = \mathbb{R},$

(b) $\quad f_n(x) = n \ln\left(1 + \dfrac{x^2}{n}\right), \quad \mathbf{A} = \mathbb{R},$

(c) $\quad f_n(x) = n \ln \dfrac{1+nx}{nx}, \quad \mathbf{A} = (0, \infty),$

(d) $\quad f_n(x) = \sqrt[2n]{1 + x^{2n}}, \quad \mathbf{A} = \mathbb{R},$

(e) $\quad f_n(x) = \sqrt[n]{2^n + |x|^n}, \quad \mathbf{A} = \mathbb{R},$

(f) $\quad f_n(x) = \sqrt{n+1} \sin^n x \cos x, \quad \mathbf{A} = \mathbb{R},$

(g) $\quad f_n(x) = n(\sqrt[n]{x} - 1), \quad \mathbf{A} = [1, a], \ a > 1.$

3.1.11. For a function f defined on $[a,b]$ set $f_n(x) = \dfrac{[nf(x)]}{n},\ x \in [a,b],\ n \in \mathbb{N}$. Show that $f_n \underset{[a,b]}{\rightrightarrows} f.$

3.1.12. Verify that the sequence $\{f_n\}$, where
$$f_n(x) = n \sin \sqrt{4\pi^2 n^2 + x^2},$$
converges uniformly on $[0, a],\ a > 0$. Does $\{f_n\}$ converge uniformly on $\mathbb{R}$?

3.1.13. Show that the sequence of polynomials $\{P_n\}$ defined inductively by

$$P_0(x) = 0, \quad P_{n+1}(x) = P_n(x) + \frac{1}{2}\left(x - P_n^2(x)\right), \quad n = 0, 1, 2, \ldots,$$

converges uniformly on the interval $[0, 1]$ to $f(x) = \sqrt{x}$.

Deduce that there exists a sequence of polynomials converging uniformly on the interval $[-1, 1]$ to the function $x \mapsto |x|$.

3.1.14. Assume that $f : \mathbb{R} \to \mathbb{R}$ is differentiable and f' is uniformly continuous on $\mathbb{R}$. Verify that

$$n\left(f\left(x + \frac{1}{n}\right) - f(x)\right) \to f'(x)$$

uniformly on $\mathbb{R}$. Show by example that the assumption of uniform continuity of f' is essential.

3.1.15. Let $\{f_n\}$ be a sequence of uniformly continuous functions converging uniformly on $\mathbb{R}$. Prove that the limit function is also uniformly continuous on $\mathbb{R}$.

3.1.16. Prove the following *Dini's Theorem*: Let $\{f_n\}$ be a sequence of continuous functions on a compact set **K** which converges pointwise to a function f that is also continuous on **K**. If $f_{n+1}(x) \leq f_n(x)$ for $x \in \mathbf{K}$ and $n \in \mathbb{N}$, then the sequence $\{f_n\}$ converges to f uniformly on **K**.

Show by example that each of the conditions in Dini's theorem (compactness of **K**, continuity of the limit function, continuity of f_n and monotonicity of the sequence $\{f_n\}$) is essential.

3.1.17. A sequence of functions $\{f_n\}$ defined on a set **A** is said to be *equicontinuous* on **A** if for every $\varepsilon > 0$ there exists a $\delta > 0$ such that $|f_n(x) - f_n(x')| < \varepsilon$ whenever $|x - x'| < \delta$, $x, x' \in \mathbf{A}$, and $n \in \mathbb{N}$. Prove that if $\{f_n\}$ is a uniformly convergent sequence of continuous functions on a compact set **K**, then $\{f_n\}$ is equicontinuous on **K**.

3.1.18. We say that a sequence of functions $\{f_n\}$, defined on a set **A**, *converges continuously* on **A** to the function f if for $x \in \mathbf{A}$ and for every sequence $\{x_n\}$ of elements of **A** converging to x the sequence

3.1. Sequences of Functions, Uniform Convergence

$\{f_n(x_n)\}$ converges to $f(x)$. Prove that if a sequence $\{f_n\}$ converges continuously on $\mathbf{A}$ to f, then for every sequence $\{x_n\}$ of elements of $\mathbf{A}$ converging to $x \in \mathbf{A}$ and for every subsequence $\{f_{n_k}\}$,
$$\lim_{k \to \infty} f_{n_k}(x_k) = f(x).$$

3.1.19. Prove that if $\{f_n\}$ converges continuously on $\mathbf{A}$ to f, then f is continuous on $\mathbf{A}$ (even if the f_n are not themselves continuous).

3.1.20. Prove that if $\{f_n\}$ converges uniformly on $\mathbf{A}$ to the continuous function f, then $\{f_n\}$ converges continuously on $\mathbf{A}$. Does the converse hold?

3.1.21. Let $\{f_n\}$ be a sequence of functions defined on a compact set $\mathbf{K}$. Prove that the following conditions are equivalent.
  (i) The sequence $\{f_n\}$ converges uniformly on $\mathbf{K}$ to $f \in C(\mathbf{K})$.
  (ii) The sequence $\{f_n\}$ converges continuously on $\mathbf{K}$ to f.

3.1.22. Assume that $\{f_n\}$ is a sequence of increasing or decreasing functions on $[a,b]$ converging pointwise to a function continuous on $[a,b]$. Prove that $\{f_n\}$ converges uniformly on $[a,b]$.

3.1.23. Let $\{f_n\}$ be a sequence of functions increasing or decreasing on $\mathbb{R}$ and uniformly bounded on $\mathbb{R}$. Prove that $\{f_n\}$ contains a subsequence pointwise convergent on $\mathbb{R}$.

3.1.24. Show that under the assumptions of the foregoing problem, if the limit function f of a pointwise convergent subsequence $\{f_{n_k}\}$ is continuous, then $\{f_n\}$ converges to f uniformly on each compact subset of $\mathbb{R}$. Must $\{f_n\}$ converge uniformly on $\mathbb{R}$?

3.1.25. Show that the limit function of a sequence of polynomials uniformly convergent on $\mathbb{R}$ is a polynomial.

3.1.26. Assume that $\{P_n\}$ is a sequence of polynomials of the form
$$P_n(x) = a_{n,p} x^p + a_{n,p-1} x^{p-1} + \cdots + a_{n,1} x + a_{n,0}.$$
Prove that the following three conditions are equivalent:
  (i) $\{P_n\}$ converges uniformly on each compact subset of $\mathbb{R}$,

(ii) there are $p+1$ distinct numbers $c_0, c_1, \ldots, c_p$ such that $\{P_n\}$ converges on $\{c_0, c_1, \ldots, c_p\}$,

(iii) the sequence of coefficients $\{a_{n,i}\}$ converges for $i = 0, 1, 2, \ldots, p$.

3.1.27. Prove that if $\{f_n\}$ is pointwise convergent and equicontinuous on a compact set **K**, then $\{f_n\}$ converges uniformly on **K**.

3.1.28. Let $\{f_n\}$ be a sequence of functions continuous on a closed interval $[a,b]$ and differentiable on the open interval (a,b). Assume that the sequence $\{f_n'\}$ is uniformly bounded on (a,b); that is, there is an $M > 0$ such that $|f_n'(x)| \leq M$ for all $n \in \mathbb{N}$ and $x \in (a,b)$. Prove that if $\{f_n\}$ is pointwise convergent on $[a,b]$, then $\{f_n\}$ is uniformly convergent on that interval.

3.1.29. Study the convergence and the uniform convergence of $\{f_n\}$ and $\{f_n'\}$ on **A**, where

(a) $\qquad f_n(x) = \dfrac{\sin nx}{\sqrt{n}}, \quad \mathbf{A} = \mathbb{R}$,

(b) $\qquad f_n(x) = \dfrac{x}{1 + n^2 x^2}, \quad \mathbf{A} = [-1, 1]$.

3.1.30. Assume that $\{f_n\}$ is uniformly convergent on **A** to the function f. Assume moreover that x_0 is a limit point of **A** and, beginning with some value of the index n, $\lim\limits_{x \to x_0} f_n(x)$ exists. Prove that

$$\lim_{n \to \infty} \lim_{x \to x_0} f_n(x) = \lim_{x \to x_0} f(x).$$

Prove also that if $\{f_n\}$ is uniformly convergent on (a, ∞) to f and, beginning with some value of the index n, $\lim\limits_{x \to \infty} f_n(x)$ exists, then

$$\lim_{n \to \infty} \lim_{x \to \infty} f_n(x) = \lim_{x \to \infty} f(x).$$

The above equalities mean that if the limit on one side of the equality exists, then the limit on the other side exists and the two are equal.

3.1.31. Let $\{f_n\}$ be a sequence of functions differentiable on $[a,b]$ and such that $\{f_n(x_0)\}$ converges for some $x_0 \in [a,b]$. Prove that if

3.2. Series of Functions, Uniform Convergence

the sequence $\{f'_n\}$ converges uniformly on $[a,b]$, then $\{f_n\}$ converges uniformly on $[a,b]$ to a function f differentiable on $[a,b]$, and

$$f'(x) = \lim_{n\to\infty} f'_n(x) \quad \text{for} \quad x \in [a,b].$$

3.1.32. For $f : [0,1] \to \mathbb{R}$, let $B_n(f,x)$ be the *Bernstein polynomial* of order n of the function f, defined by

$$B_n(f,x) = \sum_{k=0}^{n} \binom{n}{k} f\left(\frac{k}{n}\right) x^k (1-x)^{n-k}.$$

Prove that if f is continuous on $[0,1]$, then $\{B_n(f)\}$ converges uniformly on $[0,1]$ to f.

3.1.33. Use the result in the foregoing problem to prove the *approximation theorem of Weierstrass*. If $f : [a,b] \to \mathbb{R}$ is continuous on $[a,b]$, then for every $\varepsilon > 0$ there is a polynomial P such that

$$|f(x) - P(x)| < \varepsilon \quad \text{for all } x \in [a,b].$$

3.2. Series of Functions, Uniform Convergence

3.2.1. Find where the following series converge pointwise:

(a) $\sum_{n=1}^{\infty} \frac{1}{1+x^n}, \quad x \neq -1,$

(b) $\sum_{n=1}^{\infty} \frac{x^n}{1+x^n}, \quad x \neq -1,$

(c) $\sum_{n=1}^{\infty} \frac{2^n + x^n}{1+3^n x^n}, \quad x \neq -\frac{1}{3},$

(d) $\sum_{n=1}^{\infty} \frac{x^{n-1}}{(1-x^n)(1-x^{n+1})}, \quad x \neq -1, 1,$

(e) $\displaystyle\sum_{n=1}^{\infty} \frac{x^{2^{n-1}}}{1-x^{2^n}}$, $x \neq -1, 1$,

(f) $\displaystyle\sum_{n=2}^{\infty} \left(\frac{\ln n}{n}\right)^x$,

(g) $\displaystyle\sum_{n=1}^{\infty} x^{\ln n}$, $x > 0$,

(h) $\displaystyle\sum_{n=0}^{\infty} \sin^2(2\pi\sqrt{n^2+x^2})$.

3.2.2. Study the uniform convergence of the following series on the given set **A**:

(a) $\displaystyle\sum_{n=1}^{\infty} \left(\frac{\pi}{2} - \arctan(n^2(1+x^2))\right)$, $\mathbf{A} = \mathbb{R}$,

(b) $\displaystyle\sum_{n=1}^{\infty} \frac{\ln(1+nx)}{nx^n}$, $\mathbf{A} = [2, \infty)$,

(c) $\displaystyle\sum_{n=1}^{\infty} n^2 x^2 e^{-n^2 |x|}$, $\mathbf{A} = \mathbb{R}$,

(d) $\displaystyle\sum_{n=1}^{\infty} x^2(1-x^2)^{n-1}$, $\mathbf{A} = [-1, 1]$,

(e) $\displaystyle\sum_{n=1}^{\infty} \frac{n^2}{\sqrt{n!}}(x^n + x^{-n})$, $\mathbf{A} = \{x \in \mathbb{R} : 1/2 \leq |x| \leq 2\}$,

(f) $\displaystyle\sum_{n=1}^{\infty} 2^n \sin\frac{1}{3^n x}$, $\mathbf{A} = (0, \infty)$,

(g) $\displaystyle\sum_{n=2}^{\infty} \ln\left(1 + \frac{x^2}{n \ln^2 n}\right)$, $\mathbf{A} = (-a, a)$, $a > 0$.

3.2.3. Show that the series $\displaystyle\sum_{n=1}^{\infty} f_n(x)$, where f_n is defined by
$f_n(x) = 0$ if $0 \leq x \leq \frac{1}{2n+1}$ or $\frac{1}{2n-1} \leq x \leq 1$,
$f_n(x) = \frac{1}{n}$ if $x = \frac{1}{2n}$,

3.2. Series of Functions, Uniform Convergence

$f_n(x)$ is defined linearly in the intervals $[1/(2n+1), 1/(2n)]$ and $[1/(2n), 1/(2n-1)]$, is uniformly convergent on $[0,1]$ although the M-test of Weierstrass cannot be applied.

3.2.4. Study the continuity on $[0, \infty)$ of the function f defined by

$$f(x) = \sum_{n=1}^{\infty} \frac{x}{((n-1)x+1)(nx+1)}.$$

3.2.5. Study the continuity of the sum of the following series on the domain of its pointwise convergence:

(a) $\sum_{n=0}^{\infty} \frac{x^n \sin(nx)}{n!}$,   (b) $\sum_{n=0}^{\infty} x^{n^2}$,

(c) $\sum_{n=1}^{\infty} n 2^n x^n$,   (d) $\sum_{n=1}^{\infty} \ln^n(x+1)$.

3.2.6. Determine where the series $\sum_{n=1}^{\infty} |x|^{\sqrt{n}}$ converges pointwise, and study the continuity of the sum.

3.2.7. Show that the series $\sum_{n=1}^{\infty} \frac{x \sin(n^2 x)}{n^2}$ converges pointwise to a continuous function on $\mathbb{R}$.

3.2.8. Suppose that the series $\sum_{n=1}^{\infty} f_n(x)$, $x \in \mathbf{A}$, converges uniformly on $\mathbf{A}$ and that $f : \mathbf{A} \to \mathbb{R}$ is bounded. Prove that the series $\sum_{n=1}^{\infty} f(x) f_n(x)$ converges uniformly on $\mathbf{A}$.

Show by example that boundedness of f is essential. Under what assumption concerning f does the uniform convergence of the series $\sum_{n=1}^{\infty} f(x) f_n(x)$ imply the uniform convergence of $\sum_{n=1}^{\infty} f_n(x)$ on $\mathbf{A}$?

3.2.9. Assume that $\{f_n\}$ is a sequence of functions defined on $\mathbf{A}$ and such that

(1) $f_n(x) \geq 0$ for $x \in \mathbf{A}$ and $n \in \mathbb{N}$,

(2) $f_n(x) \geq f_{n+1}(x)$ for $x \in \mathbf{A}$ and $n \in \mathbb{N}$,
(3) $\sup_{x \in \mathbf{A}} f_n(x) \xrightarrow[n \to \infty]{} 0$.

Prove that $\sum_{n=1}^{\infty} (-1)^{n+1} f_n(x)$ converges uniformly on $\mathbf{A}$.

3.2.10. Prove that the following series converge uniformly on $\mathbb{R}$:

(a) $\sum_{n=1}^{\infty} \dfrac{(-1)^{n+1}}{n+x^2}$,

(b) $\sum_{n=1}^{\infty} \dfrac{(-1)^{n+1}}{\sqrt[3]{n+x^2}+x^2}$,

(c) $\sum_{n=2}^{\infty} \dfrac{(-1)^{n+1}}{\sqrt{n}+\cos x}$.

3.2.11. Show that if $\sum_{n=1}^{\infty} f_n^2(x)$ is pointwise convergent on $\mathbf{A}$, and if
$$\sup_{x \in \mathbf{A}} \left(\sum_{n=1}^{\infty} f_n^2(x) \right) < \infty,$$
and $\sum_{n=1}^{\infty} c_n^2$ converges, then $\sum_{n=1}^{\infty} c_n f_n(x)$ converges uniformly on $\mathbf{A}$.

3.2.12. Determine the domain $\mathbf{A}$ of pointwise convergence and the domain $\mathbf{B}$ of absolute convergence of the series given below. Moreover, study the uniform convergence on the indicated set $\mathbf{C}$.

(a) $\sum_{n=1}^{\infty} \dfrac{1}{n} 2^n (3x-1)^n$, $\mathbf{C} = \left[\dfrac{1}{6}, \dfrac{1}{3} \right]$,

(b) $\sum_{n=1}^{\infty} \dfrac{1}{n} \left(\dfrac{x+1}{x} \right)^n$, $\mathbf{C} = [-2, -1]$.

3.2.13. Assume that the functions $f_n, g_n : \mathbf{A} \to \mathbb{R}$, $n \in \mathbb{N}$, satisfy the following conditions:
(1) the series $\sum_{n=1}^{\infty} |f_{n+1}(x) - f_n(x)|$ is uniformly convergent on $\mathbf{A}$,
(2) $\sup_{x \in \mathbf{A}} |f_n(x)| \xrightarrow[n \to \infty]{} 0$,

3.2. Series of Functions, Uniform Convergence

(3) the sequence $\{G_n(x)\}$, where $G_n(x) = \sum_{k=1}^{n} g_n(x)$, is uniformly bounded on **A**.

Prove that the series $\sum_{n=1}^{\infty} f_n(x)g_n(x)$ converges uniformly on **A**.

Deduce the following *Dirichlet test for uniform convergence*: Assume that $f_n, g_n : \mathbf{A} \to \mathbb{R}$, $n \in \mathbb{N}$, satisfy he following conditions:
  (1′) for each fixed $x \in \mathbf{A}$ the sequence $\{f_n(x)\}$ is monotonic,
  (2′) $\{f_n(x)\}$ converges uniformly to zero on **A**,
  (3′) the sequence of partial sums of $\sum_{n=1}^{\infty} g_n(x)$ is uniformly bounded on **A**.

Then the series $\sum_{n=1}^{\infty} f_n(x)g_n(x)$ converges uniformly on **A**.

3.2.14. Show that the following series converge uniformly on the indicated set **A**:

(a) $\sum_{n=1}^{\infty} (-1)^{n+1} \dfrac{x^n}{n}$, $\mathbf{A} = [0, 1]$,

(b) $\sum_{n=1}^{\infty} \dfrac{\sin(nx)}{n}$, $\mathbf{A} = [\delta, 2\pi - \delta]$, $0 < \delta < \pi$,

(c) $\sum_{n=1}^{\infty} \dfrac{\sin(n^2 x)\sin(nx)}{n + x^2}$, $\mathbf{A} = \mathbb{R}$,

(d) $\sum_{n=1}^{\infty} \dfrac{\sin(nx)\arctan(nx)}{n}$, $\mathbf{A} = [\delta, 2\pi - \delta]$, $0 < \delta < \pi$,

(e) $\sum_{n=1}^{\infty} (-1)^{n+1} \dfrac{1}{n^x}$, $\mathbf{A} = [a, \infty)$, $a > 0$,

(f) $\sum_{n=1}^{\infty} (-1)^{n+1} \dfrac{e^{-nx}}{\sqrt{n + x^2}}$, $\mathbf{A} = [0, \infty)$.

3.2.15. Assume that the functions $f_n, g_n : \mathbf{A} \to \mathbb{R}$, $n \in \mathbb{N}$, satisfy the following conditions:
  (1) the function f_1 is bounded on **A**,

(2) the series $\sum_{n=1}^{\infty} |f_{n+1}(x) - f_n(x)|$ converges pointwise on **A** and
$$\sup_{x \in \mathbf{A}} \left(\sum_{n=1}^{\infty} |f_{n+1}(x) - f_n(x)| \right) < \infty,$$
(3) the series $\sum_{n=1}^{\infty} g_n(x)$ converges uniformly on **A**.

Prove that $\sum_{n=1}^{\infty} f_n(x) g_n(x)$ converges uniformly on **A**.

Deduce the following *Abel test for uniform convergence*: Assume that functions $f_n, g_n : \mathbf{A} \to \mathbb{R}$, $n \in \mathbb{N}$, satisfy the following conditions:
(1') for each fixed $x \in \mathbf{A}$, the sequence $\{f_n(x)\}$ is monotonic,
(2') the sequence $\{f_n\}$ is uniformly bounded on **A**,
(3') $\sum_{n=1}^{\infty} g_n(x)$ converges uniformly on **A**.

Then the series $\sum_{n=1}^{\infty} f_n(x) g_n(x)$ converges uniformly on **A**.

3.2.16. Show that the following series converge uniformly on the indicated set **A**:

(a) $\sum_{n=1}^{\infty} \dfrac{(-1)^{n+1}}{n+x^2} \arctan(nx), \quad \mathbf{A} = \mathbb{R},$

(b) $\sum_{n=2}^{\infty} \dfrac{(-1)^{n+1} \cos \frac{x}{n}}{\sqrt{n} + \cos x}, \quad \mathbf{A} = [-R, R], \ R > 0,$

(c) $\sum_{n=1}^{\infty} \dfrac{(-1)^{[\sqrt{n}]}}{\sqrt{n(n+x)}}, \quad \mathbf{A} = [0, \infty).$

3.2.17. Suppose that f_n, $n \in \mathbb{N}$, are continuous on **A** and the series $\sum_{n=1}^{\infty} f_n(x)$ converges uniformly on **A**. Show that if $x_0 \in \mathbf{A}$ is a limit point of **A**, then
$$\lim_{x \to x_0} \sum_{n=1}^{\infty} f_n(x) = \sum_{n=1}^{\infty} f_n(x_0).$$

3.2. Series of Functions, Uniform Convergence

3.2.18. Verify the following claims:

(a) $\displaystyle\lim_{x\to 1^-} \sum_{n=1}^{\infty} \frac{(-1)^{n+1}}{n} x^n = \ln 2,$

(b) $\displaystyle\lim_{x\to 1} \sum_{n=1}^{\infty} \frac{(-1)^{n+1}}{n^x} = \ln 2,$

(c) $\displaystyle\lim_{x\to 1^-} \sum_{n=1}^{\infty} (x^n - x^{n+1}) = 1,$

(d) $\displaystyle\lim_{x\to 0^+} \sum_{n=1}^{\infty} \frac{1}{2^n n^x} = 1,$

(e) $\displaystyle\lim_{x\to \infty} \sum_{n=1}^{\infty} \frac{x^2}{1+n^2 x^2} = \frac{\pi^2}{6}.$

3.2.19. Suppose that the series $\displaystyle\sum_{n=1}^{\infty} a_n$ converges. Find

$$\lim_{x\to 1^-} \sum_{n=1}^{\infty} a_n x^n.$$

3.2.20. Assume that the functions f_n, $n \in \mathbb{N}$, are continuous on $[0,1]$ and $\displaystyle\sum_{n=1}^{\infty} f_n(x)$ converges uniformly on $[0,1)$. Show that the series $\displaystyle\sum_{n=1}^{\infty} f_n(1)$ is convergent.

3.2.21. Find the domain $\mathbf{A}$ of pointwise convergence of the series $\displaystyle\sum_{n=1}^{\infty} e^{-nx} \cos(nx)$. Does the series converge uniformly on $\mathbf{A}$?

3.2.22. Assume that $f_n : [a,b] \to (0,\infty)$, $n \in \mathbb{N}$, are continuous and $f(x) = \displaystyle\sum_{n=1}^{\infty} f_n(x)$ is continuous on $[a,b]$. Show that the series $\displaystyle\sum_{n=1}^{\infty} f_n(x)$ converges uniformly on that interval.

3.2.23. Suppose that $\sum_{n=1}^{\infty} f_n(x)$ converges absolutely and uniformly on **A**. Must the series $\sum_{n=1}^{\infty} |f_n(x)|$ converge uniformly on **A**?

3.2.24. Assume that f_n, $n \in \mathbb{N}$, are monotonic on $[a,b]$. Show that if $\sum_{n=1}^{\infty} f_n(x)$ absolutely converges at the endpoints of $[a,b]$, then the series $\sum_{n=1}^{\infty} f_n(x)$ converges absolutely and uniformly on the whole $[a,b]$.

3.2.25. Suppose that $\sum_{n=1}^{\infty} \frac{1}{|a_n|}$ converges. Prove that $\sum_{n=1}^{\infty} \frac{1}{x-a_n}$ converges absolutely and uniformly on each bounded set **A** that does not contain a_n, $n \in \mathbb{N}$.

3.2.26. For a sequence of real numbers $\{a_n\}$, show that if the *Dirichlet series* $\sum_{n=1}^{\infty} \frac{a_n}{n^x}$ converges at $x = x_0$, then the series converges uniformly on $[x_0, \infty)$.

3.2.27. Study the uniform convergence on $\mathbb{R}$ of the series
$$\sum_{n=1}^{\infty} x \frac{\sin(n^2 x)}{n^2}.$$

3.2.28. Assume that f_n, $n \in \mathbb{N}$, are differentiable on $[a,b]$. Moreover, assume that $\sum_{n=1}^{\infty} f_n(x)$ converges at some $x_0 \in [a,b]$ and $\sum_{n=1}^{\infty} f'_n(x)$ converges uniformly on $[a,b]$. Show that $\sum_{n=1}^{\infty} f_n(x)$ converges uniformly on $[a,b]$ to a differentiable function, and
$$\left(\sum_{n=1}^{\infty} f_n(x)\right)' = \sum_{n=1}^{\infty} f'_n(x) \quad \text{for} \quad x \in [a,b].$$

3.2.29. Show that $f(x) = \sum_{n=1}^{\infty} \frac{1}{n^2 + x^2}$ is differentiable on $\mathbb{R}$.

3.2. Series of Functions, Uniform Convergence

3.2.30. Show that the function
$$f(x) = \sum_{n=1}^{\infty} \frac{\cos(nx)}{1+n^2}$$
is differentiable on $\left[\frac{\pi}{6}, \frac{11\pi}{6}\right]$.

3.2.31. Let $f(x) = \sum_{n=1}^{\infty} (-1)^{n+1} \ln\left(1 + \frac{x}{n}\right)$ for $x \in [0, \infty)$. Show that f is differentiable on $[0, \infty)$ and calculate $f'(0)$, $f'(1)$, and $\lim_{x \to \infty} f'(x)$.

3.2.32. Let
$$f(x) = \sum_{n=1}^{\infty} (-1)^{n+1} \frac{1}{\sqrt{n}} \arctan \frac{x}{\sqrt{n}}, \quad x \in \mathbb{R}.$$
Prove that f is continuously differentiable on $\mathbb{R}$.

3.2.33. Prove that the function
$$f(x) = \sum_{n=1}^{\infty} \frac{\sin(nx^2)}{1+n^3}, \quad x \in \mathbb{R},$$
is continuously differentiable on $\mathbb{R}$.

3.2.34. Let
$$f(x) = \sum_{n=1}^{\infty} \sqrt{n} (\tan x)^n, \quad x \in \left(-\frac{\pi}{4}, \frac{\pi}{4}\right).$$
Prove that f is continuously differentiable on $\left(-\frac{\pi}{4}, \frac{\pi}{4}\right)$.

3.2.35. Define
$$f(x) = \sum_{n=0}^{\infty} \frac{e^{-nx}}{1+n^2}, \quad x \in [0, \infty).$$
Prove that $f \in C([0, \infty))$ and $f \in C^{\infty}(0, \infty)$ and $f'(0)$ does not exist.

3.2.36. Show that the function
$$f(x) = \sum_{n=1}^{\infty} \frac{|x|}{x^2 + n^2}$$
is continuous on $\mathbb{R}$. Is it differentiable on $\mathbb{R}$?

3.2.37. Prove that the *Riemann ζ-function* defined by
$$\zeta(x) = \sum_{n=1}^{\infty} \frac{1}{n^x}$$
is in $C^{\infty}(1, \infty)$.

3.2.38. Assume that $f \in C^{\infty}([0,1])$ satisfies the following conditions:
(1) $f \not\equiv 0$,
(2) $f^{(n)}(0) = 0$ for $n = 0, 1, 2, \ldots$,
(3) for a sequence of real numbers $\{a_n\}$, the series $\sum_{n=1}^{\infty} a_n f^{(n)}(x)$ converges uniformly on $[0,1]$.

Prove that
$$\lim_{n \to \infty} n! a_n = 0.$$

3.2.39. For $x \in \mathbb{R}$ let $f_n(x)$ denote the distance from x to the nearest rational with the denominator n (the numerator and denominator do not need to be co-prime). Find all $x \in \mathbb{R}$ for which the series $\sum_{n=1}^{\infty} f_n(x)$ converges.

3.2.40. Let $g(x) = |x|$ for $x \in [-1, 1]$ and extend the definition of g to all real x by setting $g(x+2) = g(x)$. Prove that the *Weierstrass function* f defined by
$$f(x) = \sum_{n=0}^{\infty} \left(\frac{3}{4}\right)^n g(4^n x)$$
is continuous on $\mathbb{R}$ and is nowhere differentiable.

3.3. Power Series

3.3.1. Show that, given the power series $\sum_{n=0}^{\infty} a_n(x - x_0)^n$, there is $R \in [0, \infty]$ such that
(1) the power series converges absolutely for $|x - x_0| < R$ and diverges for $|x - x_0| > R$,
(2) R is the supremum of the set of the $r \in [0, \infty)$ for which $\{|a_n| r^n\}$ is a bounded sequence,

3.3. Power Series

(3) $1/R = \varlimsup_{n\to\infty} \sqrt[n]{|a_n|}$ (here $\frac{1}{0} = +\infty$ and $\frac{1}{\infty} = 0$).

R is called the *radius of convergence* of $\sum_{n=0}^{\infty} a_n(x-x_0)^n$.

3.3.2. Determine the domain of convergence of the power series given below:

(a) $\sum_{n=1}^{\infty} n^3 x^n$,

(b) $\sum_{n=1}^{\infty} \frac{2^n}{n!} x^n$,

(c) $\sum_{n=1}^{\infty} \frac{2^n}{n^2} x^n$,

(d) $\sum_{n=1}^{\infty} (2+(-1)^n)^n x^n$,

(e) $\sum_{n=1}^{\infty} \left(\frac{2+(-1)^n}{5+(-1)^{n+1}}\right)^n x^n$, (f) $\sum_{n=1}^{\infty} 2^n x^{n^2}$,

(g) $\sum_{n=1}^{\infty} 2^{n^2} x^{n!}$,

(h) $\sum_{n=1}^{\infty} \left(1+\frac{1}{n}\right)^{(-1)^n n^2} x^n$.

3.3.3. Find the domains of convergence of the following series:

(a) $\sum_{n=1}^{\infty} \frac{(x-1)^{2n}}{2^n n^3}$,

(b) $\sum_{n=1}^{\infty} \frac{n}{n+1} \left(\frac{2x+1}{x}\right)^n$,

(c) $\sum_{n=1}^{\infty} \frac{n 4^n}{3^n} x^n (1-x)^n$,

(d) $\sum_{n=1}^{\infty} \frac{(n!)^2}{(2n)!} (x-1)^n$,

(e) $\sum_{n=1}^{\infty} \sqrt{n}(\tan x)^n$,

(f) $\sum_{n=1}^{\infty} \left(\arctan \frac{1}{x}\right)^{n^2}$.

3.3.4. Show that if the radii of convergence of $\sum_{n=0}^{\infty} a_n x^n$ and $\sum_{n=0}^{\infty} b_n x^n$ are R_1 and R_2, respectively, then

(a) the radius of convergence R of $\sum_{n=0}^{\infty} (a_n+b_n)x^n$ is $\min\{R_1, R_2\}$, if $R_1 \neq R_2$. What can be said about R if $R_1 = R_2$?

(b) the radius of convergence R of $\sum_{n=0}^{\infty} a_n b_n x^n$ satisfies $R \geq R_1 R_2$. Show by example that the inequality may be strict.

3.3.5. Let R_1 and R_2 be the radii of convergence of $\sum_{n=0}^{\infty} a_n x^n$ and $\sum_{n=0}^{\infty} b_n x^n$, respectively. Show that

(a) if $R_1, R_2 \in (0, \infty)$, then the radius of convergence R of the power series
$$\sum_{n=0}^{\infty} \frac{a_n}{b_n} x^n, \quad b_n \neq 0, \; n = 0, 1, 2, \ldots,$$
satisfies $R \leq \frac{R_1}{R_2}$,

(b) the radius of convergence R of the Cauchy product (see, e.g., I, 3.6.1) of the given series satisfies $R \geq \min\{R_1, R_2\}$.

Show, by example, that the inequalities in (a) and (b) can be strict.

3.3.6. Find the radius of convergence R of $\sum_{n=0}^{\infty} a_n x^n$, if

(a) there are α and $L > 0$ such that $\lim\limits_{n \to \infty} |a_n n^\alpha| = L$,

(b) there exist positive α and L such that $\lim\limits_{n \to \infty} |a_n \alpha^n| = L$,

(c) $\lim\limits_{n \to \infty} |a_n n!| = L, \; L \in (0, \infty)$.

3.3.7. Suppose that the radius of convergence of $\sum_{n=0}^{\infty} a_n x^n$ is R and $0 < R < \infty$. Evaluate the radius of convergence of:

(a) $\sum_{n=0}^{\infty} 2^n a_n x^n$, (b) $\sum_{n=0}^{\infty} n^n a_n x^n$,

(c) $\sum_{n=0}^{\infty} \frac{n^n}{n!} a_n x^n$, (d) $\sum_{n=0}^{\infty} a_n^2 x^n$.

3.3.8. Find all power series uniformly convergent on $\mathbb{R}$.

3.3.9. Find the radius of convergence R of the power series
$$\sum_{n=0}^{\infty} \frac{x^{2n+1}}{(2n+1)!!}$$
and show that its sum f satisfies the equation $f'(x) = 1 + x f(x)$, $x \in (-R, R)$.

3.3. Power Series

3.3.10. Show that the series $\sum_{n=0}^{\infty} \frac{x^{3n}}{(3n)!}$ converges on $\mathbb{R}$ and the sum f satisfies the equation $f''(x) + f'(x) + f(x) = e^x$, $x \in \mathbb{R}$.

3.3.11. Let $R > 0$ be the radius of convergence of the power series $\sum_{n=0}^{\infty} a_n x^n$ and let $S_n(x) = \sum_{k=0}^{n} a_k x^k$, $n = 0, 1, 2, \ldots$. Show that if f is the sum of the series and if $x_0 \in (-R, R)$ is such that $S_n(x_0) < f(x_0)$, $n = 0, 1, 2, \ldots$, then $f'(x_0) \neq 0$.

3.3.12. Let $\{S_n\}$ be the sequence of partial sums of $\sum_{n=0}^{\infty} a_n$ and let $T_n = \frac{S_0 + S_1 + \cdots + S_n}{n+1}$. Prove that if $\{T_n\}$ is bounded, then the power series $\sum_{n=0}^{\infty} a_n x^n$, $\sum_{n=0}^{\infty} S_n x^n$ and $\sum_{n=0}^{\infty} (n+1) T_n x^n$ converge for $|x| < 1$, and

$$\sum_{n=0}^{\infty} a_n x^n = (1-x) \sum_{n=0}^{\infty} S_n x^n = (1-x)^2 \sum_{n=0}^{\infty} (n+1) T_n x^n.$$

3.3.13. Let $f(x) = \sum_{n=0}^{\infty} x^{2^n}$, $|x| < 1$. Prove that there is an $M > 0$ such that

$$|f'(x)| < \frac{M}{1 - |x|}, \quad |x| < 1.$$

3.3.14. Prove the following *Abel theorem*. If $\sum_{n=0}^{\infty} a_n$ converges to L, then

(1) $\sum_{n=0}^{\infty} a_n x^n$ converges uniformly on $[0, 1]$,

(2) $\lim_{x \to 1^-} \sum_{n=0}^{\infty} a_n x^n = L$.

3.3.15. Prove the following generalization of the Abel theorem. If $\{S_n\}$ is the sequence of partial sums of $\sum_{n=0}^{\infty} a_n$ and the radius of convergence of the power series $f(x) = \sum_{n=0}^{\infty} a_n x^n$ is 1, then

$$\varliminf_{n \to \infty} S_n \leq \varliminf_{x \to 1^-} f(x) \leq \varlimsup_{x \to 1^-} f(x) \leq \varlimsup_{n \to \infty} S_n.$$

3.3.16. Prove the *Tauber theorem*. Assume that the radius of convergence of the power series $f(x) = \sum_{n=0}^{\infty} a_n x^n$ is 1. If $\lim_{n\to\infty} na_n = 0$ and $\lim_{x\to 1^-} f(x) = L$, $L \in \mathbb{R}$, then $\sum_{n=0}^{\infty} a_n$ converges to L.

3.3.17. Show by example that in the Tauber theorem the assumption $\lim_{n\to\infty} na_n = 0$ is essential.

3.3.18. Suppose that $\{a_n\}$ is a positive sequence and the radius of convergence of $f(x) = \sum_{n=1}^{\infty} a_n x^n$ is 1. Prove that $\lim_{x\to 1^-} f(x)$ exists and is finite if and only if $\sum_{n=1}^{\infty} a_n$ converges.

3.3.19. Prove the following generalization of the Tauber theorem. Assume that the radius of convergence of $f(x) = \sum_{n=0}^{\infty} a_n x^n$ is 1. If

$$\lim_{n\to\infty} \frac{a_1 + 2a_2 + \cdots + na_n}{n} = 0 \quad \text{and} \quad \lim_{x\to 1^-} f(x) = L, \ L \in \mathbb{R},$$

then the series $\sum_{n=0}^{\infty} a_n$ converges to L.

3.3.20. Assume that the radius of convergence of $f(x) = \sum_{n=0}^{\infty} a_n x^n$ is 1. Prove that if $\sum_{n=1}^{\infty} na_n^2$ converges and $\lim_{x\to 1^-} f(x) = L$, $L \in \mathbb{R}$, then $\sum_{n=0}^{\infty} a_n$ converges and has sum L.

3.3.21. Assume that $a_n, b_n > 0$, $n = 0, 1, 2, \ldots$, and the power series $f(x) = \sum_{n=0}^{\infty} a_n x^n$ and $g(x) = \sum_{n=0}^{\infty} b_n x^n$ both have the same radius of convergence, equal to 1. Moreover, assume that $\lim_{x\to 1^-} f(x) = \lim_{x\to 1^-} g(x) = +\infty$. Prove that if $\lim_{n\to\infty} \frac{a_n}{b_n} = A \in [0, \infty)$, then also $\lim_{x\to 1^-} \frac{f(x)}{g(x)} = A$.

3.3.22. Prove the following generalization of the result in the foregoing problem. Assume that the power series $f(x) = \sum_{n=0}^{\infty} a_n x^n$ and

3.3. Power Series

$g(x) = \sum_{n=0}^{\infty} b_n x^n$ both have the same radius of convergence, equal to 1. Moreover, assume that $S_n = a_0 + a_1 + \cdots + a_n$ and $T_n = b_0 + b_1 + \cdots + b_n$, $n \in \mathbb{N}$, are positive and both series $\sum_{n=0}^{\infty} S_n$ and $\sum_{n=0}^{\infty} T_n$ diverge. If $\lim_{n \to \infty} \frac{S_n}{T_n} = A \in [0, \infty)$, then $\lim_{x \to 1^-} \frac{f(x)}{g(x)} = A$.

3.3.23. Show by example that the converse of the above theorem fails to hold. Namely, the fact that $\lim_{x \to 1^-} \frac{f(x)}{g(x)} = A$ does not imply the existence of $\lim_{n \to \infty} \frac{S_n}{T_n}$.

3.3.24. Let the radius of convergence of the power series $f(x) = \sum_{n=0}^{\infty} a_n x^n$ with nonnegative coefficients be 1 and let $\lim_{x \to 1^-} f(x)(1-x) = A \in (0, \infty)$. Prove that there are positive A_1 and A_2 such that

$$A_1 n \leq S_n = a_0 + a_1 + \cdots + a_n \leq A_2 n, \quad n \in \mathbb{N}.$$

3.3.25. Prove the following *theorem of Hardy and Littlewood*. Let the radius of convergence of the power series $f(x) = \sum_{n=0}^{\infty} a_n x^n$ with nonnegative coefficients be 1 and let $\lim_{x \to 1^-} f(x)(1-x) = A \in (0, \infty)$. Then

$$\lim_{n \to \infty} \frac{S_n}{n} = A,$$

where $S_n = a_0 + a_1 + \cdots + a_n$.

3.3.26. Let the radius of convergence of $f(x) = \sum_{n=0}^{\infty} a_n x^n$ be equal to 1. Prove that if the sequence $\{na_n\}$ is bounded and $\lim_{x \to 1^-} f(x) = L$, $L \in \mathbb{R}$, then the series $\sum_{n=0}^{\infty} a_n$ converges and has sum L.

3.3.27. Let the radius of convergence of $f(x) = \sum_{n=0}^{\infty} a_n x^n$ be equal to 1. Prove that if $\lim_{x \to 1^-} (1-x) f(x)$ exists and is different from zero, then $\{a_n\}$ cannot converge to zero.

3.4. Taylor Series

3.4.1. Assume that f is in $C^\infty([a,b])$. Show that if all derivatives $f^{(n)}$ are uniformly bounded on $[a,b]$, then for x and x_0 in $[a,b]$,

$$f(x) = \sum_{n=0}^{\infty} \frac{f^{(n)}(x_0)}{n!}(x-x_0)^n.$$

3.4.2. Define

$$f(x) = \begin{cases} e^{-\frac{1}{x^2}} & \text{if } x \neq 0, \\ 0 & \text{if } x = 0. \end{cases}$$

Does the equality

$$f(x) = \sum_{n=0}^{\infty} \frac{f^{(n)}(0)}{n!} x^n$$

hold for $x \neq 0$?

3.4.3. Define $f(x) = \sum_{n=0}^{\infty} \frac{\cos(n^2 x)}{e^n}$, $x \in \mathbb{R}$. Show that f is in $C^\infty(\mathbb{R})$ and the equality

$$f(x) = \sum_{n=0}^{\infty} \frac{f^{(n)}(0)}{n!} x^n$$

holds only at $x = 0$.

3.4.4. Show that if $\alpha \in \mathbb{R} \setminus \mathbb{N}$ and $|x| < 1$, then

$$(1+x)^\alpha = 1 + \sum_{n=1}^{\infty} \frac{\alpha(\alpha-1)\cdots(\alpha-n+1)}{n!} x^n.$$

This is called *Newton's binomial formula*.

3.4.5. Show that for $|x| \leq 1$,

$$|x| = 1 - \frac{1}{2}(1-x^2) - \sum_{n=2}^{\infty} \frac{(2n-3)!!}{(2n)!!}(1-x^2)^n.$$

3.4. Taylor Series

3.4.6. Show that if the power series $\sum_{n=1}^{\infty} a_n x^n$ has positive radius of convergence R and $f(x) = \sum_{n=1}^{\infty} a_n x^n$ for $x \in (-R, R)$, then f is in $C^{\infty}(-R, R)$, and
$$a_n = \frac{f^{(n)}(0)}{n!}, \quad n = 0, 1, 2, \ldots.$$

3.4.7. Prove that if x_0 is in the interval of convergence $(-R, R)$, $R > 0$, of the power series $f(x) = \sum_{n=0}^{\infty} a_n x^n$, then
$$f(x) = \sum_{n=0}^{\infty} \frac{f^{(n)}(x_0)}{n!}(x - x_0)^n \quad \text{for} \quad |x - x_0| < R - |x_0|.$$

3.4.8. Assume that $\sum_{n=0}^{\infty} a_n x^n$ and $\sum_{n=0}^{\infty} b_n x^n$ converge in the same interval $(-R, R)$. Let $\mathbf{A}$ be the set of all $x \in (-R, R)$ for which
$$\sum_{n=0}^{\infty} a_n x^n = \sum_{n=0}^{\infty} b_n x^n.$$
Prove that if $\mathbf{A}$ has a limit point in $(-R, R)$, then $a_n = b_n$ for $n = 0, 1, 2, \ldots$.

3.4.9. Find the Taylor series of f about zero when

(a) $f(x) = \sin x^3, \quad x \in \mathbb{R}$,
(b) $f(x) = \sin^3 x, \quad x \in \mathbb{R}$,
(c) $f(x) = \sin x \cos 3x, \quad x \in \mathbb{R}$,
(d) $f(x) = \sin^6 x + \cos^6 x, \quad x \in \mathbb{R}$,
(e) $f(x) = \frac{1}{2} \ln \frac{1+x}{1-x}, \quad x \in (-1, 1)$,
(f) $f(x) = \ln(1 + x + x^2), \quad x \in (-1, 1)$,
(g) $f(x) = \frac{1}{1 - 5x + 6x^2}, \quad x \in (-1/3, 1/3)$,
(h) $f(x) = \frac{e^x}{1 - x}, \quad x \in (-1, 1)$.

3.4.10. Find the Taylor series of the following functions f about the point $x = 1$:

(a) $f(x) = (x+1)e^x$, $x \in \mathbb{R}$,

(b) $f(x) = \dfrac{e^x}{x}$, $x \neq 0$,

(c) $f(x) = \dfrac{\cos x}{x}$, $x \neq 0$,

(d) $f(x) = \dfrac{\ln x}{x}$, $x > 0$.

3.4.11. For $|x| < 1$, establish the following equalities:

(a) $$\arcsin x = x + \sum_{n=1}^{\infty} \frac{(2n-1)!!}{(2n)!!(2n+1)} x^{2n+1},$$

(b) $$\arctan x = \sum_{n=0}^{\infty} (-1)^n \frac{1}{2n+1} x^{2n+1}.$$

Using the above identities, show that

$$\frac{\pi}{6} = \frac{1}{2} + \sum_{n=1}^{\infty} \frac{(2n-1)!!}{2^{2n+1}(2n)!!(2n+1)} \quad \text{and} \quad \frac{\pi}{4} = \sum_{n=0}^{\infty} (-1)^n \frac{1}{2n+1}.$$

3.4.12. Find the Taylor series for f about zero when

(a) $f(x) = x \arctan x - \dfrac{1}{2} \ln(1 + x^2)$, $x \in (-1, 1)$,

(b) $f(x) = x \arcsin x + \sqrt{1 - x^2}$, $x \in (-1, 1)$.

3.4.13. Find the sum of the series

(a) $\displaystyle\sum_{n=1}^{\infty} \frac{(-1)^{n+1}}{n(n+1)}$,

(b) $\displaystyle\sum_{n=0}^{\infty} \frac{(-1)^n n}{(2n+1)!}$,

(c) $\displaystyle\sum_{n=2}^{\infty} \frac{(-1)^n}{n^2 + n - 2}$,

(d) $\displaystyle\sum_{n=1}^{\infty} \frac{(-1)^{n-1}}{n(2n-1)}$,

(e) $\displaystyle\sum_{n=1}^{\infty} \frac{(-1)^n (2n-1)!!}{(2n)!!}$,

(f) $\displaystyle\sum_{n=0}^{\infty} \frac{3^n(n+1)}{n!}$.

3.4. Taylor Series

3.4.14. Find the sum of the series $\sum_{n=1}^{\infty} \frac{((n-1)!)^2}{(2n)!}(2x)^{2n}$ for $|x| \leq 1$.

3.4.15. Using the Taylor formula with integral remainder (see, e.g., 2.3.4) prove the following *theorem of Bernstein*. Assume that f is infinitely differentiable on an open interval **I** and all its derivatives $f^{(n)}$ are nonnegative in **I**. Then f is *real analytic* on **I**; that is, for each $x_0 \in \mathbf{I}$ there is a neighborhood $(x_0 - r, x_0 + r) \subset \mathbf{I}$ such that

$$f(x) = \sum_{n=0}^{\infty} \frac{f^{(n)}(x_0)}{n!}(x-x_0)^n \quad \text{for} \quad |x-x_0| < r.$$

3.4.16. Suppose that f is infinitely differentiable on an open interval **I**. Prove that if for every $x_0 \in \mathbf{I}$ there are an open interval $\mathbf{J} \subset \mathbf{I}$ with $x_0 \in \mathbf{J}$, and constants $C > 0$ and $\rho > 0$ such that

$$|f^{(n)}(x)| \leq C \frac{n!}{\rho^n} \quad \text{for} \quad x \in \mathbf{J},$$

then

$$f(x) = \sum_{n=0}^{\infty} \frac{f^{(n)}(x_0)}{n!}(x-x_0)^n \quad \text{for} \quad x \in (x_0 - \rho, x_0 + \rho) \cap \mathbf{J}.$$

3.4.17. Assume that f is real analytic on an open interval **I**. Show that for every $x_0 \in \mathbf{I}$ there are an open interval **J**, with $x_0 \in \mathbf{J} \subset \mathbf{I}$, and positive constants A, B such that

$$|f^{(n)}(x)| \leq A \frac{n!}{B^n} \quad \text{for} \quad x \in \mathbf{J}.$$

3.4.18. Apply the formula of Faà di Bruno (see, e.g., 2.1.38) to show that for each positive integer n and for $A > 0$,

$$\sum \frac{k!}{k_1! k_2! \cdots k_n!} A^k = A(1+A)^{n-1},$$

where $k = k_1 + k_2 + \cdots + k_n$ and the sum is taken over all $k_1, k_2, \ldots, k_n$ such that $k_1 + 2k_2 + \cdots + nk_n = n$.

3.4.19. Let **I**, **J** be open intervals and let $f : \mathbf{I} \to \mathbf{J}$ and $g : \mathbf{J} \to \mathbb{R}$ be real analytic on **I** and **J**, respectively. Show that $h = g \circ f$ is real analytic on **I**.

3.4.20. Let f be in C^∞ on an open interval **I** and $(-1)^n f^{(n)}(x) \geq 0$ for $x \in \mathbf{I}$ and $n \in \mathbb{N}$. Prove that f is real analytic on **I**.

3.4.21. Apply the formula of Faà di Bruno to prove that, for each positive integer n,

$$\sum \frac{(-1)^k k!}{k_1! k_2! \cdots k_n!} \left(\frac{\frac{1}{2}}{1}\right)^{k_1} \left(\frac{\frac{1}{2}}{2}\right)^{k_2} \cdots \left(\frac{\frac{1}{2}}{n}\right)^{k_n} = 2(n+1)\left(\frac{\frac{1}{2}}{n+1}\right),$$

where $k = k_1 + k_2 + \cdots + k_n$ and the sum is taken over all $k_1, k_2, \ldots, k_n$ such that $k_1 + 2k_2 + \cdots + nk_n = n$, and $\binom{\alpha}{k} = \frac{\alpha(\alpha-1)\cdots(\alpha-k+1)}{k!}$.

3.4.22. Assume that f is real analytic on an open interval **I**. Prove that if $f'(x_0) \neq 0$ for an $x_0 \in \mathbf{I}$, then there are an open interval **J** containing x_0 and a real analytic function g defined on an open interval **K** containing $f(x_0)$ and such that $(g \circ f)(x) = x$ for $x \in \mathbf{J}$ and $(f \circ g)(x) = x$ for $x \in \mathbf{K}$.

3.4.23. Prove that if f is differentiable on $(0, \infty)$ and such that $f^{-1} = f'$, then f is real analytic on $(0, \infty)$.

3.4.24. Prove that there is only one function f differentiable on $(0, \infty)$ and such that $f^{-1} = f'$.

3.4.25. Prove that the only function satisfying the assumptions of the foregoing problem is $f(x) = ax^c$, where $c = \frac{1+\sqrt{5}}{2}$ and $a = c^{1-c}$.

3.4.26. Apply the result in 2.3.10 to show that for $x \in (0, 2)$,

$$\ln(1+x) = 2 \sum_{n=0}^{\infty} \frac{1}{2n+1} \left(\frac{x}{2+x}\right)^{2n+1}.$$

3.4.27. Let $M_p(x, y)$ and $L(x, y)$ denote the power mean and the logarithmic mean of positive numbers x and y (see, e.g., 2.5.41 and 2.5.42 for the definitions). Show that if $p \geq \frac{1}{3}$, then

$$L(x, y) < M_p(x, y) \quad \text{for} \quad x, y > 0, \ x \neq y.$$

3.4. Taylor Series

3.4.28. Show that, in the notation of 3.4.27, if $p < \frac{1}{3}$, then there exist positive numbers x and y for which $L(x,y) > M_p(x,y)$.

3.4.29. Show that, in the notation of 3.4.27, if $p \leq 0$, then
$$L(x,y) > M_p(x,y) \quad \text{for} \quad x,y > 0, \ x \neq y.$$

3.4.30. Show that, in the notation of 3.4.27, if $p > 0$, then there exist positive numbers x and y for which $L(x,y) < M_p(x,y)$.

Solutions

Chapter 1

Limits and Continuity

1.1. The Limit of a Function

1.1.1.
(a) Since $\left|x\cos\frac{1}{x}\right| \leq |x|$, the limit is 0.
(b) For $x > 0$, $1-x < x\left[\frac{1}{x}\right] \leq 1$ and for $x < 0$, $1 \leq x\left[\frac{1}{x}\right] < 1-x$. Therefore $\lim\limits_{x \to 0} x\left[\frac{1}{x}\right] = 1$.
(c) As in (b), one can show that the limit is equal to $\frac{b}{a}$.
(d) The limit does not exist because the one-sided limits are different.
(e) The limit is equal to $\frac{1}{2}$ (compare with the solution of I, 3.2.1).
(f) We have

$$\lim_{x \to 0} \frac{\cos\left(\frac{\pi}{2}\cos x\right)}{\sin(\sin x)} = \lim_{x \to 0} \frac{\sin\left(\frac{\pi}{2}(1+\cos x)\right)}{\sin(\sin x)}$$

$$= \lim_{x \to 0} \frac{\sin\left(\pi \cos^2 \frac{x}{2}\right)}{\sin(\sin x)}$$

$$= \lim_{x \to 0} \frac{\sin\left(\pi \sin^2 \frac{x}{2}\right)}{\sin(\sin x)}$$

$$= \lim_{x \to 0} \pi \frac{\sin\frac{x}{2}}{2\cos\frac{x}{2}} \cdot \frac{2\sin\frac{x}{2}\cos\frac{x}{2}}{\sin\left(2\sin\frac{x}{2}\cos\frac{x}{2}\right)} \cdot \frac{\sin\left(\pi \sin^2\frac{x}{2}\right)}{\pi \sin^2 \frac{x}{2}}$$

$$= 0.$$

111

1.1.2.

(a) Suppose that $\lim_{x \to 0} f(x) = l$. Then, given $\varepsilon > 0$, there is $0 < \delta < \frac{\pi}{2}$ such that

(1) $\qquad |f(y) - l| < \varepsilon \quad \text{if} \quad 0 < |y| < \delta.$

Note also that if $0 < |x| < \delta$, then $0 < |y| = |\sin x| < |x| < \delta$. Thus, by (1), $|f(\sin x) - l| < \varepsilon$, which gives $\lim_{x \to 0} f(\sin x) = l$.
Now suppose that $\lim_{x \to 0} f(\sin x) = l$. Given $\varepsilon > 0$, there exists $0 < \delta < \frac{\pi}{2}$ such that

(2) $\qquad |f(\sin x) - l| < \varepsilon \quad \text{if} \quad 0 < |x| < \delta.$

Now if $0 < |y| < \sin \delta$, then $0 < |x| = |\arcsin y| < \delta$ and by (2) we get $|f(y) - l| = |f(\sin x) - l| < \varepsilon$. This means that $\lim_{x \to 0} f(x) = l$.

(b) The implication follows immediately from the definition of a limit. To show that the other implication does not hold, observe that, e.g., $\lim_{x \to 0}[|x|] = 0$ but $\lim_{x \to 0}[x]$ does not exist.

1.1.3. Clearly, $f(x) + \frac{1}{f(x)} \geq 2$. Hence, by assumption, given $\varepsilon > 0$, there is $\delta > 0$ such that

$$0 \leq f(x) + \frac{1}{f(x)} - 2 < \varepsilon \quad \text{for} \quad 0 < |x| < \delta.$$

This condition can be rewritten equivalently as

(1) $\qquad 0 \leq (f(x) - 1) + \left(\frac{1}{f(x)} - 1\right) < \varepsilon$

or

(2) $\qquad 0 \leq (f(x) - 1)\left(1 - \frac{1}{f(x)}\right) < \varepsilon.$

Squaring both sides of (1) and using (2), we get

$$(f(x) - 1)^2 + \left(\frac{1}{f(x)} - 1\right)^2 \leq \varepsilon^2 + 2\varepsilon.$$

Consequently, $(f(x) - 1)^2 \leq \varepsilon^2 + 2\varepsilon$.

1.1. The Limit of a Function.

1.1.4. Suppose that $\lim_{x \to a} f(x)$ exists and is equal to l. Then on account of our condition, we get $l + \frac{1}{|l|} = 0$, which implies $l = -1$. Now we show that $\lim_{x \to a} f(x) = -1$. To this end observe that there is $\delta > 0$ such that $f(x) < 0$ for $x \in (a-\delta, a+\delta) \setminus \{a\}$. Indeed, if in every deleted neighborhood of a there were an x_0 such that $f(x_0) > 0$, then we would get $f(x_0) + \frac{1}{f(x_0)} \geq 2$, contrary to our assumption. Since $f(x) < 0$, the following inequality holds:

$$|f(x) + 1| \leq \left| f(x) + \frac{1}{|f(x)|} \right|.$$

1.1.5. There is $M \geq 0$ such that $|f(x)| \leq M$ for $x \in [0,1]$. Since $f(ax) = bf(x)$ for $x \in [0, \frac{1}{a}]$, $f(a^2 x) = b^2 f(x)$ for $x \in [0, \frac{1}{a^2}]$. One can show by induction that

$$f(a^n x) = b^n f(x) \quad \text{for} \quad x \in \left[0, \frac{1}{a^n}\right], \quad n \in \mathbb{N}.$$

Therefore

$$(*) \qquad |f(x)| \leq M \frac{1}{b^n} \quad \text{for} \quad x \in \left[0, \frac{1}{a^n}\right], \quad n \in \mathbb{N}.$$

On the other hand, the equality $f(ax) = bf(x)$ implies $f(0) = 0$, which together with $(*)$ gives the desired result.

1.1.6.

(a) We have

$$x^2 \left(1 + 2 + 3 + \cdots + \left[\frac{1}{|x|} \right] \right) = x^2 \frac{1 + \left[\frac{1}{|x|} \right]}{2} \left[\frac{1}{|x|} \right].$$

It follows from the definition of the greatest integer function that if $0 < |x| < 1$, then

$$\frac{1}{2}(1 - |x|) < x^2 \left(1 + 2 + 3 + \cdots + \left[\frac{1}{|x|} \right] \right) \leq \frac{1}{2}(1 + |x|).$$

Consequently, the limit is $\frac{1}{2}$.

(b) As in (a), one can show that the limit is $\frac{k(k+1)}{2}$.

1.1.7. Since P is a polynomial with positive coefficients, for $x > 1$ we get
$$\frac{P(x)-1}{P(x)} \leq \frac{[P(x)]}{P([x])} \leq \frac{P(x)}{P(x-1)}.$$
Thus $\lim\limits_{x\to\infty} \frac{[P(x)]}{P([x])} = 1$.

1.1.8. Consider $f : \mathbb{R} \to \mathbb{R}$ defined by setting
$$f(x) = \begin{cases} (-1)^n & \text{if } x = \frac{1}{2^n},\ n = 0,1,2,3,\ldots, \\ 0 & \text{otherwise.} \end{cases}$$
Now if $f(x) \geq \varphi(x)$, then
$$\varphi(x) \leq f(x) = (f(x)+f(2x)) - f(2x) \leq (f(x)+f(2x)) - \varphi(2x),$$
which gives $\lim\limits_{x\to 0} f(x) = 0$.

1.1.9.

(a) Consider, for example, $f : \mathbb{R} \to \mathbb{R}$ defined as follows:
$$f(x) = \begin{cases} (-1)^n & \text{if } x = \frac{1}{2^{2^n}},\ n = 1,2,3,\ldots, \\ 0 & \text{otherwise.} \end{cases}$$

(b) If $f(x) \geq |x|^\alpha$ and $f(x)f(2x) \leq |x|$, then
$$|x|^\alpha \leq f(x) \leq \frac{|x|}{f(2x)} \leq \frac{|x|}{|2x|^\alpha}.$$
Since $\frac{1}{2} < \alpha < 1$, we see that $\lim\limits_{x\to 0} f(x) = 0$.

1.1.10. We have $\frac{g(a)}{a^\alpha} = \lim\limits_{x\to\infty} \frac{f(ax)}{a^\alpha x^\alpha} = \lim\limits_{t\to\infty} \frac{f(t)}{t^\alpha} = g(1)$.

1.1.11. It follows from $\lim\limits_{x\to\infty} \frac{f(2x)}{f(x)} = 1$ that for any $n \in \mathbb{N}$,
$$\lim_{x\to\infty} \frac{f(2^n x)}{f(x)} = \lim_{x\to\infty} \left(\frac{f(2^n x)}{f(2^{n-1}x)} \frac{f(2^{n-1}x)}{f(2^{n-2}x)} \cdots \frac{f(2x)}{f(x)} \right) = 1.$$
Assume that f is increasing and $c \geq 1$. Clearly, there is $n \in \mathbb{N} \cup \{0\}$ such that $2^n \leq c < 2^{n+1}$. Hence by the monotonicity of f we obtain $f(2^n x) \leq f(cx) \leq f(2^{n+1}x)$, which gives
$$\lim_{x\to\infty} \frac{f(cx)}{f(x)} = 1 \quad \text{for} \quad c \geq 1.$$

1.1. The Limit of a Function.

In view of the above, if $0 < c < 1$, then

$$\lim_{x\to\infty} \frac{f(cx)}{f(x)} = \lim_{t\to\infty} \frac{f(t)}{f(\frac{1}{c}t)} = 1.$$

1.1.12.

(a) Note first that if $a > 1$, then $\lim_{x\to\infty} a^x = +\infty$. Indeed, given $M > 0$, $a^x > M$ if and only if $x > \frac{\ln M}{\ln a}$. To see that $\lim_{n\to\infty} \frac{a^n}{n+1} = +\infty$, write $\frac{a^n}{n+1} = \frac{(1+(a-1))^n}{n+1}$ and observe that by the binomial formula $(1+(a-1))^n > \frac{n(n-1)}{2}(a-1)^2$. Thus, given N, there is n_0 such that $\frac{a^n}{n+1} > N$ whenever $n > n_0$. Now for $x > n_0+1$, set $n = [x]$. Then $\frac{a^x}{x} > \frac{a^n}{n+1} > N$, which gives $\lim_{x\to\infty} \frac{a^x}{x} = +\infty$.

(b) Clearly, $\lim_{x\to\infty} \frac{a^x}{x^\alpha} = +\infty$ for $\alpha \leq 0$. In the case when $\alpha > 0$, we get

$$\frac{a^x}{x^\alpha} = \left(\frac{a^{\frac{x}{\alpha}}}{x}\right)^\alpha = \left(\frac{b^x}{x}\right)^\alpha,$$

where $b = a^{\frac{1}{\alpha}} > 1$. By (a), $\lim_{x\to\infty} \frac{b^x}{x} = +\infty$. Consequently,

$$\lim_{x\to\infty} \frac{a^x}{x^\alpha} = \lim_{x\to\infty} \left(\frac{b^x}{x}\right)^\alpha = +\infty$$

for positive α.

1.1.13. It follows from the foregoing problem that $\lim_{y\to\infty} \frac{\alpha y}{e^{\alpha y}} = 0$. Substituting $y = \ln x$ yields $\lim_{x\to\infty} \frac{\ln x}{x^\alpha} = 0$.

1.1.14. We know that $\lim_{n\to\infty} a^{\frac{1}{n}} = \lim_{n\to\infty} a^{-\frac{1}{n}} = 1$. Suppose first that $a > 1$. Let $\varepsilon > 0$ be given. There exists an integer n_0 such that $n > n_0$ implies

$$1 - \varepsilon < a^{-\frac{1}{n}} < a^x < a^{\frac{1}{n}} < 1 + \varepsilon \quad \text{for} \quad |x| < \frac{1}{n}.$$

Therefore $\lim_{x\to 0} a^x = 1$ for $a > 1$. If $0 < a < 1$, it follows from the above that

$$\lim_{x\to 0} a^x = \lim_{x\to 0} \frac{1}{(1/a)^x} = 1.$$

The case $a = 1$ is obvious. To show the continuity of the exponential function $x \mapsto a^x$, choose $x_0 \in \mathbb{R}$ arbitrarily. Then

$$\lim_{x \to x_0} a^x = \lim_{x \to x_0} a^{x_0} a^{x-x_0} = a^{x_0} \lim_{y \to 0} a^y = a^{x_0}.$$

1.1.15.

(a) Since (see, e.g., I, 2.1.38) $\lim_{n \to \infty} \left(1 + \frac{1}{n}\right)^n = e$, given $\varepsilon > 0$, there is n_0 such that if $x > n_0 + 1$, and if $n = [x]$, then

$$e - \varepsilon < \left(1 + \frac{1}{n+1}\right)^n < \left(1 + \frac{1}{x}\right)^x < \left(1 + \frac{1}{n}\right)^{n+1} < e + \varepsilon.$$

(b) We have

$$\lim_{x \to -\infty} \left(1 + \frac{1}{x}\right)^x = \lim_{y \to +\infty} \left(1 - \frac{1}{y}\right)^{-y}$$

$$= \lim_{y \to +\infty} \left(1 + \frac{1}{y-1}\right)^{y-1} \left(1 + \frac{1}{y-1}\right).$$

Hence the required equality follows from (a).

(c) In view of (a) and (b) we get $\lim_{x \to 0^+} (1+x)^{\frac{1}{x}} = \lim_{y \to +\infty} \left(1 + \frac{1}{y}\right)^y = e$
and $\lim_{x \to 0^-} (1+x)^{\frac{1}{x}} = \lim_{y \to -\infty} \left(1 + \frac{1}{y}\right)^y = e.$

1.1.16. It is known that (see, e.g., I, 2.1.38) $0 < \ln\left(1 + \frac{1}{n}\right) < \frac{1}{n}$, $n \in \mathbb{N}$. Moreover, given $\varepsilon > 0$, there is n_0 such that $\frac{1}{n_0-1} < \varepsilon$. Consequently, if $|x| < \frac{1}{n_0}$, then

$$-\varepsilon < -\frac{1}{n_0 - 1} < \ln\left(1 - \frac{1}{n_0}\right) < \ln(1+x) < \ln\left(1 + \frac{1}{n_0}\right) < \frac{1}{n_0} < \varepsilon.$$

Hence $\lim_{x \to 0} \ln(1+x) = 0$. To prove the continuity of the logarithmic function take an $x_0 \in (0, \infty)$. Then

$$\lim_{x \to x_0} \ln x = \lim_{x \to x_0} \left(\ln x_0 + \ln \frac{x}{x_0}\right) = \ln x_0 + \lim_{y \to 1} \ln y$$
$$= \ln x_0 + \lim_{t \to 0} \ln(1+t) = \ln x_0.$$

1.1. The Limit of a Function.

1.1.17.

(a) By the result in 1.1.15 and by the continuity of the logarithmic function (see 1.1.16),

$$\lim_{x \to 0} \frac{\ln(1+x)}{x} = \lim_{x \to 0} \ln(1+x)^{\frac{1}{x}} = \ln e = 1.$$

(b) Note first that the continuity of the logarithmic function with base a, $a > 0$, $a \neq 1$, follows from the continuity of the natural logarithm function and from the equality $\log_a x = \frac{\ln x}{\ln a}$. So, by (a),

$$\lim_{x \to 0} \frac{\log_a(1+x)}{x} = \log_a e.$$

Set $y = a^x - 1$. Then

$$\lim_{x \to 0} \frac{a^x - 1}{x} = \lim_{y \to 0} \frac{y}{\log_a(y+1)} = \frac{1}{\log_a e} = \ln a.$$

(c) Put $y = (1+x)^\alpha - 1$. Clearly, x tends to zero if and only if y tends to zero. Moreover,

$$\frac{(1+x)^\alpha - 1}{x} = \frac{y}{\ln(1+y)} \cdot \frac{\ln(1+y)}{x} = \frac{y}{\ln(1+y)} \cdot \frac{\alpha \ln(1+x)}{x}.$$

This and (a) give $\lim\limits_{x \to 0} \frac{(1+x)^\alpha - 1}{x} = \alpha$.

1.1.18.

(a) Set $y = (\ln x)^{\frac{1}{x}}$. Then $\ln y = \frac{\ln(\ln x)}{\ln x} \cdot \frac{\ln x}{x}$. Hence, by 1.1.13 and by the continuity of the exponential function, $\lim\limits_{x \to \infty} (\ln x)^{\frac{1}{x}} = 1$.

(b) Set $y = x^{\sin x}$. Then $\ln y = \frac{\sin x}{x} \cdot x \ln x$. By 1.1.13,

$$\lim_{x \to 0^+} x \ln x = \lim_{t \to +\infty} \frac{-\ln t}{t} = 0.$$

Again by the continuity of the exponential function, we obtain $\lim\limits_{x \to 0^+} x^{\sin x} = 1$.

(c) Setting $y = (\cos x)^{\frac{1}{\sin^2 x}}$, we see that

$$\ln y = \frac{\ln(\cos x)}{\cos x - 1} \cdot \frac{\cos x - 1}{\sin^2 x}.$$

Now, by 1.1.17 (a), $\lim_{x\to 0}(\cos x)^{\frac{1}{\sin^2 x}} = e^{-\frac{1}{2}}$.

(d) For sufficiently large x,

$$\frac{e}{2^{\frac{1}{x}}} \le (e^x - 1)^{\frac{1}{x}} \le e.$$

Since $\lim_{x\to\infty} 2^{\frac{1}{x}} = 1$ (see 1.1.14), the limit is e.

(e) We have $\lim_{x\to 0^+}(\sin x)^{\frac{1}{\ln x}} = e^a$, where

$$a = \lim_{x\to 0^+} \frac{\ln \sin x}{\ln x} = \lim_{x\to 0^+} \frac{\ln \frac{\sin x}{x} + \ln x}{\ln x} = 1.$$

The last equality follows from the continuity of the logarithmic function (see 1.1.16).

1.1.19.

(a) We have

$$\lim_{x\to 0} \frac{\sin 2x + 2\arctan 3x + 3x^2}{\ln(1+3x+\sin^2 x) + xe^x} = \lim_{x\to 0} \frac{\frac{\sin 2x + 2\arctan 3x + 3x^2}{x}}{\frac{\ln(1+3x+\sin^2 x)}{x} + e^x} = 2,$$

because, by 1.1.17(a), $\lim_{x\to 0} \frac{\ln(1+3x+\sin^2 x)}{x} = 3$.

(b) By 1.1.17(a) we get

$$\lim_{x\to 0} \frac{2\ln\cos x}{-x^2} = \lim_{x\to 0} \frac{\ln(1-\sin^2 x)}{-x^2} = 1.$$

Hence $\lim_{x\to 0} \frac{\ln\cos x}{\tan x^2} = -\frac{1}{2}$.

(c) We have

$$\lim_{x\to 0^+} \frac{\sqrt{1-e^{-x}} - \sqrt{1-\cos x}}{\sqrt{\sin x}} = \lim_{x\to 0^+} \frac{\frac{\sqrt{1-e^{-x}} - \sqrt{1-\cos x}}{\sqrt{x}}}{\sqrt{\frac{\sin x}{x}}} = 1.$$

(d) We have $\lim_{x\to 0}(1+x^2)^{\cot x} = e^a$, where

$$a = \lim_{x\to 0} \frac{\ln(1+x^2)}{\tan x} = \lim_{x\to 0} \frac{x^2}{x} = 0,$$

because, in view of 1.1.17(a), $\lim_{x\to 0} \frac{\ln(1+x^2)}{x^2} = 1$.

1.1. The Limit of a Function. 119

1.1.20.

(a) Observe first that

(1) $$\frac{2\ln\tan\frac{\pi x}{2x+1}}{x} = \frac{\ln\left(\frac{1}{\cos^2\frac{\pi x}{2x+1}} - 1\right)}{x}.$$

By 1.1.16 and 1.1.18 (d),

$$\lim_{x\to\infty}\frac{\ln(x-1)}{\ln x} = \lim_{x\to\infty}\ln(x-1)^{\frac{1}{\ln x}} = \lim_{y\to\infty}\ln(e^y-1)^{\frac{1}{y}} = 1.$$

Hence

(2) $$\lim_{x\to\infty}\frac{\ln\left(\frac{1}{\cos^2\frac{\pi x}{2x+1}} - 1\right)}{x} = \lim_{x\to\infty}\frac{\ln\frac{1}{\cos^2\frac{\pi x}{2x+1}}}{x}$$
$$= \lim_{x\to\infty}\frac{-2\ln\cos\frac{\pi x}{2x+1}}{x}.$$

Next, by 1.1.18 (e),

$$\lim_{x\to\infty}\frac{-2\ln\cos\frac{\pi x}{2x+1}}{x} = \lim_{x\to\infty}\frac{-2\ln\sin\frac{\pi}{2(2x+1)}}{x} = \lim_{x\to\infty}\frac{2\ln\frac{2(2x+1)}{\pi}}{x}.$$

The last limit is 0 (see 1.1.13). This combined with (1) and (2) implies that the limit is 1.

(b) We have

$$\lim_{x\to\infty}x\left(\ln\left(1+\frac{x}{2}\right) - \ln\frac{x}{2}\right) = \lim_{x\to\infty}\frac{\ln\left(1+\frac{2}{x}\right)}{\frac{1}{x}}$$
$$= \lim_{y\to 0}\frac{\ln(1+y)}{\frac{1}{2}y} = 2,$$

where the last equality is a consequence of 1.1.17 (a).

1.1.21. Put $b(x) = \frac{f(x)}{x^\alpha}$. Then

$$\lim_{x\to 0^+}g(x)\ln f(x) = \lim_{x\to 0^+}(\alpha g(x)\ln x + g(x)\ln b(x))$$
$$= \lim_{x\to 0^+}\alpha g(x)\ln x = \gamma.$$

1.1.22. By 1.1.17(a),

$$\lim_{x\to 0} g(x)\ln f(x) = \lim_{x\to 0} g(x)\frac{\ln(f(x)-1+1)}{f(x)-1}(f(x)-1) = \gamma.$$

1.1.23.

(a) Apply the result in 1.1.21 with

$$g(x) = x, \quad \alpha = 1/2 \quad \text{and} \quad f(x) = 2\sin\sqrt{x} + \sqrt{x}\sin\frac{1}{x}$$

and use the equality $\lim_{x\to 0^+} x\ln\sqrt{x} = 0$ (see, e.g., 1.1.13). The limit is 1.

(b) Put

$$f(x) = 1 + xe^{-\frac{1}{x^2}}\sin\frac{1}{x^4} \quad \text{and} \quad g(x) = e^{\frac{1}{x^2}},$$

and note that $\lim_{x\to 0} g(x)(f(x)-1) = 0$. Thus, by 1.1.22, the limit is 1.

(c) As in (b) one can show that the limit equals $e^{\frac{\pi}{2}}$.

1.1.24. No. For a fixed positive and irrational α, consider the function defined by

$$f(x) = \begin{cases} 1 & \text{if } x = n\alpha,\ n\in\mathbb{N}, \\ 0 & \text{otherwise.} \end{cases}$$

This function satisfies our assumption. Indeed, if $a \geq 0$ and $a+k = n\alpha$ for some $k, n \in \mathbb{N}$, then there are no other $k', n' \in \mathbb{N}$ such that $a + k' = n'\alpha$. If there were, we would get $k - k' = (n - n')\alpha$, a contradiction. Clearly, $\lim_{x\to\infty} f(x)$ does not exist.

1.1.25. No. Consider the function defined by

$$f(x) = \begin{cases} 1 & \text{if } x = n\sqrt[n]{2},\ n\in\mathbb{N}, \\ 0 & \text{otherwise.} \end{cases}$$

The limit $\lim_{x\to\infty} f(x)$ does not exist, although the function enjoys the property given in the problem. In fact, if $a > 0$, and for some $k, n \in \mathbb{N}$

1.1. The Limit of a Function. 121

we have $ak = n\sqrt[n]{2}$, then there are no other $k', n' \in \mathbb{N}$ for which $ak' = n' \sqrt[n']{2}$. If there were, we would get

$$\frac{k}{k'} = \frac{n}{n'} 2^{\frac{n'-n}{nn'}},$$

a contradiction.

1.1.26. No. Consider the function defined in the solution of the foregoing problem. To see that the function satisfies the given condition, suppose that a and b are positive and $a + bn = m\sqrt[m]{2}$, $a + bk = l\sqrt[l]{2}$ for some $n, m, k, l \in \mathbb{N}$ such that $n \neq k$, $m \neq l$. Then

$$(1) \qquad a = \frac{nl\sqrt[l]{2} - mk\sqrt[m]{2}}{n - k}, \quad b = \frac{m\sqrt[m]{2} - l\sqrt[l]{2}}{n - k}.$$

If there were $p, q \in \mathbb{N}$ such that $p \neq n$, $p \neq k$ and $q \neq m$, $q \neq l$, and $a + bp = q\sqrt[q]{2}$, then in view of (1) we would get

$$m(p - k)\sqrt[m]{2} + l(n - p)\sqrt[l]{2} = q(n - k)\sqrt[q]{2},$$

a contradiction.

1.1.27. Fix $\varepsilon > 0$ arbitrarily. By assumption there is $\delta > 0$ such that

$$\frac{|f(x) - f(\frac{1}{2}x)|}{|x|} < \varepsilon, \quad \text{whenever} \quad 0 < |x| < \delta.$$

Hence, for $0 < |x| < \delta$,

$$\left|\frac{f(x)}{x}\right| = \lim_{n \to \infty} \left|\frac{f(x) - f\left(\frac{1}{2^{n+1}}x\right)}{x}\right|$$

$$\leq \varlimsup_{n \to \infty} \sum_{k=1}^{n+1} \frac{\frac{1}{2^{k-1}} \left|f\left(\frac{x}{2^{k-1}}\right) - f\left(\frac{x}{2^k}\right)\right|}{\frac{1}{2^{k-1}}|x|}$$

$$\leq \varlimsup_{n \to \infty} \sum_{k=1}^{n+1} \frac{1}{2^{k-1}} \varepsilon = 2\varepsilon.$$

1.1.28. Put $\lim\limits_{x\to\infty} (f(x+1) - f(x)) = l$ and set

$$M_n = \sup_{x\in[n,n+1)} f(x) \quad \text{and} \quad m_n = \inf_{x\in[n,n+1)} f(x).$$

The sequences $\{M_n\}$ and $\{m_n\}$ are well defined for $n \geq [a] + 1$. By the definition of the supremum, given $\varepsilon > 0$ there is $\{x_n\}$ such that $x_n \in [n, n+1)$ and $f(x_n) > M_n - \varepsilon$. Then

$$f(x_n + 1) - f(x_n) - \varepsilon < M_{n+1} - M_n < f(x_{n+1}) - f(x_{n+1} - 1) + \varepsilon,$$

and consequently,

$$l - \varepsilon \leq \varliminf_{n\to\infty} (M_{n+1} - M_n) \leq \varlimsup_{n\to\infty} (M_{n+1} - M_n) \leq l + \varepsilon.$$

Since $\varepsilon > 0$ can be arbitrarily chosen, $\lim\limits_{n\to\infty} (M_{n+1} - M_n) = l$. In much the same way one can show that $\lim\limits_{n\to\infty} (m_{n+1} - m_n) = l$. It follows from the Stolz theorem that (see also, e.g., I, 2.3.2)

$$\lim_{n\to\infty} \frac{M_n}{n} = \lim_{n\to\infty} \frac{m_n}{n+1} = l.$$

Hence, given $\varepsilon > 0$, there is n_0 such that for $n > n_0$,

$$(*) \qquad -\varepsilon < \frac{m_n}{n+1} - l < \varepsilon \quad \text{and} \quad -\varepsilon < \frac{M_n}{n} - l < \varepsilon.$$

It follows from the above that if $l > 0$, then $f(x) > 0$ for sufficiently large x. Therefore if $n_x = [x]$, then

$$\frac{m_{n_x}}{n_x + 1} \leq \frac{f(x)}{x} \leq \frac{M_{n_x}}{n_x}.$$

Now, by $(*)$, we see that for $x > n_0 + 1$,

$$-\varepsilon < \frac{m_{n_x}}{n_x + 1} - l \leq \frac{f(x)}{x} - l \leq \frac{M_{n_x}}{n_x} - l < \varepsilon.$$

For $l < 0$, one can show that

$$\frac{m_{n_x}}{n_x} \leq \frac{f(x)}{x} \leq \frac{M_{n_x}}{n_x + 1}$$

1.1. The Limit of a Function.

and proceed analogously. In this way the assertion is proved for $l \neq 0$. To show that our assertion is also true for $l = 0$, put $M_n = \sup_{x \in [n,n+1]} |f(x)|$. As above, one can find a sequence $\{x_n\}$ such that

$$|f(x_n+1)| - |f(x_n)| - \varepsilon < M_{n+1} - M_n < |f(x_{n+1})| - |f(x_{n+1}-1)| + \varepsilon$$

and show that $\lim_{n \to \infty} \frac{M_n}{n} = 0$. Since $\left|\frac{f(x)}{x}\right| \leq \frac{M_n}{n}$ for $x \in [n, n+1)$, we get $\lim_{x \to \infty} \frac{f(x)}{x} = 0$.

1.1.29. For $n \geq [a] + 1$ set $m_n = \inf_{x \in [n,n+1]} f(x)$. By the definition of the infimum, given $\varepsilon > 0$, there is a sequence $\{x_n\}$ such that $x_n \in [n, n+1)$ and $m_n \leq f(x_n) < m_n + \varepsilon$. Then

$$f(x_{n+1}) - f(x_{n+1} - 1) < m_{n+1} - m_n + \varepsilon.$$

This, in turn, implies $\lim_{n \to \infty} (m_{n+1} - m_n) = \infty$. By the Stolz theorem (see also I, 2.3.4), $\lim_{n \to \infty} \frac{m_n}{n} = +\infty$. If $x \in [n, n+1)$, then $\frac{f(x)}{x} \geq \frac{m_n}{n+1}$, which gives $\lim_{x \to \infty} \frac{f(x)}{x} = +\infty$.

1.1.30. Using notation introduced in the solution of Problem 1.1.28, one can show that

$$\lim_{n \to \infty} \frac{M_{n+1} - M_n}{n^k} = \lim_{n \to \infty} \frac{m_{n+1} - m_n}{n^k} = l.$$

Now by the Stolz theorem (see, e.g., I, 2.3.11),

$$\lim_{n \to \infty} \frac{M_n}{n^{k+1}} = \frac{1}{k+1} \lim_{n \to \infty} \frac{M_{n+1} - M_n}{n^k}$$

and

$$\lim_{n \to \infty} \frac{m_n}{n^{k+1}} = \frac{1}{k+1} \lim_{n \to \infty} \frac{m_{n+1} - m_n}{n^k}.$$

To prove our assertion it is enough to apply the reasoning analogous to that used in the solutions of the two preceding problems.

1.1.31. Set $\lim_{x \to +\infty} \frac{f(x+1)}{f(x)} = l$ and note that the function $x \mapsto \ln(f(x))$ satisfies the assumptions of Problem 1.1.28. Therefore we obtain $\lim_{x \to \infty} \frac{\ln(f(x))}{x} = \ln l$. Hence

$$\lim_{x \to +\infty} (f(x))^{\frac{1}{x}} = e^{\ln l} = l.$$

1.1.32. No. Consider the function defined by
$$f(x) = \begin{cases} 0 & \text{if } x = \frac{1}{n}, \ n = 1, 2, \ldots, \\ 1 & \text{otherwise.} \end{cases}$$

1.1.33. No. Let us take the function defined as follows:
$$f(x) = \begin{cases} 1 & \text{if } x = \frac{1}{n\sqrt[n]{2}}, \ n = 1, 2, \ldots, \\ 0 & \text{otherwise,} \end{cases}$$
and proceed as in the solution of 1.1.25.

1.1.34. Given $\varepsilon > 0$, there is δ, $0 < \delta < 1$, such that if $0 < |x| < \delta$, then
$$\left| f\left(x\left(\frac{1}{x} - \left[\frac{1}{x}\right]\right)\right)\right| < \varepsilon.$$
Now take $n \in \mathbb{N}$ so large that $\frac{1}{n} < \delta$. For $0 < s < \frac{1}{n+1}$, set $x = \frac{1-s}{n}$. Then
$$\frac{1}{n+1} = \frac{1 - \frac{1}{n+1}}{n} < \frac{1-s}{n} = x < \frac{1}{n}.$$
Thus $n < \frac{1}{x} < n+1$ and $\left[\frac{1}{x}\right] = n$. Consequently,
$$x\left(\frac{1}{x} - \left[\frac{1}{x}\right]\right) = x\left(\frac{1}{x} - n\right) = 1 - \frac{1-s}{n}n = s.$$
Finally, if $0 < s < \frac{1}{n+1}$, then $|f(s)| = \left|f\left(x\left(\frac{1}{x} - \left[\frac{1}{x}\right]\right)\right)\right| < \varepsilon$. For $s < 0$, one can proceed analogously.

1.1.35.

(a) Assume that f is monotonically increasing on (a, b). If $\{x_n\}$ is a decreasing sequence convergent to x_0, then $\{f(x_n)\}$ is also monotonically decreasing and bounded below by $f(x_0)$. Thus (see, e.g., I, 2.1.1), $\lim_{n\to\infty} f(x_n) = \inf_{n\in\mathbb{N}} f(x_n)$. Clearly,
$$\inf_{n\in\mathbb{N}} f(x_n) \geq \inf_{x>x_0} f(x).$$
Moreover, given $x > x_0$, there is an n such that $x_n < x$, and consequently, $f(x_n) \leq f(x)$. Hence
$$\inf_{n\in\mathbb{N}} f(x_n) \leq \inf_{x>x_0} f(x).$$

1.1. The Limit of a Function.

In this way we have proved that if $\{x_n\}$ is monotonically decreasing to x_0, then
$$\lim_{n\to\infty} f(x_n) = \inf_{x>x_0} f(x).$$
Now assume that $\{x_n\}$ is a sequence convergent to x_0 and such that $x_n > x_0$. Then (see, e.g., I, 2.4.29) $\{x_n\}$ contains a monotonically decreasing subsequence $\{x_{n_k}\}$. In view of the above,
$$\lim_{k\to\infty} f(x_{n_k}) = \inf_{x>x_0} f(x).$$
If the sequence $\{x_n\}$ contained a subsequence $\{z_{n_k}\}$ such that $\lim_{k\to\infty} f(z_{n_k}) \neq \inf_{x>x_0} f(x)$, then we could find a monotonic subsequence of it not convergent to $\inf_{x>x_0} f(x)$, a contradiction. This implies that $\lim_{x\to x_0^+} f(x) = \inf_{x>x_0} f(x)$.

It is worth noting here that the above analysis shows that to determine a one-sided limit it is enough to consider only monotonic sequences.

The same reasoning applies to the other equalities in (a) and (b).

(c) Assume that f is monotonically increasing. Since $f(x) \geq f(x_0)$ for $x \geq x_0$, $f(x_0^+) = \inf_{x>x_0} f(x) \geq f(x_0)$. Likewise, one can show that $f(x_0^-) = \sup_{x<x_0} f(x) \leq f(x_0)$.

1.1.36.

(a) It follows from the solution of the foregoing problem that
$$f(t) \leq f(x^-) \leq f(x) \quad \text{whenever} \quad a < x_0 < t < x.$$
If $x \to x_0^+$, then $t \to x_0^+$, and therefore
$$f(x_0^+) = \lim_{t\to x_0^+} f(t) \leq \lim_{x\to x_0^+} f(x^-)$$
and
$$\overline{\lim_{x\to x_0^+}} f(x^-) \leq f(x_0^+) = \lim_{x\to x_0^+} f(x).$$
Consequently, $\lim_{x\to x_0^+} f(x^-) = f(x_0^+)$.

(b) This follows by the same reasoning as in (a).

1.1.37. Necessity of the condition follows immediately from the definition of a limit. Indeed, if $\lim_{x \to a} f(x) = l$, then given $\varepsilon > 0$, there is $\delta > 0$ such that the relation $0 < |x - a| < \delta$ implies $|f(x) - l| < \frac{\varepsilon}{2}$. Consequently,

$$|f(x) - f(x')| \leq |f(x) - l| + |f(x') - l| < \varepsilon.$$

Now we show that the condition is sufficient. Suppose that it is satisfied and f does not have a limit at a. Take $\{x_n\}$ such that $\lim_{n \to \infty} x_n = a$, $x_n \neq a$ and $\{f(x_n)\}$ does not converge. Consequently, $\{f(x_n)\}$ is not a Cauchy sequence. On the other hand, since $\lim_{n \to \infty} x_n = a$, there is n_0 such that if $n, k \geq n_0$, then $0 < |x_n - a| < \delta$ and $0 < |x_k - a| < \delta$. It follows from the assumption that $|f(x_n) - f(x_k)| < \varepsilon$, a contradiction.

In an entirely similar manner one can show that in order that the finite limit $\lim_{x \to \infty} f(x)$ exist the following condition is necessary and sufficient: for every $\varepsilon > 0$ there exists $M > 0$ such that $x, x' > M$ implies $|f(x) - f(x')| < \varepsilon$.

1.1.38. Let $\{x_n\}$, $x_n \neq a$, be an arbitrary sequence converging to a. It follows from the definition of the limit of a function at a that $\lim_{n \to \infty} f(x_n) = A$. Set $y_n = f(x_n)$. Since $f(x) \neq A$ in a deleted neighborhood of a, $f(x_n) \neq A$ for sufficiently large n. Hence $\lim_{n \to \infty} g(y_n) = B$, or equivalently, $\lim_{n \to \infty} g(f(x_n)) = B$. This means that $\lim_{x \to a} g(f(x)) = B$.

1.1.39. Consider the functions f and g defined as follows:

$$f(x) = \begin{cases} 0 & \text{if } x = \frac{1}{n}, \ n = 1, 2, \ldots, \\ \sin x & \text{otherwise}, \end{cases}$$

$$g(y) = \begin{cases} 0 & \text{if } y = 0, \\ \frac{\sin y}{y} & \text{otherwise}. \end{cases}$$

Then

$$g(f(x)) = \begin{cases} 0 & \text{if } x = \frac{1}{n}, \ n \in \mathbb{N}, \text{ or } x = k\pi, \ k \in \mathbb{Z}, \\ \frac{\sin(\sin x)}{\sin x} & \text{otherwise}, \end{cases}$$

1.1. The Limit of a Function. 127

and $\lim_{x\to 0} f(x) = 0$ and $\lim_{y\to 0} g(y) = 1$, but $\lim_{x\to 0} g(f(x))$ does not exist.

1.1.40. By the periodicity of $x \mapsto f(x) - x$, $f(x+1) = f(x) + 1$. Consequently, for any integer n, $f(x+n) = f(x) + n$, $x \in \mathbb{R}$. Since each real x can be written as the sum of its integral and fractional parts (that is $x = [x] + r$, where $0 \leq r < 1$), we get

$$(*) \qquad f(x) = f(r) + [x].$$

The monotonicity of f gives

$$f(0) \leq f(r) \leq f(1) = f(0) + 1 \quad \text{for} \quad 0 \leq r < 1.$$

One can prove by induction that

$$f^n(0) \leq f^n(r) \leq f^n(0) + 1 \quad \text{for} \quad 0 \leq r < 1 \quad \text{and} \quad n \in \mathbb{N}.$$

Therefore,

$$\frac{f^n(0)}{n} \leq \frac{f^n(r)}{n} \leq \frac{f^n(0)}{n} + \frac{1}{n}.$$

These inequalities prove our assertion in the case when $0 \leq x < 1$. Moreover, by $(*)$, $f^n(x) = f^n(r) + [x]$, which implies that the assertion holds for the other $x \in \mathbb{R}$.

1.1.41 [6, page 47]. Observe first that

$$\begin{aligned}
x + f(0) - 1 &\leq [x] + f(0) = f([x]) \leq f(x) \\
&\leq f(1 + [x]) = f(0) + [x] + 1 \\
&\leq x + f(0) + 1.
\end{aligned}$$

Now we show by induction that for $n \in \mathbb{N}$,

$$(1) \qquad x + n(f(0) - 1) \leq f^n(x) \leq x + n(f(0) + 1).$$

Fix n arbitrarily and assume that (1) is true. Then, as in the solution of 1.1.40, we get

$$\begin{aligned}
f^{n+1}(x) &= f(f^n(x)) = f([f^n(x)] + r) \\
&= [f^n(x)] + f(r) \leq f^n(x) + f(1) \\
&\leq x + n(f(0) + 1) + f(0) + 1 \\
&= x + (n+1)(f(0) + 1),
\end{aligned}$$

where $r = f^n(x) - [f^n(x)]$. This proves the right inequality in (1). In much the same way one can prove the left inequality. Again by induction, we will prove that

(2) $$f^{n(m_p-1)}(0) \le np \le f^{nm_p}(0), \quad n \in \mathbb{N}.$$

For $n = 1$ the inequalities follow from the definition of m_p. Suppose that they hold for an arbitrarily fixed n. Then

$$f^{(n+1)m_p}(0) = f^{m_p}(f^{nm_p}(0))$$
$$\ge f^{m_p}(0+np) = f^{m_p}(0) + np$$
$$\ge p + np.$$

Likewise,

$$f^{(n+1)(m_p-1)}(0) = f^{m_p-1}(f^{n(m_p-1)}(0)) \le f^{m_p-1}(0+np)$$
$$= np + f^{m_p-1}(0)$$
$$\le np + p.$$

Thus the inequalities (2) are proved.

Any positive integer n can be written as $n = km_p + q$, where $0 \le q < m_p$. By (1) and (2) we obtain

$$kp + q(f(0) - 1) \le f^q(kp) \le f^q(f^{km_p}(0))$$
$$= f^n(0) = f^{q+k}(f^{k(m_p-1)}(0))$$
$$\le f^{q+k}(kp) \le kp + (q+k)(1+f(0)),$$

which implies

(3) $$\frac{kp}{n} + \frac{q(f(0)-1)}{n} \le \frac{f^n(0)}{n} \le \frac{kp}{n} + \frac{k+q}{n}(1+f(0)).$$

Since $\lim_{n\to\infty} \frac{k}{n} = \frac{1}{m_p}$ and $\lim_{n\to\infty} \frac{q}{n} = 0$, the required inequality is a consequence of (3).

1.1.42 [6, page 47]. Note that by 1.1.40 it is enough to show that $\lim_{n\to\infty} \frac{f^n(0)}{n}$ exists. If $f(0) = 0$, then the limit is 0. Assume now that $f(0) > 0$. Then either for any positive integer p there is an integer m such that $f^m(0) > p$, or there is a positive integer p such that $f^m(0) \le p$ for all $m \in \mathbb{N}$. In the latter case $\{f^n(0)\}$ is a bounded sequence

1.2. Properties of Continuous Functions

and, consequently, $\lim_{n\to\infty} \frac{f^n(0)}{n} = 0$. In the first case $\lim_{p\to\infty} m_p = \infty$, where m_p is defined as in 1.1.41. Passage to the limit as $p \to \infty$ in the inequalities given in 1.1.41 shows that $\lim_{p\to\infty} \frac{p}{m_p}$ exists, and consequently $\lim_{n\to\infty} \frac{f^n(0)}{n}$ also exists.

In the case where $f(0) < 0$ one can prove an inequality similar to (2) in the solution of the foregoing problem, and proceed analogously.

1.2. Properties of Continuous Functions

1.2.1. The function is discontinuous at each $x_0 \neq \pi k$, where $k \in \mathbb{Z}$. Indeed, if $\{x_n\}$ is a sequence of irrationals converging to x_0, then $\lim_{n\to\infty} f(x_n) = 0$. On the other hand, if $\{z_n\}$ is a sequence of rationals converging to x_0, then, by the continuity of the sine function, $\lim_{n\to\infty} f(z_n) = \lim_{n\to\infty} \sin|z_n| = \sin|x_0| \neq 0$. Similarly, one can show that f is continuous at $k\pi$ with $k \in \mathbb{Z}$.

1.2.2. As in the solution of the foregoing problem we can show that f is continuous at -1 and at 1 only.

1.2.3.

(a) Observe first that if $\{x_n\}$ converges to x, with $x_n = \frac{p_n}{q_n}$, where $p_n \in \mathbb{Z}$ and $q_n \in \mathbb{N}$ are relatively prime, and $x_n \neq x$, $n \in \mathbb{N}$, then $\lim_{n\to\infty} q_n = \infty$. So, if x is irrational and $\{x_n\}$ is as above, then $\lim_{n\to\infty} f(x_n) = \lim_{n\to\infty} \frac{1}{q_n} = 0 = f(x)$. If $\{z_n\}$ is a sequence of irrationals converging to x, then $\lim_{n\to\infty} f(z_n) = 0 = f(x)$. This means that f is continuous at every irrational. Likewise, one can show that 0 is a point of continuity of f. Suppose now that $x \neq 0$ and $x = \frac{p}{q}$, where p and q are co-prime. If $\{x_n\}$ is a sequence of irrationals converging to x, then $\lim_{n\to\infty} f(x_n) = 0 \neq f(x)$. Consequently, f is discontinuous at every rational different from zero.

(b) Suppose $x \in \mathbb{R} \setminus \mathbb{Q}$ and let $\{z_n\}$ be a sequence of irrationals different from x approaching x. Then $\lim_{n\to\infty} f(z_n) = \lim_{n\to\infty} |z_n| = |x|$. If $\{x_n\}$ is a sequence of rationals approaching x, then, by the

remark at the beginning of the solution of (a),
$$\lim_{n\to\infty} f(x_n) = \lim_{n\to\infty} \frac{x_n q_n}{q_n + 1} = x.$$
This means that f is continuous at every positive irrational and discontinuous at every negative irrational. Similarly, one can show that f is continuous at zero. Now let $0 \neq x = \frac{p}{q}$ (p,q co-prime). Then
$$x_n = \frac{p}{q} \cdot \frac{(np+1)q+1}{(np+1)q}$$
converges to $\frac{p}{q}$. Note that the numerator and denominator of x_n are relatively prime. Therefore,
$$\lim_{n\to\infty} f(x_n) = \lim_{n\to\infty} \frac{(np+1)pq+p}{(np+1)q^2+1} = \frac{p}{q} \neq \frac{p}{q+1}.$$
Thus the function is discontinuous at every rational different from zero.

1.2.4. Let $f \in C([a,b])$ and let x_0 be a point in $[a,b]$. Given $\varepsilon > 0$, there is $\delta > 0$ such that if $x \in [a,b]$ and $0 < |x - x_0| < \delta$, then $|f(x) - f(x_0)| < \varepsilon$. Now the continuity of $|f|$ at x_0 follows from the obvious inequality $||f(x)| - |f(x_0)|| \leq |f(x) - f(x_0)|$.
The function given by
$$f(x) = \begin{cases} 1 & \text{for } x \in \mathbb{Q} \cap [a,b], \\ -1 & \text{for } x \in [a,b] \setminus \mathbb{Q}, \end{cases}$$
is discontinuous at each point in $[a,b]$, although $|f|$ is a constant function and therefore continuous on $[a,b]$.

1.2.5. In order that f be continuous on $\mathbb{R}$, a necessary and sufficient condition is that
$$\lim_{x\to 2n^-} f(x) = \lim_{x\to 2n^+} f(x) \quad \text{and} \quad \lim_{x\to (2n-1)^-} f(x) = \lim_{x\to (2n-1)^+} f(x)$$
for each $n \in \mathbb{Z}$. This gives
$$b_n + 1 = a_n \quad \text{and} \quad a_{n-1} = b_n - 1.$$
Consequently, by induction, $a_n = 2n + a_0$ and $b_n = 2n - 1 + a_0$, $a_0 \in \mathbb{R}$.

1.2. Properties of Continuous Functions

1.2.6. Since the function is odd, we will study its continuity only for $x \geq 0$. Clearly, f is continuous at each $x \neq \sqrt{n}$, $n = 1, 2, \ldots$. Now suppose that $n = k^2$, where k is a positive integer. Then

$$\lim_{x \to k^+} f(x) = n \lim_{x \to k^+} \sin \pi x = 0$$

and

$$\lim_{x \to k^-} f(x) = (n-1) \lim_{x \to k^-} \sin \pi x = 0.$$

Hence the function is also continuous at every $n = k^2$. If $n \in \mathbb{N}$ is not a square of an integer, then

$$\lim_{x \to \sqrt{n}^+} f(x) = n \lim_{x \to \sqrt{n}^+} \sin \pi x = n \sin(\pi\sqrt{n})$$

and

$$\lim_{x \to \sqrt{n}^-} f(x) = (n-1) \sin(\pi\sqrt{n}).$$

We conclude that f is discontinuous at $x = \pm\sqrt{n}$, where $n \neq k^2$.

1.2.7. We get

$$f(x) = \begin{cases} 1 & \text{if } x \in [\frac{1}{2}, 1), \\ n + (x-n)^n & \text{if } x \in [n, n+1), \ n \in \mathbb{N}. \end{cases}$$

Consequently, the function is continuous at each $x \neq n$, $n \in \mathbb{N}$. Moreover, $\lim_{x \to n^+} f(x) = \lim_{x \to n^-} f(x) = n = f(n)$. So, f is continuous on $[\frac{1}{2}, \infty)$.

Now we show that f is strictly increasing on $[1, \infty)$. Clearly, f is strictly increasing on each interval $[n, n+1)$. If $x_1 \in [n-1, n)$ and $x_2 \in [n, n+1)$, then

$$f(x_2) - f(x_1) = (x_2 - n)^n + 1 - (x_1 - n + 1)^{n-1} > (x_2 - n)^n \geq 0.$$

It then follows that $f(x_2) - f(x_1) > 0$ for $x_2 \in [m, m+1)$ and $x_1 \in [n, n+1)$, if $m > n+1$.

1.2.8.

(a) We have

$$f(x) = \begin{cases} 1 & \text{if } x > 0, \\ 0 & \text{if } x = 0, \\ -1 & \text{if } x < 0. \end{cases}$$

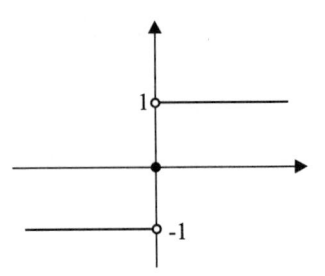

The function is discontinuous only at zero.

(b) By the definition of f,
$$f(x) = \begin{cases} x^2 & \text{if } x \geq 0, \\ x & \text{if } x < 0. \end{cases}$$

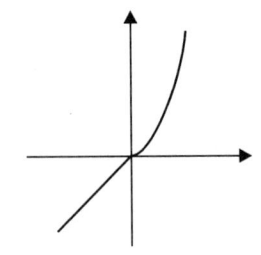

The function is continuous on $\mathbb{R}$.

(c) We get
$$f(x) = \lim_{n \to \infty} \frac{\ln(e^n + x^n)}{n} = \lim_{n \to \infty} \frac{n + \ln\left(1 + (x/e)^n\right)}{n}.$$

Consequently,
$$f(x) = \begin{cases} 1 & \text{if } 0 \leq x \leq e, \\ \ln x & \text{if } x > e. \end{cases}$$

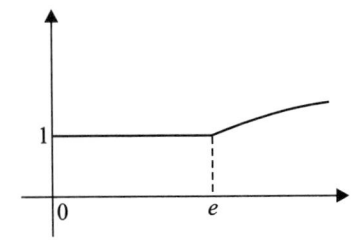

The function is continuous on $[0, \infty)$.

1.2. Properties of Continuous Functions

(d) $f(x) = \max\{4, x^2, \frac{1}{x^2}\}$. The function is continuous on $\mathbb{R} \setminus \{0\}$.

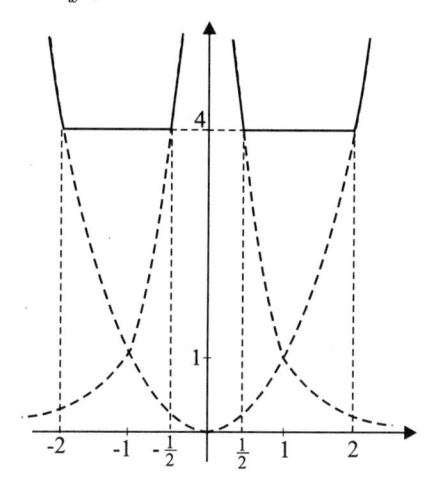

(e) $f(x) = \max\{|\cos x|, |\sin x|\}$. Clearly, f is continuous on $\mathbb{R}$.

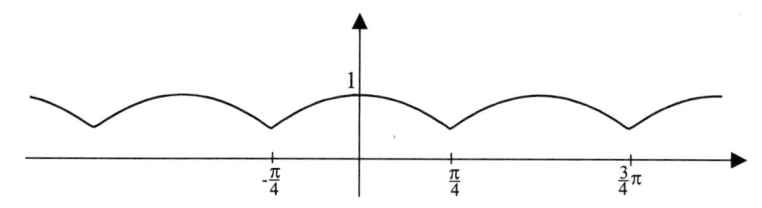

1.2.9. Let $T > 0$ be a period of f. By the continuity of f on $[0, T]$ there are $x_\star \in [0, T]$ and $x^\star \in [0, T]$ such that $f(x_\star) = \inf\limits_{x \in [0,T]} f(x)$ and $f(x^\star) = \sup\limits_{x \in [0,T]} f(x)$. The desired result follows from the periodicity of f.

1.2.10. Since P is a polynomial of even degree, we obtain $\lim\limits_{x \to \infty} P(x) = \lim\limits_{x \to -\infty} P(x) = +\infty$. Therefore for any $M > 0$ there is $a > 0$ such that if $|x| > a$, then $P(x) > M$. Let $x_0 \in [-a, a]$ be such that
$$P(x_0) = \inf\limits_{x \in [-a,a]} P(x).$$
If $P(x_0) \leq M$, then we can set $x_\star = x_0$. If $P(x_0) > M$, take $b > 0$ such that $P(x) > P(x_0)$ whenever $|x| > b$. By continuity there exists

$x_\star \in [-b, b]$ such that $P(x_\star) = \inf\limits_{x \in [-b,b]} P(x)$.
To prove the second assertion, observe that

$$\lim_{x \to -\infty} |P(x)| = \lim_{x \to \infty} |P(x)| = +\infty$$

and proceed analogously.

1.2.11.

(a) Set
$$f(x) = \begin{cases} 2x - 1 & \text{if } x \in (0,1), \\ 0 & \text{if } x = 0 \text{ or } x = 1. \end{cases}$$

(b) For $n \in \mathbb{N}$, put
$$\mathbf{A}_n = \left\{ 0, \frac{1}{2^n}, \frac{2}{2^n}, \frac{3}{2^n}, \dots, \frac{2^n - 1}{2^n} \right\}$$

and $\mathbf{B}_1 = \mathbf{A}_1$, $\mathbf{B}_n = \mathbf{A}_n \setminus \bigcup\limits_{k=1}^{n-1} \mathbf{A}_k = \mathbf{A}_n \setminus \mathbf{A}_{n-1}$. Clearly, $\bigcup\limits_{k=1}^{\infty} \mathbf{A}_k = \bigcup\limits_{k=1}^{\infty} \mathbf{B}_k$. Define f as follows:

$$f(x) = \begin{cases} 0 & \text{if } x \in [0,1] \setminus \bigcup\limits_{k=1}^{\infty} \mathbf{A}_k, \\ \frac{1}{2^n} - 1 & \text{if } x \in \mathbf{B}_n,\ n \in \mathbb{N}. \end{cases}$$

For any a and b, $0 \leq a < b \leq 1$, $\inf\limits_{x \in [a,b]} f(x) = -1$; -1 is not attained by f on $[a, b]$.

1.2.12. Observe first that

(1) $\qquad \omega_f(x_0, \delta_1) \leq \omega_f(x_0, \delta_2) \quad$ whenever $\quad 0 < \delta_1 < \delta_2$.

Assume that $\lim\limits_{\delta \to 0^+} \omega_f(x_0, \delta) = 0$. Then, given $\varepsilon > 0$, there is $\delta_0 > 0$ such that $\omega_f(x_0, \delta) < \varepsilon$ if $\delta < \delta_0$. Consequently, if $|x - x_0| < \delta < \delta_0$, then $|f(x) - f(x_0)| < \varepsilon$, which proves the continuity of f at x_0.

Assume now that f is continuous at x_0. Then, given $\varepsilon > 0$, there is $\delta_0 > 0$ such that $|x - x_0| < \delta_0$ implies $|f(x_0) - f(x)| < \frac{\varepsilon}{2}$. Hence, in view of (1), if $0 < \delta < \delta_0$, then

$$\omega_f(x_0, \delta) \leq \omega_f(x, \delta_0) < \varepsilon,$$

1.2. Properties of Continuous Functions 135

and consequently, $\lim_{\delta\to 0^+} \omega_f(x_o,\delta) = 0$.

1.2.13.

(a) Let $x_0 \in [a,b]$ and $\varepsilon > 0$ be chosen arbitrarily. It follows from the continuity of f and g that there is $\delta > 0$ such that if $x \in [a,b]$ and $|x - x_0| < \delta$, then

$$f(x_0)-\varepsilon < f(x) < f(x_0)+\varepsilon \quad \text{and} \quad g(x_0)-\varepsilon < g(x) < g(x_0)+\varepsilon.$$

Hence

(1) $\quad h(x) < \min\{f(x_0) + \varepsilon,\ g(x_0) + \varepsilon\}$
$\quad\quad\quad = \min\{f(x_0), g(x_0)\} + \varepsilon = h(x_0) + \varepsilon$

and

$$f(x) > f(x_0)-\varepsilon \geq h(x_0)-\varepsilon \quad \text{and} \quad g(x) > g(x_0)-\varepsilon \geq h(x_0)-\varepsilon.$$

Consequently,

(2) $\quad\quad\quad\quad\quad h(x) > h(x_0) - \varepsilon.$

The continuity of h at x_0 follows from (1) and (2). In much the same way one can prove that H is continuous on $[a,b]$.

(b) As in (a) one can show that $\max\{f_1, f_2, f_3\}$ and $\min\{f_1, f_2, f_3\}$ are continuous on $[a,b]$. Continuity of f follows from

$$f(x) = f_1(x) + f_2(x) + f_3(x) - \max\{f_1(x), f_2(x), f_3(x)\}$$
$$- \min\{f_1(x), f_2(x), f_3(x)\}.$$

1.2.14. Since f is continuous, the functions m and M are well defined. Let x_0 be in $[a,b]$ and let $\varepsilon > 0$. By the continuity of f, there is $\delta > 0$ such that

$$\sup_{|h|<\delta} |f(x_0+h) - f(x_0)| < \varepsilon.$$

It follows from the definition of m that

(1) $\quad m(x_0+h) - m(x_0) = \inf_{\zeta\in[a,x_0+h]} f(\zeta) - \inf_{\zeta\in[a,x_0]} f(\zeta) \leq 0.$

Observe that if the first infimum is attained at a point in $[a, x_0]$, then the equality holds in (1). So, suppose that $x_h \in [x_0, x_0 + h]$ and
$$m(x_0 + h) = \inf_{\zeta \in [a, x_0 + h]} f(\zeta) = f(x_h).$$
Then for $|h| < \delta$,
$$m(x_0 + h) - m(x_0) = f(x_h) - \inf_{\zeta \in [a, x_0]} f(\zeta) \geq f(x_h) - f(x_0) > -\varepsilon,$$
because $|x_h - x_0| \leq |h| < \delta$. We therefore have shown that m is continuous at each x_0 in $[a, b]$. The same argument can be applied to show that M is continuous on $[a, b]$.

1.2.15. Since f is bounded, the functions m and M are well defined and bounded. Moreover m is decreasing on $(a, b]$ and M is increasing on $[a, b)$. For $x_0 \in (a, b)$, by 1.1.35,
$$\lim_{x \to x_0^-} m(x) = \inf_{\zeta \in (a, x_0)} m(\zeta) \geq m(x_0).$$
If $\inf_{\zeta \in (a, x_0)} m(\zeta) > m(x_0)$, then there is a positive d such that
$$\inf_{\zeta \in (a, x_0)} m(\zeta) = m(x_0) + d.$$
Thus, for each $\zeta \in (a, x_0)$,
$$m(\zeta) = \inf_{a \leq x < \zeta} f(x) \geq m(x_0) + d,$$
and consequently, $f(x) \geq m(x_0) + d$ for every $x \in [a, x_0)$, a contradiction. So, we have proved that $\lim_{x \to x_0^-} m(x) = m(x_0)$. The continuity from the left of M can be proved in precisely the same manner.

1.2.16. No. Consider the following function:
$$f(x) = \begin{cases} 2 & \text{if } x \in [0, 1), \\ 1 & \text{if } x \in [1, 2), \\ 3 & \text{if } x \in [2, 3]. \end{cases}$$
Then $m^\star$ is not continuous from the left at $x_0 = 1$, and $M^\star$ is not continuous from the left at $x_1 = 2$.

1.2. Properties of Continuous Functions

1.2.17. Set $\lim\limits_{x\to\infty} f(x) = l$. Then, given $\varepsilon > 0$, there is $M > a$ such that $|f(x)-l| < \varepsilon$ for $x > M$. Thus if $x > M$, then $l-\varepsilon < f(x) < l+\varepsilon$. Obviously, since f is continuous, it is bounded on $[a, M]$.

1.2.18. Suppose that $\lim\limits_{n\to\infty} x_n = a$. By the continuity of f, for every $\varepsilon > 0$ there exists $\delta > 0$ such that

(1) $\qquad |f(x) - f(a)| < \varepsilon \quad \text{for} \quad |x - a| < \delta.$

It follows from the definition of limit inferior that there is a sequence $\{x_{n_k}\}$ for which $|x_{n_k} - a| < \delta$ beginning with some value k_0 of the index k. Now by (1) we get $|f(x_{n_k}) - f(a)| < \varepsilon$ for $k > k_0$. We therefore have shown that

$$\varliminf_{n\to\infty} f(x_n) \leq f(\varliminf_{n\to\infty} x_n).$$

We now show by example that this inequality may be strict. Take $f(x) = -x$, $x \in \mathbb{R}$, and $x_n = (-1)^n$, $n \in \mathbb{N}$. Then

$$-1 = \varliminf_{n\to\infty} f(x_n) < f(\varliminf_{n\to\infty} x_n) = 1.$$

In an entirely similar manner one can show that

$$\varlimsup_{n\to\infty} f(x_n) \geq f(\varlimsup_{n\to\infty} x_n).$$

The same example can be used to show that this inequality may also be strict.

1.2.19.

(a) It has been shown in the solution of the preceding problem that for any bounded sequence $\{x_n\}$ and for any continuous function f the following inequalities hold:

$$\varliminf_{n\to\infty} f(x_n) \leq f(\varliminf_{n\to\infty} x_n)$$

and

$$\varlimsup_{n\to\infty} f(x_n) \geq f(\varlimsup_{n\to\infty} x_n).$$

Put $\varliminf\limits_{n\to\infty} x_n = a$. Then there exists a sequence $\{x_{n_k}\}$ such that

(1) $\qquad f(x_{n_k}) \leq f(a) + \varepsilon$

(see the solution of the foregoing problem). Clearly, for sufficiently large n we have $x_n > a - \frac{\delta}{2}$. Hence by monotonicity and continuity of f we obtain

$$f(x_n) \geq f\left(a - \frac{\delta}{2}\right) > f(a) - \varepsilon.$$

Combined with (1), this gives $\lim\limits_{n \to \infty} f(x_n) = f(\lim\limits_{n \to \infty} x_n)$.
(b) The proof of this equality runs as in (a).

1.2.20. Apply 1.2.19 to $-f$.

1.2.21. Note that g is well defined and increasing on $\mathbb{R}$.
(a) By Problem 1.1.35,

(1) $$g(x_0^-) = \sup_{x < x_0} g(x) \leq g(x_0).$$

Suppose that $g(x_0^-) < g(x_0)$. Then there is a positive d such that $g(x_0^-) = g(x_0) - d$. Consequently, for every $x < x_0$,

$$\sup\{t : f(t) < x\} \leq g(x_0) - d,$$

or equivalently, $t \leq g(x_0) - d$ if $f(t) < x$. This implies that $t \leq g(x_0) - d$ if $f(t) < x_0$, which gives $g(x_0) = \sup\{t : f(t) < x_0\} \leq g(x_0) - d$, a contradiction.
(b) The function g may be discontinuous, as the following example shows. If

$$f(x) = \begin{cases} x & \text{for } x < 1, \\ -x + 2 & \text{for } 1 \leq x \leq 2, \\ x - 2 & \text{for } x > 2, \end{cases}$$

then

$$g(x) = \begin{cases} x & \text{for } x \leq 0, \\ 2 + x & \text{for } x > 0. \end{cases}$$

1.2. Properties of Continuous Functions

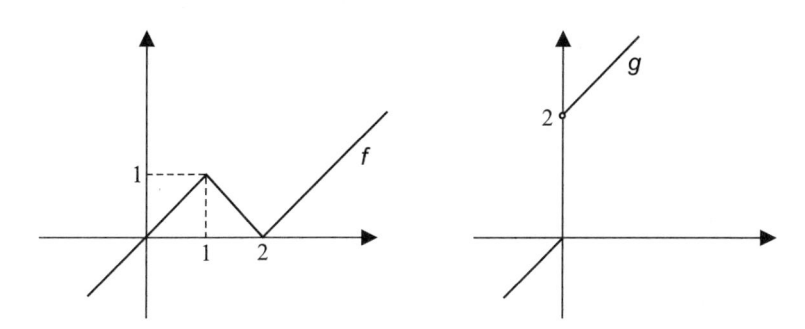

1.2.22. We know that the set $\left\{m + n\frac{T_1}{T_2} : m, n \in \mathbb{Z}\right\}$ is dense in $\mathbb{R}$ (see, e.g., I, 1.1.15). Thus, given an $x \in \mathbb{R}$, there is a sequence $\left\{m_k + n_k\frac{T_1}{T_2}\right\}$ convergent to $\frac{x}{T_2}$. By the periodicity and continuity of f we get
$$f(0) = \lim_{k \to \infty} f(m_k T_2 + n_k T_1) = f(x).$$
Let T_1 and T_2 be two incommensurate numbers and let
$$\mathbb{W} = \{x \in \mathbb{R} : x = rT_1 + sT_2, \; s, t \in \mathbb{Q}\}.$$
Define f by setting
$$f(x) = \begin{cases} 1 & \text{for} \quad x \in \mathbb{W}, \\ 0 & \text{for} \quad x \in \mathbb{R} \setminus \mathbb{W}. \end{cases}$$
Then T_1 and T_2 are periods of f.

1.2.23.

(a) Assume that T_n, $n \in \mathbb{N}$, are positive periods of f such that $\lim_{n \to \infty} T_n = 0$. By the continuity of f, given $x_0 \in \mathbb{R}$ and $\varepsilon > 0$, there exists $\delta > 0$ such
$$|f(x) - f(x_0)| < \varepsilon \quad \text{whenever} \quad |x - x_0| < \delta.$$
Since $\lim_{n \to \infty} T_n = 0$, there exists n_0 for which $0 < T_{n_0} < \frac{\delta}{2}$. Then at least one of the numbers kT_{n_0} with $k \in \mathbb{Z}$ belongs to the interval $(x_0 - \delta, x_0 + \delta)$. Consequently,
$$|f(x_0) - f(0)| = |f(x_0) - f(kT_{n_0})| < \varepsilon.$$
It then follows from the arbitrariness of $\varepsilon > 0$ and $x_0 \in \mathbb{R}$ that f is constant, contrary to our assumption.

(b) The Dirichlet function defined by setting

$$f(x) = \begin{cases} 1 & \text{if } x \in \mathbb{Q}, \\ 0 & \text{if } x \in \mathbb{R} \setminus \mathbb{Q}, \end{cases}$$

is periodic. Every rational number is its period. Therefore a fundamental period does not exist.

(c) Assume that the set of all periods of f is not dense in $\mathbb{R}$. Then there exists an interval (a,b) which does not contain any period of f. As in (a), one can show that there is a period T and an integer k such that $kT \in (a,b)$. A contradiction.

1.2.24.

(a) Let $x_0 \in \mathbb{R}$ be a point of continuity of f. Since f is not constant, there is $x_1 \neq x_0$ such that $f(x_1) \neq f(x_0)$. If there were no minimal positive period of f, there would exist a sequence $\{T_n\}$ of positive periods of f converging to zero. Take $0 < \varepsilon < |f(x_1) - f(x_0)|$. It follows from the continuity of f at x_0 that there is $\delta > 0$ such that

(1) $\qquad |f(x) - f(x_0)| < \varepsilon \quad \text{whenever} \quad |x - x_0| < \delta.$

Since $\lim\limits_{n \to \infty} T_n = 0$, there exists an index n_0 for which $0 < T_{n_0} < \frac{\delta}{2}$. Thus at least one of the numbers kT_{n_0}, $k \in \mathbb{Z}$, belongs to the interval $(x_0 - x_1 - \delta, x_0 - x_1 + \delta)$. Consequently, $x_1 + kT_{n_0} \in (x_0 - \delta, x_0 + \delta)$ and, by (1),

$$|f(x_1) - f(x_0)| = |f(x_1 + kT_{n_0}) - f(x_0)| < \varepsilon.$$

A contradiction.

(b) This is an immediate consequence of (a).

1.2.25. Let T_1 and T_2 be positive periods of f and g, respectively. Suppose that $f \neq g$. Then there is x_0 such that $f(x_0) \neq g(x_0)$, or in other words,

(1) $\qquad |f(x_0) - g(x_0)| = M > 0.$

For $0 < \varepsilon < \frac{M}{2}$, there is $\delta > 0$ such that

(2) $\qquad |f(x_0 + h) - f(x_0)| < \varepsilon \quad \text{whenever} \quad |h| < \delta.$

1.2. Properties of Continuous Functions   141

By the assumption $\lim_{x\to\infty}(f(x)-g(x))=0$, there is a positive integer k such that, if $x \geq x_0 + kT_2$, then
$$|f(x) - g(x)| < \varepsilon.$$
Consequently, for any positive integer m,

(3) $\qquad |f(x_0 + kmT_2) - g(x_0 + kmT_2)| < \varepsilon.$

By (2), (3) and the periodicity of f and g we get

$$\begin{aligned}&|f(x_0) - g(x_0)|\\ &= |f(x_0) - f(x_0 + kmT_2) + f(x_0 + kmT_2) - g(x_0 + kmT_2)|\\ (4)\quad &\leq |f(x_0) - f(x_0 + kmT_2)| + |f(x_0 + kmT_2) - g(x_0 + kmT_2)|\\ &= |f(x_0) - f(x_0 + kmT_2 - nT_1)|\\ &\quad + |f(x_0 + kmT_2) - g(x_0 + kmT_2)| < \varepsilon + \varepsilon = 2\varepsilon,\end{aligned}$$

whenever

(5) $\qquad |mkT_2 - nT_1| < \delta.$

However, since $2\varepsilon < M$, (4) would contradict (1) if there were $m \in \mathbb{N}$ and $n \in \mathbb{Z}$ satisfying (5). On the other hand, if $\frac{T_1}{T_2}$ is rational, (5) is obviously satisfied for some integers m and n. If $\frac{T_1}{T_2}$ is irrational, then (5) is also satisfied (see, e.g., I, 1.1.14).

1.2.26.

(a) Set $f(x) = \sin x$ and $g(x) = x - [x]$ for $x \in \mathbb{R}$. Then f and g are periodic with fundamental periods 2π and 1, respectively. Therefore no period of f is commensurate with any period of g. Put $h = f + g$. If h were periodic with period T, then we would get
$$\sin T + T - [T] = 0, \quad \sin(-T) - T - [-T] = 0.$$
Consequently, $(T - [T]) + (-T - [-T]) = 0$, which would imply $T - [T] = 0$. This would mean that T is an integer, a contradiction with $\sin T = 0$.

(b) [A. D. Kudriašov, A. S. Meşeriakov, Mathematics in School, 6 (1969), 19-21 (Russian)] Let α, β and γ be such real numbers that the equality $a\alpha + b\beta + c\gamma = 0$, where $a, b, c \in \mathbb{Q}$, holds if and only if $a = b = c = 0$. Such numbers do exist. One can take, for example, $\alpha = 1, \beta = \sqrt{2}$ and $\gamma = \sqrt{3}$. Define

$$\mathbb{W} = \{a\alpha + b\beta + c\gamma : a, b, c \in \mathbb{Q}\}.$$

Consider two functions f and g defined by setting

$$f(x) = \begin{cases} -b - c - b^2 + c^2 & \text{if } x = a\alpha + b\beta + c\gamma \in \mathbb{W}, \\ 0 & \text{if } x \notin \mathbb{W}, \end{cases}$$

$$g(x) = \begin{cases} a + c + a^2 - c^2 & \text{if } x = a\alpha + b\beta + c\gamma \in \mathbb{W}, \\ 0 & \text{if } x \notin \mathbb{W}. \end{cases}$$

Note that each number $r\alpha$, $r \in \mathbb{Q} \setminus \{0\}$, is a period of f and each number $s\beta$, $s \in \mathbb{Q} \setminus \{0\}$, is a period of g. We will show that these functions do not have any other periods. If T is a period of f, then $f(\beta + T) = f(\beta)$, and because $f(\beta) = -2$, we get $\beta + T \in \mathbb{W}$. Consequently, $T \in \mathbb{W}$. Therefore $T = r\alpha + s\beta + t\gamma$ with some $r, s, t \in \mathbb{Q}$. Since $f(T) = f(0)$, we obtain $-s - t - s^2 + t^2 = 0$, or equivalently, $(s+t)(1+s-t) = 0$. We now show that $1+s-t \neq 0$. Indeed, if $1 + s - t = 0$, then $T = r\alpha + s\beta + (1+s)\gamma$. Using

(1) $$f(x + T) = f(x),$$

with $x = -\gamma$, we get $-s - s - s^2 + s^2 = 1 + 1$, or $s = -1$. Therefore $T = r\alpha - \beta$. Now substituting $x = \beta$ into (1) yields $f(r\alpha) = f(\beta)$, and consequently, $0 = -1 - 1$, a contradiction. Thus we have proved that $1 + s - t \neq 0$. It then follows that $s + t = 0$. Hence $T = r\alpha + s\beta - s\gamma$. The task is now to show that $s = 0$. To this end we take $x = \gamma$ in (1), and we get

$$-s + s - 1 - s^2 + (s-1)^2 = -1 + 1,$$

which implies $s = 0$. In an entirely similar way one can show that the only periods of g are those mentioned above. So no period of f is commensurate with any period of g. Note now that $h = f + g$ is given by the formula

$$h(x) = \begin{cases} a - b + a^2 - b^2 & \text{if } x = a\alpha + b\beta + c\gamma \in \mathbb{W}, \\ 0 & \text{if } x \notin \mathbb{W}. \end{cases}$$

As above, one can show that the only periods of h are the numbers $t\gamma$, where $t \in \mathbb{Q}$ and $t \neq 0$.

1.2. Properties of Continuous Functions

1.2.27. Suppose that $h = f + g$ is periodic with period T. Since $\frac{T_1}{T_2} \notin \mathbb{Q}$, we see that $\frac{T}{T_1} \notin \mathbb{Q}$ or $\frac{T}{T_2} \notin \mathbb{Q}$. Assume, for example, that $\frac{T}{T_1} \notin \mathbb{Q}$. By the periodicity of h we get $f(x+T) + g(x+T) = h(x+T) = h(x) = f(x) + g(x)$ for $x \in \mathbb{R}$. Therefore the function H defined by setting $H(x) = f(x+T) - f(x) = g(x) - g(x+T)$ is continuous and periodic with two incommensurate periods T_1 and T_2. By the result in 1.2.22, H is constant. This means that there is $c \in \mathbb{R}$ such that $f(x+T) = f(x) + c$ for $x \in \mathbb{R}$. Suppose that $c \neq 0$. Substituting $x = 0$ and then $x = T$ in the last equality, we get

$$f(2T) = f(T) + c = f(0) + 2c.$$

One can show by induction that $f(nT) = f(0) + nc$, which contradicts the boundedness of f (see, e.g., 1.2.9). Hence $c = 0$ and T is a period of f. Consequently, $T = nT_1$ with some $n \in \mathbb{Z}$, a contradiction.

1.2.28. The proof of this result is a modification of that presented in the solution of the preceding problem. Assume that T_1 is the fundamental period of f. As in the solution of the foregoing problem one can show that the function H given by the formula

$$H(x) = f(x+T) - f(x) = g(x) - g(x+T)$$

is identically equal to zero. Therefore T is a common period of f and g, a contradiction.

1.2.29. Suppose, for example, that f is monotonically increasing. Let x_0 be a point of discontinuity of f. By the result in 1.1.35, $f(x_0^+) - f(x_0^-) > 0$. This means that f has a simple discontinuity at x_0. With each such point x_0 we can associate an interval $(f(x_0^-), f(x_0^+))$. It follows from the monotonicity of f and from the result in 1.1.35 that the intervals associated with different points of discontinuity of f are disjoint. Taking one rational number from each interval, we obtain a one-to-one correspondence between the set of points of discontinuity of f and a subset of $\mathbb{Q}$.

1.2.30. Since f is uniformly continuous on $[0,1]$, given $\varepsilon > 0$ there exists $n_0 \in \mathbb{N}$ such that for $2n > n_0$ and for $k = 1, 2, \ldots, 2n$ we have

$$\left| f\left(\frac{k}{2n}\right) - f\left(\frac{k-1}{2n}\right) \right| < \varepsilon.$$

Thus if $2n > n_0$, then

$$|S_{2n}| = \left|\frac{1}{2n}\sum_{k=1}^{2n}(-1)^k f\left(\frac{k}{2n}\right)\right| \leq \frac{\varepsilon}{2}.$$

Moreover,

$$|S_{2n+1}| = \left|\frac{1}{2n+1}\sum_{k=1}^{2n+1}(-1)^k f\left(\frac{k}{2n+1}\right)\right| \leq \frac{n}{2n+1}\varepsilon + \frac{1}{2n+1}|f(1)|.$$

It then follows that

$$\lim_{n\to\infty}\frac{1}{n}\sum_{k=1}^{n}(-1)^k f\left(\frac{k}{n}\right) = 0.$$

1.2.31. As in the solution of the foregoing problem, note first that f is uniformly continuous on $[0,1]$. Hence, given $\varepsilon > 0$, there is $n_0 \in \mathbb{N}$ such that if $n > n_0$ and $k = 0, 1, 2, \ldots n$, then

$$\left|f\left(\frac{k}{n}\right) - f\left(\frac{k+1}{n}\right)\right| < \varepsilon.$$

Consequently, for $n > n_0$,

$$S_n = \frac{1}{2^n}\sum_{k=0}^{n}(-1)^k \binom{n}{k} f\left(\frac{k}{n}\right)$$

$$= \frac{1}{2^n}\sum_{k=0}^{n-1}(-1)^k \binom{n-1}{k}\left(f\left(\frac{k}{n}\right) - f\left(\frac{k+1}{n}\right)\right).$$

Therefore

$$|S_n| < \frac{\varepsilon}{2^n}\sum_{k=0}^{n-1}\binom{n-1}{k} = \frac{\varepsilon}{2}.$$

1.2.32. Put $M = \lim\sup_{r\to\infty}_{x\geq r} f(x)$ and $m = \lim\inf_{r\to\infty}_{x\geq r} f(x)$. Suppose that $M > m$. Then there is a real number k such that $M > k > m$, and there exists a satisfying $f(a) > k$. By the continuity of f there exists $b > a$ such that $f(t) > k$ for all $t \in [a,b]$.

1.2. Properties of Continuous Functions 145

Take $p = \frac{ab}{b-a}$. Then $\frac{x}{a} \geq \frac{x}{b} + 1$ whenever $x \geq p$. Indeed,

$$\frac{x}{a} - \frac{x}{b} = x\left(\frac{1}{a} - \frac{1}{b}\right) = \frac{x}{p} \geq 1.$$

Therefore there is a positive integer n_0 between $\frac{x}{b}$ and $\frac{x}{a}$; that is, $\frac{x}{a} \geq n_0 \geq \frac{x}{b}$, or equivalently, $a \leq \frac{x}{n_0} \leq b$. By assumption,

$$f(x) = f\left(n_0 \frac{x}{n_0}\right) \geq f\left(\frac{x}{n_0}\right) > k$$

for all $x \geq p$, which contradicts the definition of m. Consequently, $m = M$, which means that $\lim\limits_{x \to \infty} f(x)$ exists and is finite or infinite.

1.2.33. Let f be convex on (a,b) and $a < s < u < v < t < b$. It follows from the geometric interpretation of convexity that the point $(u, f(u))$ lies below the line passing through $(s, f(s))$ and $(v, f(v))$. This means that

(1) $$f(u) \leq f(s) + \frac{f(v) - f(s)}{v - s}(u - s).$$

Similarly, the point $(v, f(v))$ is below the line passing through $(u, f(u))$ and $(t, f(t))$. Thus

(2) $$f(v) \leq f(u) + \frac{f(t) - f(u)}{t - u}(v - u).$$

Inequalities (1) and (2) give

$$f(s) + \frac{f(u) - f(s)}{u - s}(v - s) \leq f(v) \leq f(u) + \frac{f(t) - f(u)}{t - u}(v - u).$$

It follows from these inequalities and the squeeze law that, if $\{v_n\}$ is a sequence converging to u from the right, then $\lim\limits_{n \to \infty} f(v_n) = f(u)$, which means that $\lim\limits_{x \to u^+} f(x) = f(u)$. Likewise, $\lim\limits_{x \to u^-} f(x) = f(u)$. Thus the continuity of f at any point u in (a,b) is proved.

The following example shows that the assertion is not true if an interval is not open:

$$f(x) = \begin{cases} x^2 & \text{if } x \in [0, 1), \\ 2 & \text{if } x = 1. \end{cases}$$

1.2.34. It follows from the uniform convergence of $\{f_n\}$ that, given $\varepsilon > 0$, there exists n_0 such that
$$|f_n(x) - f(x)| < \frac{1}{3}\varepsilon \quad \text{for} \quad n \geq n_0,\ x \in \mathbf{A}.$$
Fix $a \in \mathbf{A}$. By the continuity of f_{n_0} at a there is $\delta > 0$ such that
$$|f_{n_0}(x) - f_{n_0}(a)| < \frac{1}{3}\varepsilon \quad \text{whenever} \quad |x - a| < \delta.$$
Thus
$$|f(x) - f(a)| \leq |f_{n_0}(x) - f(x)| + |f_{n_0}(x) - f_{n_0}(a)| + |f_{n_0}(a) - f(a)| < \varepsilon.$$

1.3. Intermediate Value Property

1.3.1. Let f be defined on $[a, b]$ by setting
$$f(x) = \begin{cases} \sin \frac{1}{x-a} & \text{if} \quad a < x \leq b, \\ 0 & \text{if} \quad x = a. \end{cases}$$

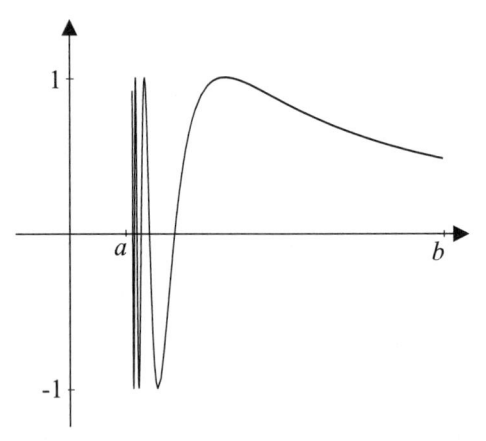

Clearly, f enjoys the intermediate value property on $[a, b]$ but it is discontinuous at a.

We now construct a function enjoying the intermediate value property and having infinitely many points of discontinuity. Let $\mathbf{C}$ denote the *Cantor set*. Recall that the Cantor set is defined as follows. We divide the interval $[0, 1]$ into three equal parts, remove the

1.3. Intermediate Value Property   147

interval $\left(\frac{1}{3},\frac{2}{3}\right)$, and denote by $\mathbf{E}_1$ the union of the intervals $[0,\frac{1}{3}]$ and $[\frac{2}{3},1]$. At the second stage we remove the open middle thirds of the remaining two intervals and set

$$\mathbf{E}_2 = \left[0,\frac{1}{9}\right] \cup \left[\frac{2}{9},\frac{3}{9}\right] \cup \left[\frac{6}{9},\frac{7}{9}\right] \cup \left[\frac{8}{9},1\right].$$

Proceeding analogously, we remove at the nth stage the union of the open middle thirds of the remaining 2^{n-1} intervals and we denote by $\mathbf{E}_n$ the union of 2^n closed intervals, each of length 3^{-n}. Then

$$\mathbf{C} = \bigcap_{n=1}^{\infty} \mathbf{E}_n.$$

Note that if (a_i, b_i), $i = 1, 2, \ldots$, is the sequence of removed intervals, then

$$\mathbf{C} = [0,1] \setminus \bigcup_{i=1}^{\infty} (a_i, b_i).$$

Define the function g by setting

$$g(x) = \begin{cases} 0 & \text{if } x \in \mathbf{C}, \\ \frac{2(x-a_i)}{b_i - a_i} - 1 & \text{if } x \in (a_i, b_i),\ i = 1, 2, \ldots. \end{cases}$$

It follows from the construction of the Cantor set that each interval $[a,b] \subset [0,1]$ contains an open subinterval disjoint with $\mathbf{C}$. Indeed, if (a,b) is free of points of $\mathbf{C}$, then (a,b) is one of the removed intervals (a_i, b_i) or its subinterval. If there is $x \in (a,b) \cap \mathbf{C}$, then there are $n \in \mathbb{N}$ and $k \in \{0,1,2,\ldots,3^n - 1\}$ such that $x \in \left[\frac{k}{3^n}, \frac{k+1}{3^n}\right] \subset (a,b)$. Then the open middle third of $\left[\frac{k}{3^n}, \frac{k+1}{3^n}\right]$, which in fact is one of the intervals (a_i, b_i), is an open subinterval free of points of $\mathbf{C}$.

The function g is discontinuous at each $x \in \mathbf{C}$, and it follows from the above that g enjoys the intermediate value property.

1.3.2. Let $x_0 \in (a,b)$ be arbitrarily fixed. It follows from the monotonicity of f that

$$\sup_{a \leq x < x_0} f(x) = f(x_0^-) \leq f(x_0) \leq f(x_0^+) = \inf_{x_0 < x \leq b} f(x)$$

(see, e.g., 1.1.35). Suppose now that

$$f(x_0) < f(x_0^+).$$

Then there is a strictly decreasing sequence $\{x_n\}$, $x_n \in (x_0, b]$, converging to x_0 and such that $\lim_{n\to\infty} f(x_n) = f(x_0^+)$. Since f is strictly increasing, $f(x_n) > f(x_0^+) > f(x_0)$. By the intermediate value property, there is $x' \in (x_0, x_n)$ such that $f(x') = f(x_0^+)$. Then

$$\inf_{x_0 < x < x'} f(x) \geq \inf_{x_0 < x \leq b} f(x) = f(x').$$

On the other hand, by the strict monotonicity of f, $\inf_{x_0 < x < x'} f(x) < f(x')$, a contradiction. So we have proved that $f(x_0) = f(x_0^+)$. The equalities $f(x_0^-) = f(x_0)$, $f(a) = f(a^+)$, and $f(b) = f(b^-)$ can be proved in an entirely similar manner.

1.3.3. The function g defined by $g(x) = f(x) - x$, $x \in [0,1]$, is continuous, and $g(0) = f(0) \geq 0$, and $g(1) = f(1) - 1 \leq 0$. Since g has the intermediate value property, there exists $x_0 \in [0,1]$ such that $g(x_0) = 0$.

1.3.4. Consider the function $h(x) = f(x) - g(x)$, $x \in [a,b]$, and observe that $h(a) < 0$ and $h(b) > 0$. By the intermediate value property there is $x_0 \in (a,b)$ such that $h(x_0) = 0$.

1.3.5. Define the function g by setting

$$g(x) = f\left(x + \frac{T}{2}\right) - f(x).$$

Then g is continuous on $\mathbb{R}$, $g(0) = f(\frac{T}{2}) - f(0)$, and $g(\frac{T}{2}) = f(0) - f(\frac{T}{2})$. Thus there is $x_0 \in [0, \frac{T}{2}]$ for which $g(x_0) = 0$.

1.3.6. Put

$$m = \min\{f(x_1), \ldots, f(x_n)\} \quad \text{and} \quad M = \max\{f(x_1), \ldots, f(x_n)\}.$$

Then

$$m \leq \frac{1}{n}(f(x_1) + f(x_2) + \cdots + f(x_n)) \leq M.$$

Consequently, there is $x_0 \in (a,b)$ such that

$$f(x_0) = \frac{1}{n}(f(x_1) + f(x_2) + \cdots + f(x_n)).$$

1.3. Intermediate Value Property

1.3.7.

(a) Set $f(x) = (1-x)\cos x - \sin x$. Then $f(0) = 1$ and $f(1) = -\sin 1 < 0$. Therefore there is $x_0 \in (0,1)$ satisfying $f(x_0) = 0$.

(b) It is well known that (see, e.g., 1.1.12)

$$\lim_{x \to \infty} e^{-x}|P(x)| = 0 \quad \text{and} \quad \lim_{x \to -\infty} e^{-x}|P(x)| = +\infty.$$

Consequently, there is an $x_0 \in \mathbb{R}$ such that $e^{-x_0}|P(x_0)| = 1$.

1.3.8. Let us observe that

$$\operatorname{sgn} P(-a_l) = (-1)^l \quad \text{and} \quad \operatorname{sgn} P(-b_l) = (-1)^{l+1}, \quad l = 0, 1, \ldots, n.$$

By the intermediate value property, there is a root of the polynomial P in every interval $(-b_l, -a_l)$, $l = 0, 1, \ldots, n$.

1.3.9. No. Consider, for example, f and g defined as follows:

$$f(x) = \begin{cases} \sin \frac{1}{x-a} & \text{if } a < x \leq b, \\ 0 & \text{if } x = a, \end{cases}$$

and

$$g(x) = \begin{cases} -\sin \frac{1}{x-a} & \text{if } a < x \leq b, \\ 1 & \text{if } x = a. \end{cases}$$

1.3.10. Set

$$g(x) = f(x+1) - f(x), \quad x \in [0,1].$$

Then $g(1) = f(2) - f(1) = -g(0)$. Hence there exists $x_0 \in [0,1]$ such that $f(x_0 + 1) = f(x_0)$. So, we can take $x_2 = x_0 + 1$ and $x_1 = x_0$.

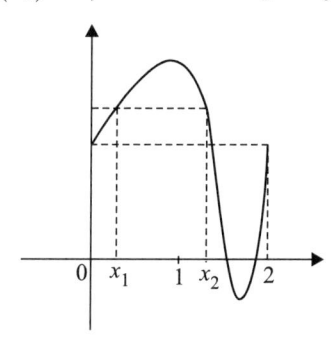

1.3.11. Consider the function
$$g(x) = f(x+1) - f(x) - \frac{1}{2}(f(2) - f(0)), \quad x \in [0,1],$$
and apply the reasoning analogous to that used in the solution of the preceding problem.

1.3.12. Define the function g by the formula
$$g(x) = f(x+1) - f(x) \quad \text{for} \quad x \in [0, n-1].$$
If $g(0) = 0$, then $f(1) = f(0)$. So suppose, for example, that $g(0) > 0$. Then $f(1) > f(0)$. If also $f(k+1) > f(k)$ for $k = 1, 2, \ldots, n-1$, then we would get
$$f(0) < f(1) < f(2) < \cdots < f(n) = f(0).$$
A contradiction. It then follows that there is a k_0 such that $g(k_0) > 0$ and $g(k_0 + 1) \leq 0$. Since g is continuous, there is $x_0 \in (k_0, k_0 + 1]$ for which $g(x_0) = 0$. Consequently, $f(x_0 + 1) = f(x_0)$. Analogous reasoning can be applied when $g(0) < 0$.

1.3.13. The function f can be extended on $[0, \infty)$ so as to have period n. The extended function is denoted also by f. For an arbitrarily fixed $k \in \{1, 2, \ldots, n-1\}$, define
$$g(x) = f(x+k) - f(x), \quad x \geq 0.$$
Now we show that there is $x_0 \in [0, kn]$ such that $g(x_0) = 0$. Indeed, if $g(0) = 0$, then $x_0 = 0$. So suppose, for example, that $g(0) > 0$. If also $g(j) > 0$ for all $j = 0, 1, 2, \ldots, kn - k$, then we would get
$$f(0) < f(k) < f(2k) < \cdots < f(kn) = f(0).$$
A contradiction. It then follows that there is j_0 such that $g(j_0) > 0$ and $g(j_0 + 1) \leq 0$. Since g is continuous, there is $x_0 \in (j_0, j_0 + 1]$ for which $g(x_0) = 0$. Consequently, $f(x_0 + k) = f(x_0)$. Suppose first that $x_0 \in [(l-1)n, ln - k]$ for some $1 \leq l \leq k$. It then follows from the periodicity of f that $f(x_0) = f(x_0 - (l-1)n)$ and $f(x_0 + k) = f(x_0 - (l-1)n + k)$. Therefore we can take $x_k = x_0 - (l-1)n$ and $x'_k = x_0 - (l-1)n + k$. If $x_0 \in [ln - k, ln]$, then $x_0 + k \in [ln, (l+1)n]$,

1.3. Intermediate Value Property

and we have $f(x_0 - (l-1)n) = f(x_0) = f(x_0 + k) = f(x_0 - ln + k)$. So we can take $x_k = x_0 - (l-1)n$ and $x'_k = x_0 - ln + k$.

It is not true that for any $k \in \{1, 2, \ldots, n-1\}$ there are x_k and x'_k such that $x_k - x'_k = k$ and $f(x_k) = f(x'_k)$. In fact, it is enough to consider the function

$$f(x) = \sin\left(\frac{\pi}{2}x\right) \quad \text{for} \quad x \in [0, 4].$$

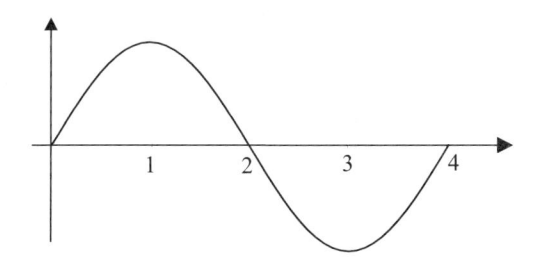

It is easy to see that $f(x+3) \neq f(x)$ for all $x \in [0, 1]$.

1.3.14. The following solution of this problem is due to our student Grzegorz Michalak.

Without loss of generality we can assume that $f(0) = f(n) = 0$. The case $n = 1$ is obvious. So suppose that $n > 1$. We will consider first the case where $f(1) > 0, f(2) > 0, \ldots, f(n-1) > 0$. For $k = 1, 2, \ldots, n-1$, we set $g_k(x) = f(x+k) - f(x)$. The function g_k is continuous on $[0, n-k]$, and by assumption $g_k(0) > 0$ and $g_k(n-k) < 0$. Consequently, there is $x_k \in [0, n-k]$ such that $g_k(x_k) = 0$, or in other words, $f(x_k + k) = f(x_k)$. This shows that the assertion is true in this case. In an entirely similar manner we can see that it is also true if $f(1) < 0, f(2) < 0, \ldots, f(n-1) < 0$. Suppose now that $f(1) > 0$ (resp. $f(1) < 0$), the numbers $f(1), f(2), \ldots, f(n-1)$ are distinct and different from zero, and there is m, $2 \leq m \leq n-1$, with $f(m) < 0$ (resp. $f(m) > 0$). Then there are integers $k_1, k_2, \ldots, k_s$

between 1 and $n-2$ such that
$$f(1) > 0, f(2) > 0, \ldots, f(k_1) > 0,$$
$$f(k_1 + 1) < 0, f(k_1 + 2) < 0, \ldots, f(k_2) < 0,$$
$$\ldots$$
$$f(k_s + 1) < 0, f(k_s + 2) < 0, \ldots, f(n-1) < 0$$
(or $f(k_s + 1) > 0, f(k_s + 2) > 0, \ldots, f(n-1) > 0$)

(resp. $f(1) < 0, f(2) < 0, \ldots, f(k_1) < 0, \ldots$). Now reasoning similar to that in the proof of the first case shows that there are k_1 solutions in $[0, k_1 + 1]$, $k_2 - k_1$ solutions in $[k_1, k_2 + 1]$, and so on. Clearly, in this case all these solutions must be distinct and therefore the assertion is proved. Finally, consider the case where there are integers k and m, $0 \leq k < m \leq n$, with $f(k) = f(m)$. Suppose also that the numbers $f(k), f(k+1), \ldots, f(m-1)$ are distinct. It follows from the above that there are $m - k$ solutions in the interval $[k, m]$. Next define

$$f_1(x) = \begin{cases} f(x) & \text{if } 0 \leq x \leq k, \\ f(x + m - k) & \text{if } k < x \leq n - (m-k). \end{cases}$$

Clearly, f_1 is continuous on $[0, n - (m-k)]$ and $f_1(n - (m-k)) = f_1(0) = 0$. If $f_1(0), f_1(1), \ldots, f_1(n-(m-k)-1)$ are distinct, then by the first part of the proof we get $n - (m-k)$ solutions which together with the $m - k$ solutions obtained above give the desired result. If some of the numbers $f_1(0), f_1(1), \ldots, f_1(n - (m-k) - 1)$ coincide, the procedure can be repeated.

1.3.15. Suppose, contrary to our claim, that the equation $f(x) = g(x)$ has no solutions. Then the function $h(x) = f(x) - g(x)$ would be either positive or negative. Hence

$$0 \neq h(f(x)) + h(g(x))$$
$$= f(f(x)) - g(f(x)) + f(g(x)) - g(g(x))$$
$$= f^2(x) - g^2(x).$$

A contradiction.

The following example shows that the assumption of continuity is essential:
$$f(x) = \begin{cases} \sqrt{2} & \text{if } x \in \mathbb{R} \setminus \mathbb{Q}, \\ 0 & \text{if } x \in \mathbb{Q}, \end{cases}$$

1.3. Intermediate Value Property

$$g(x) = \begin{cases} 0 & \text{if } x \in \mathbb{R} \setminus \mathbb{Q}, \\ \sqrt{2} & \text{if } x \in \mathbb{Q}. \end{cases}$$

1.3.16. Assume, contrary to our claim, that there are x_1, x_2 and x_3 such that $x_1 < x_2 < x_3$ and, for example, $f(x_1) > f(x_2)$ and $f(x_2) < f(x_3)$. By the intermediate value property, for every u such that $f(x_2) < u < \min\{f(x_1), f(x_3)\}$ there are $s \in (x_1, x_2)$ and $t \in (x_2, x_3)$ satisfying $f(s) = u = f(t)$. Since f is injective, $s = t$, contrary to the fact that $x_1 < s < x_2 < t < x_3$.

1.3.17. It follows from the result in the foregoing problem that f is either strictly decreasing or strictly increasing.

(a) Suppose that f is strictly increasing and there is x_0 such that $f(x_0) \neq x_0$. Let, for example, $f(x_0) > x_0$. Then $f^n(x_0) > x_0$, contrary to our assumption. Similar arguments apply to the case $f(x_0) < x_0$.

(b) If f is strictly decreasing, then f^2 is strictly increasing. Since $f^n(x) = x$, we get $f^{2n}(x) = x$, which means that the nth iteration of f^2 is the identity. Therefore, by (a), $f^2(x) = x$.

1.3.18. Note that f is injective. Indeed, if $f(x_1) = f(x_2)$, then $-x_1 = f^2(x_1) = f^2(x_2) = -x_2$. Hence $x_1 = x_2$. It follows from 1.3.16 that if f were continuous, then it would be either strictly increasing or strictly decreasing. In both cases f^2 would be strictly increasing. A contradiction.

1.3.19. As in the solution of the foregoing problem, one can show that f is injective on $\mathbb{R}$. Analysis similar to that in the solution of 1.3.16 shows that f is either strictly increasing or strictly decreasing. In both cases f^{2k}, $k \in \mathbb{N}$, is strictly increasing. Consequently, the integer n in the condition $f^n(x) = -x$ has to be odd. If f were strictly increasing, f^n also would be strictly increasing, which would contradict our condition. So, f is strictly decreasing. Moreover, since

$$f(-x) = f(f^n(x)) = f^n(f(x)) = -f(x),$$

we see that f is an odd function (and so is every iteration of f).

Now we will show that $f(x) = -x$, $x \in \mathbb{R}$. Suppose that there is an x_0 such that $x_1 = f(x_0) > -x_0$, or in other words, $-x_1 < x_0$.

It then follows that $x_2 = f(x_1) < f(-x_0) = -f(x_0) = -x_1 < x_0$. One can show by induction that if $x_k = f(x_{k-1})$, then $(-1)^n x_n < x_0$, which contradicts our assumption that $x_n = f^n(x_0) = -x_0$. Similar reasoning applies to the case where $f(x_0) < -x_0$. Hence $f(x) = -x$ for all $x \in \mathbb{R}$.

1.3.20. Suppose that f has a discontinuity at x. Then there exists a sequence $\{x_n\}$ convergent to x such that $\{f(x_n)\}$ does not converge to $f(x)$. This means that there exists $\varepsilon > 0$ such that for every $k \in \mathbb{N}$ there is $n_k > k$ for which

$$|f(x_{n_k}) - f(x)| \geq \varepsilon.$$

So $f(x_{n_k}) \geq f(x) + \varepsilon > f(x)$ or $f(x_{n_k}) \leq f(x) - \varepsilon < f(x)$. Assume, for example, that the first inequality holds. There exists a rational number q such that $f(x) + \varepsilon > q > f(x)$. Thus $f(x_{n_k}) > q > f(x)$ for $k \in \mathbb{N}$. By the intermediate value property of f there is z_k between x and x_{n_k} such that $f(z_k) = q$, which means $z_k \in f^{-1}(\{q\})$. Clearly, $\lim\limits_{k \to \infty} z_k = x$. Since $f^{-1}(\{q\})$ is closed, $x \in f^{-1}(\{q\})$, and therefore $f(x) = q$. A contradiction.

1.3.21. To prove our theorem it is enough to consider the case when $T > 0$. Set $g(x) = f(x+T) - f(x)$. Then there are two possibilities.
  (1) There exists $x_0 > a$ such that $g(x)$ is positive (or negative) for all $x > x_0$.
  (2) There is no such x_0.

In case (1), if, for example, g is positive on (x_0, ∞), then the sequence $\{f(x_0 + nT)\}$ is monotonically increasing. Since f is bounded, the following limit exists and is finite:

$$\lim_{n \to \infty} f(x_0 + nT) = \lim_{n \to \infty} f(x_0 + (n+1)T).$$

Therefore one can take $x_n = x_0 + nT$. In case (2), by the intermediate value property of g, for every positive integer $n > a$ there is $x_n > n$ such that $g(x_n) = 0$.

1.3. Intermediate Value Property

1.3.22. Set
$$g(x) = \begin{cases} x+2 & \text{if } -3 \leq x \leq -1, \\ -x & \text{if } -1 < x \leq 1, \\ x-2 & \text{if } 1 < x \leq 3, \end{cases}$$

and define f by the formula

$$f(x) = g(x - 6n) + 2n \quad \text{for} \quad 6n - 3 \leq x \leq 6n + 3, \ n \in \mathbb{Z}.$$

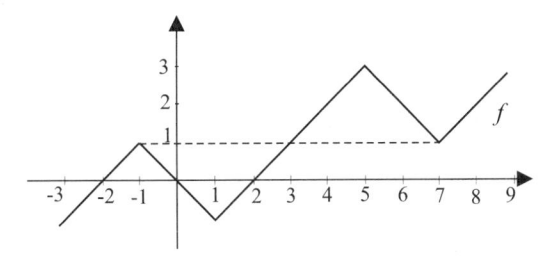

The function f has the desired property.

There is no continuous function on $\mathbb{R}$ that attains each of its values exactly two times. Suppose, contrary to this claim, that f is such a function. Let x_1, x_2 be such that $f(x_1) = f(x_2) = b$. Then $f(x) \neq b$ for $x \neq x_1, x_2$. So either $f(x) > b$ for all $x \in (x_1, x_2)$ or $f(x) < b$ for all $x \in (x_1, x_2)$. In the former case there is one $x_0 \in (x_1, x_2)$ such that $f(x_0) = \max\{f(x) : x \in [x_1, x_2]\}$. Indeed, if there were more points at which f attains its maximum on $[x_1, x_2]$, then there would be values of f assumed more than two times in $[x_1, x_2]$. Consequently, there is exactly one point x'_0 (outside the interval $[x_1, x_2]$) such that $c = f(x_0) = f(x'_0) > b$. Then the intermediate value property of f implies that every value in (b, c) is attained at least three times. A contradiction. Analogous reasoning can be applied to the case when $f(x) < b$ for $x \in (x_1, x_2)$.

1.3.23. Assume that f is strictly monotone on each interval $[t_{i-1}, t_i]$, where $i = 1, 2, \ldots, n$ and $0 = t_0 < t_1 < \cdots < t_n = 1$. The set $\mathbf{Y} = \{f(t_i) : 0 \leq i \leq n\}$ consists of at most $n+1$ elements $y_0, y_1, \ldots, y_m$. We can assume that $y_0 < y_1 < \cdots < y_m$. Put $z_{2i} = y_i$, $0 \leq i \leq m$, and choose $z_1, z_3, \ldots, z_{2m-1}$ so that

$z_0 < z_1 < z_2 < z_3 < \cdots < z_{2m-1} < z_{2m}$. Let

$$\mathbf{X}_k = \{x \in [0,1] : f(x) = z_k\},$$
$$\mathbf{X} = \mathbf{X}_0 \cup \mathbf{X}_1 \cup \cdots \cup \mathbf{X}_{2m} = \{x_1, x_2, \ldots, x_N\},$$

and let $0 = x_1 < x_2 < \cdots < x_N = 1$. For $1 \leq j \leq N$, let k_j denote the only element of the set $\{0, 1, 2, \ldots, 2m\}$ for which $f(x_j) = z_{k_j}$. Then k_1 and k_N are even and $k_j - k_{j+1} = \pm 1$, $1 \leq j < N$. It then follows that the number N of elements of the set $\mathbf{X}$ is odd. Consequently, one of the sets $\mathbf{X}_k = f^{-1}(z_k)$ has an odd number of elements.

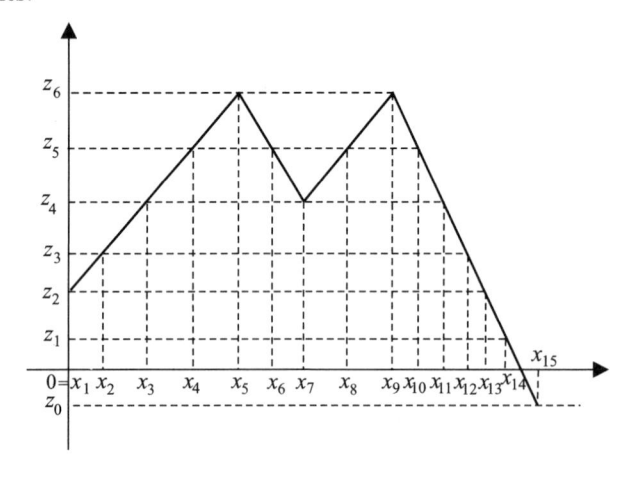

1.3.24. We first show that there are at most countably many proper local extrema of f. Indeed, if $x_0 \in (0,1)$ and $f(x_0)$ is a proper local maximum (minimum) of f, then there exists an interval $(p,q) \subset [0,1]$ with rational endpoints such that $f(x) < f(x_0)$ ($f(x) > f(x_0)$) for $x \neq x_0$ and $x \in (p,q)$. Consequently, our assertion follows from the fact that there are countably many intervals with rational endpoints.

Since there are at most countably many proper local extrema of f, there is a y between $f(0)$ and $f(1)$ which is not an extremal value of f. Assume that $f(0) < f(1)$ and put $f^{-1}(y) = \{x_1, x_2, \ldots, x_n\}$, where $x_1 < x_2 < \cdots < x_n$. Moreover, set $x_0 = 0$ and $x_{n+1} = 1$. Then the function $x \mapsto f(x) - y$ is either positive or negative on each interval (x_i, x_{i+1}), and signs are different in the adjoint intervals. Note that

1.3. Intermediate Value Property

the function is negative in the first interval and positive in the last one. Therefore the number of these intervals is even. Consequently, n is odd.

1.3.25. Define the sequence $\{x_n\}$ by setting $x_n = f^n(x_0)$. If there is a term of this sequence which is a fixed point of f, then $\{x_n\}$ is constant beginning with some value of the index n. Thus it converges. If there is a term of this sequence which is its limit point, then by assumption, the sequence is as above. So it is enough to consider the case where no term of the sequence $\{x_n\}$ is its limit point. Suppose, contrary to our claim, that the sequence is not convergent. Then

$$a = \varliminf_{n\to\infty} x_n < b = \varlimsup_{n\to\infty} x_n.$$

Let $x_{k_0} \in (a,b)$. Since x_{k_0} is not a limit point of $\{x_n\}$, there is an interval $(c,d) \subset (a,b)$ which does not contain any other term of the sequence. Moreover, there are infinitely many terms of the sequence in each of the intervals $(-\infty, c)$ and (d, ∞). If there are no terms of the sequence in (a,b), then we can take $c = a$ and $d = b$. Now we define a subsequence $\{x_{n_k}\}$ of $\{x_n\}$ in such a way that $x_{n_k} < c$ and $x_{n_k+1} > d$ for $k \in \mathbb{N}$. Therefore, if g is a limit point of $\{x_{n_k}\}$, then $g \leq c$ and $f(g) \geq d$. This contradicts our assumption that each limit point of the sequence is a fixed point of f.

1.3.26 [6]. By the result in 1.1.42, we know that $\lim\limits_{n\to\infty} \frac{f^n(0)}{n} = \alpha(f)$ exists. We will now show that there is $x_0 \in [0,1]$ such that $f(x_0) = x_0 + \alpha(f)$. If $f(x) \geq x + \alpha(f) + \varepsilon$ for all $x \in [0,1]$ and for some $\varepsilon > 0$, then, in particular, $f(0) \geq \alpha(f) + \varepsilon$. We shall show by induction that for $n \in \mathbb{N}$, $f^n(0) \geq n(\alpha(f) + \varepsilon)$. Indeed, setting $r = f(0) - [f(0)]$, we get

$$\begin{aligned} f^2(0) = f(f(0)) &= f([f(0)] + r) = [f(0)] + f(r) \\ &\geq [f(0)] + r + \alpha(f) + \varepsilon = f(0) + \alpha(f) + \varepsilon \\ &\geq 2(\alpha(f) + \varepsilon). \end{aligned}$$

Similar arguments can be applied to prove that $f^n(0) \geq n(\alpha(f) + \varepsilon)$ implies $f^{n+1}(0) \geq (n+1)(\alpha(f) + \varepsilon)$. Now observe that if $f^n(0) \geq n(\alpha(f) + \varepsilon)$, then $\alpha(f) \geq \alpha(f) + \varepsilon$, a contradiction. In an entirely similar manner one can show that if $f(x) \leq x + \alpha(f) - \varepsilon$ for all $x \in [0,1]$ and for some $\varepsilon > 0$, then $\alpha(f) \leq \alpha(f) - \varepsilon$. A contradiction again. Consequently, by the intermediate value property there is $x_0 \in [0,1]$ such that $F(x_0) = f(x_0) - x_0 = \alpha(f)$. In particular, if $\alpha(f) = 0$, then x_0 is a fixed point of f. On the other hand, if x_0 is a fixed point of f, then $\alpha(f) = \lim\limits_{n\to\infty} \frac{f^n(x_0)}{n} = 0$.

1.3.27. Let $\mathbf{A} = \{x \in [0,1] : f(x) \geq 0\}$, $s = \inf \mathbf{A}$, and $h = f + g$. Since h is decreasing, we get $h(s) \geq h(x) \geq g(x)$ for $x \in \mathbf{A}$. Since g is continuous, this implies that $h(s) \geq g(s)$. Consequently, $f(s) \geq 0$. It follows from our assumption that $g(0) > h(0) \geq h(s) \geq g(s)$. By the intermediate value property of g, there exists $t \in (0, s]$ such that $g(t) = h(s)$. Then $h(t) \geq h(s) = g(t)$, which gives $f(t) \geq 0$. By the definition of s, we have $t = s$, which implies that $g(s) = h(s)$, or equivalently, $f(s) = 0$.

1.3.28. Note first that f is not continuous on $\mathbb{R}$. If f were continuous on $\mathbb{R}$, then by the result in 1.3.16, it would be strictly monotone, for example, strictly increasing. In this case, if $f(x_0) = 0$, we would get $f(x) > 0$ for $x > x_0$, and $f(x) < 0$ for $x < x_0$, which would contradict the assumption that f maps $\mathbb{R}$ onto $[0, \infty)$. Similar analysis may be used to show that f cannot be strictly decreasing. Consequently, f is not continuous on $\mathbb{R}$.

Suppose now, contrary to our claim, that f has a finite numbers of points of discontinuity, say, $x_1 < x_2 < \cdots < x_n$. Then f is strictly monotone on every interval $(-\infty, x_1), (x_1, x_2), \ldots, (x_n, \infty)$. Consequently, by the intermediate value property of f,

$$f((-\infty, x_1)), f((x_1, x_2)), \ldots, f((x_n, \infty))$$

are pairwise disjoint open intervals. Hence

$$[0, \infty) \setminus \left(f((-\infty, x_1)) \cup \bigcup_{k=1}^{n-1} f((x_k, x_{k+1})) \cup f((x_n, \infty)) \right)$$

has at least $n+1$ elements. On the other hand, the only elements of

$$\mathbb{R} \setminus \left((-\infty, x_1) \cup \bigcup_{k=1}^{n-1} (x_k, x_{k+1}) \cup (x_n, \infty) \right)$$

1.3. Intermediate Value Property 159

are $x_1, x_2, \ldots, x_n$. Therefore f cannot be bijective, a contradiction. So, we have proved that f has infinitely many points of discontinuity.

1.3.29. We show that if $\mathbf{I}$ is a subinterval of $(0,1)$ with nonempty interior, then $f(\mathbf{I}) = [0,1]$. To this end, note that such an $\mathbf{I}$ contains a subinterval $\left(\frac{k}{2^{n_0}}, \frac{k+1}{2^{n_0}}\right)$. So it is enough to show that $f\left(\left(\frac{k}{2^{n_0}}, \frac{k+1}{2^{n_0}}\right)\right) = [0,1]$. Now observe that if $x \in (0,1)$, then either $x = \frac{m}{2^{n_0}}$ with some m and n_0, or $x \in \left(\frac{j}{2^{n_0}}, \frac{j+1}{2^{n_0}}\right)$ with some j, $j = 0, 1, \ldots, 2^{n_0}-1$. If $x = \frac{m}{2^{n_0}}$, then $f(x) = 1$ and the value of f at the middle point of $\left(\frac{k}{2^{n_0}}, \frac{k+1}{2^{n_0}}\right)$ is also 1. Next if $x \in \left(\frac{j}{2^{n_0}}, \frac{j+1}{2^{n_0}}\right)$ for some j, then there is $x' \in \left(\frac{k}{2^{n_0}}, \frac{k+1}{2^{n_0}}\right)$ such that $f(x) = f(x')$. Indeed, all numbers in $\left(\frac{k}{2^{n_0}}, \frac{k+1}{2^{n_0}}\right)$ have the same first n_0 digits, and we can find x' in this interval for which all the remaining digits are as in the binary expansion of x. Since

$$\lim_{n \to \infty} \frac{\sum_{i=1}^{n} a_i}{n} = \lim_{n \to \infty} \frac{\sum_{i=n_0+1}^{n} a_i}{n - n_0},$$

we get $f(x) = f(x')$. Consequently, it is enough to show that $f((0,1)) = [0,1]$, or in other words, that for every $y \in [0,1]$ there is $x \in (0,1)$ such that $f(x) = y$. It follows from the above that 1 is attained, for example, at $x = \frac{1}{2}$. To show that 0 is also attained, take $x = .a_1 a_2 \ldots$, where

$$a_i = \begin{cases} 1 & \text{if } i = 2^k, \; k = 1, 2, \ldots, \\ 0 & \text{otherwise.} \end{cases}$$

Then

$$f(x) = \lim_{k \to \infty} \frac{k}{2^k} = 0.$$

To obtain the value $y = \frac{p}{q}$, where p and q are co-prime positive integers, take

$$x = .\underbrace{00\ldots0}_{q-p}\underbrace{11\ldots1}_{p}\underbrace{00\ldots0}_{q-p}\ldots,$$

where blocks of $q - p$ zeros alternate with blocks of p ones. Then $f(x) = \lim\limits_{k \to \infty} \frac{kp}{kq} = \frac{p}{q}$. Now our task is to show that every irrational $y \in [0,1]$ is also attained. It is well known (see, e.g., I, 1.1.14) that there is a sequence of rationals $\frac{p_n}{q_n}$, where each pair of positive integers

p_n and q_n is co-prime, converging to y. Let

$$x = .\underbrace{00\ldots 0}_{q_1-p_1}\underbrace{11\ldots 1}_{p_1}\underbrace{00\ldots 0}_{q_2-p_2}\ldots,$$

where $q_1 - p_1$ zeros are followed by p_1 ones, then $q_2 - p_2$ zeros are followed by p_2 ones, and so on. Then

$$f(x) = \lim_{n\to\infty} \frac{p_1 + p_2 + \cdots + p_n}{q_1 + q_2 + \cdots + q_n} = \lim_{n\to\infty} \frac{p_n}{q_n} = y.$$

Since $\lim\limits_{n\to\infty} q_n = +\infty$, the second equality follows immediately from the result in I, 2.3.9 or from the Stolz theorem (see, e.g., I, 2.3.11).

1.4. Semicontinuous Functions

1.4.1.

(a) Set $\sup\limits_{\delta>0}\inf\{f(x) : x \in \mathbf{A},\ 0 < |x - x_0| < \delta\} = a$. Assume first that a is a real number. We shall show that $a = \varliminf\limits_{x\to x_0} f(x)$. By the definition of supremum, for every $\delta > 0$

(i) $\qquad \inf\{f(x) : x \in \mathbf{A},\ 0 < |x - x_0| < \delta\} \leq a,$

and for every $\varepsilon > 0$ there is δ^* such that

(ii) $\qquad \inf\{f(x) : x \in \mathbf{A},\ 0 < |x - x_0| < \delta^*\} > a - \varepsilon.$

By (ii),

(iii) $\qquad f(x) > a - \varepsilon \quad \text{if} \quad 0 < |x - x_0| < \delta^*.$

Now let $\{x_n\}$ be a sequence of points of $\mathbf{A}$ different from x_0. If the sequence converges to x_0, then, beginning with some value of the index n, $0 < |x_n - x_0| < \delta^*$. Therefore $f(x_n) > a - \varepsilon$. If $\{f(x_n)\}$ converges, say to y, then we get $y \geq a - \varepsilon$, and consequently, $\varliminf\limits_{x\to x_0} f(x) \geq a$. To show that also $\varliminf\limits_{x\to x_0} f(x) \leq a$, we will use (i). It follows from the definition of infimum that, given $\varepsilon_1 > 0$, there

1.4. Semicontinuous Functions

exists $x^* \in \mathbf{A}$ such that $0 < |x^* - x_0| < \delta$ and $f(x^*) < a + \varepsilon_1$. Taking $\delta = \frac{1}{n}$, we get a sequence $\{x_n^*\}$ such that

$$0 < |x_n^* - x_0| < \frac{1}{n} \quad \text{and} \quad f(x_n^*) < a + \varepsilon_1.$$

Combined with (iii), this gives $a - \varepsilon < f(x_n^*) < a + \varepsilon_1$. Without loss of generality we may assume $\{f(x_n^*)\}$ is convergent. Then its limit is less than or equal to $a+\varepsilon_1$. It follows from the arbitrariness of $\varepsilon_1 > 0$ that

$$\varlimsup_{x \to x_0} f(x) \leq a.$$

If $a = +\infty$, then, given $M > 0$, there is δ^* such that

$$\inf\{f(x) : x \in \mathbf{A},\ 0 < |x - x_0| < \delta^*\} > M.$$

Hence if $0 < |x - x_0| < \delta^*$, then $f(x) > M$. Consequently, if $\{x_n\}$ converges to x_0, then, beginning with some value of the index n, $f(x_n) > M$. Thus $\lim_{n \to \infty} f(x_n) = +\infty$, which means that $\varliminf_{x \to x_0} f(x) = \varlimsup_{x \to x_0} f(x) = +\infty$. Finally, if $a = -\infty$, then for any $\delta > 0$,

$$\inf\{f(x) : x \in \mathbf{A},\ 0 < |x - x_0| < \delta\} = -\infty.$$

Therefore there is a sequence $\{x_n^*\}$ convergent to x_0 such that $\lim_{n \to \infty} f(x_n^*) = -\infty$, which gives $\varlimsup_{x \to x_0} f(x) = -\infty$.

(b) The proof runs as in (a).

1.4.2. The result is an immediate consequence of 1.1.35 and the preceding problem.

1.4.3. It follows from the result in the preceding problem that, given $\varepsilon > 0$, there is $\delta > 0$ such that

$$0 \leq y_0 - \inf\{f(x) : x \in \mathbf{A},\ 0 < |x - x_0| < \delta\} < \varepsilon.$$

By the definition of infimum this is equivalent to the conditions (i) and (ii).

By 1.4.2 (b), $\tilde{y} = \varlimsup_{x \to x_0} f(x)$ if and only if for every $\varepsilon > 0$ the following two conditions are satisfied:
 (1) There is $\delta > 0$ such that $f(x) < \tilde{y} + \varepsilon$ for all $x \in \mathbf{A}$ from the deleted neighborhood $0 < |x - x_0| < \delta$.

(2) For every $\delta > 0$ there is $x' \in \mathbf{A}$ from the deleted neighborhood $0 < |x' - x_0| < \delta$ for which $f(x') > \tilde{y} - \varepsilon$.

1.4.4.
(a) By 1.4.2(a), $\lim\limits_{x \to x_0} f(x) = -\infty$ if and only if for any $\delta > 0$

$$\inf\{f(x) : x \in \mathbf{A},\ 0 < |x - x_0| < \delta\} = -\infty.$$

This means that for any $\delta > 0$ the set

$$\{f(x) : x \in \mathbf{A},\ 0 < |x - x_0| < \delta\}$$

is unbounded below, which gives the desired result.

(b) The proof runs as in (a).

1.4.5. Let $\{\delta_n\}$ be a monotonically decreasing sequence of positive numbers converging to zero. It follows from 1.4.2(a) that

$$l = \lim\limits_{n \to \infty} \inf\{f(x) : x \in \mathbf{A},\ 0 < |x - x_0| < \delta_n\}.$$

For a real l, this is equivalent to the following two conditions
  (1) For $n \in \mathbb{N}$ there exists $k_n \in \mathbb{N}$ such that $0 < |x - x_0| < \delta_k$ implies $f(x) > l - \frac{1}{n}$ for $k > k_n$.
  (2) For $n \in \mathbb{N}$ there exist $k_n > n$ and $x_{k_n} \in \mathbf{A}$ such that $0 < |x_{k_n} - x_0| < \delta_{k_n}$ and $f(x_{k_n}) < l + \frac{1}{n}$.

Consequently, there exists a sequence $\{x_{k_n}\}$ convergent to x_0 such that $\lim\limits_{n \to \infty} f(x_{k_n}) = l$.

If $\lim\limits_{x \to x_0} f(x) = -\infty$, then, by 1.4.4(a), for any $n \in \mathbb{N}$ and $\delta > 0$ there is $x_n \in \mathbf{A}$ such that $0 < |x_n - x_0| < \delta$ and $f(x_n) < -n$. Therefore $\lim\limits_{n \to \infty} x_n = x_0$ and $\lim\limits_{n \to \infty} f(x_n) = -\infty$.

If $\lim\limits_{x \to x_0} f(x) = +\infty$, then the existence of $\{x_n\}$ follows immediately from the definition.

1.4.6. The result follows immediately from I, 1.1.2 and from 1.4.1.

1.4.7. It is enough to apply I, 1.1.4 and 1.4.1.

1.4. Semicontinuous Functions

1.4.8. Note that

(1) $\quad\inf_{x\in \mathbf{A}}(f(x)+g(x)) \geq \inf_{x\in \mathbf{A}} f(x) + \inf_{x\in \mathbf{A}} g(x),$

(2) $\quad\sup_{x\in \mathbf{A}}(f(x)+g(x)) \leq \sup_{x\in \mathbf{A}} f(x) + \sup_{x\in \mathbf{A}} g(x).$

Indeed, for $x \in \mathbf{A}$,

$$f(x)+g(x) \geq \inf_{x\in \mathbf{A}} f(x) + \inf_{x\in \mathbf{A}} g(x),$$

which implies (1). Inequality (2) can be proved analogously.

We first show that

(3) $\quad \varliminf_{x\to x_0} f(x) + \varliminf_{x\to x_0} g(x) \leq \varliminf_{x\to x_0} (f(x)+g(x)).$

By (1), we get

$$\inf\{f(x)+g(x) : x\in \mathbf{A},\ 0<|x-x_0|<\delta\}$$
$$\geq \inf\{f(x) : x\in \mathbf{A},\ 0<|x-x_0|<\delta\}$$
$$+ \inf\{g(x) : x\in \mathbf{A},\ 0<|x-x_0|<\delta\}.$$

Passage to the limit as $\delta \to 0^+$ and the result in 1.4.2(a) give (3). The inequality

(4) $\quad \varlimsup_{x\to x_0}(f(x)+g(x)) \leq \varlimsup_{x\to x_0} f(x) + \varlimsup_{x\to x_0} g(x)$

can be proved in an entirely similar manner. Furthermore, it follows from Problem 1.4.6 and (3) that

$$\varliminf_{x\to x_0} f(x) = \varliminf_{x\to x_0}(f(x)+g(x)-g(x))$$
$$\geq \varliminf_{x\to x_0}(f(x)+g(x)) + \varliminf_{x\to x_0}(-g(x))$$
$$= \varliminf_{x\to x_0}(f(x)+g(x)) - \varlimsup_{x\to x_0} g(x).$$

One can prove, in much the same way, that

$$\varliminf_{x\to x_0} f(x) + \varlimsup_{x\to x_0} g(x) \leq \varlimsup_{x\to x_0}(f(x)+g(x)).$$

To show that the inequalities can be strict, consider the functions defined as follows:

$$f(x) = \begin{cases} \sin\frac{1}{x} & \text{if } x > 0, \\ 0 & \text{if } x \leq 0, \end{cases}$$

$$g(x) = \begin{cases} 0 & \text{if } x \geq 0, \\ \sin\frac{1}{x} & \text{if } x < 0. \end{cases}$$

For $x_0 = 0$ the given inequalities are of the form $-2 < -1 < 0 < 1 < 2$.

1.4.9. Observe first that if f and g are nonnegative on $\mathbf{A}$, then

(1) $\quad \inf_{x \in \mathbf{A}} (f(x) \cdot g(x)) \geq \inf_{x \in \mathbf{A}} f(x) \cdot \inf_{x \in \mathbf{A}} g(x),$

(2) $\quad \sup_{x \in \mathbf{A}} (f(x) \cdot g(x)) \leq \sup_{x \in \mathbf{A}} f(x) \cdot \sup_{x \in \mathbf{A}} g(x).$

The rest of the proof runs as in the solution of the foregoing problem. To see that the given inequalities can be strict, consider the functions given by setting

$$f(x) = \begin{cases} \frac{1}{\sin^2\frac{1}{x}+1} & \text{if } x > 0, \\ 2 & \text{if } x \leq 0, \end{cases}$$

$$g(x) = \begin{cases} 3 & \text{if } x \geq 0, \\ \frac{1}{\sin^2\frac{1}{x}+1} & \text{if } x < 0. \end{cases}$$

For $x_0 = 0$ the given inequalities are of the form $\frac{1}{4} < 1 < \frac{3}{2} < 3 < 6$.

1.4.10. We have $\varliminf_{x \to x_0} f(x) = \varlimsup_{x \to x_0} f(x) = \lim_{x \to x_0} f(x)$. So, by 1.4.8,

$$\lim_{x \to x_0} f(x) + \varliminf_{x \to x_0} g(x) \leq \varliminf_{x \to x_0} (f(x) + g(x)) \leq \varlimsup_{x \to x_0} f(x) + \varliminf_{x \to x_0} g(x).$$

Therefore

$$\varliminf_{x \to x_0} (f(x) + g(x)) = \lim_{x \to x_0} f(x) + \varliminf_{x \to x_0} g(x).$$

The other equalities can be proved analogously.

1.4. Semicontinuous Functions

1.4.11. If $\lambda = l$ or $\lambda = L$, then the assertion follows immediately from 1.4.5. So assume that $\lambda \in (l, L)$. Then by 1.4.5, there exist sequences $\{x'_n\}$ and $\{x''_n\}$ both converging to a and such that

$$\lim_{n\to\infty} f(x'_n) = l \quad \text{and} \quad \lim_{n\to\infty} f(x''_n) = L.$$

It then follows that $f(x'_n) < \lambda < f(x''_n)$ beginning with some value of the index n. Since f is continuous, it enjoys the intermediate value property. Hence there is x_n in the interval with the endpoints x'_n and x''_n for which $f(x_n) = \lambda$. Since $\{x'_n\}$ and $\{x''_n\}$ converge to a, the sequence $\{x_n\}$ does also.

1.4.12. The function is continuous at every $k\pi$ with $k \in \mathbb{Z}$ (see, e.g., 1.2.1). Clearly,

$$\varlimsup_{x\to x_0} f(x) = \begin{cases} \sin x_0 & \text{if } \sin x_0 > 0, \\ 0 & \text{if } \sin x_0 \le 0, \end{cases}$$

and

$$\varliminf_{x\to x_0} f(x) = \begin{cases} 0 & \text{if } \sin x_0 > 0, \\ \sin x_0 & \text{if } \sin x_0 \le 0. \end{cases}$$

Consequently, f is upper semicontinuous on the set

$$\left(\mathbb{Q} \cap \bigcup_{k\in\mathbb{Z}}(2k\pi,(2k+1)\pi)\right) \cup \left((\mathbb{R}\setminus\mathbb{Q}) \cap \bigcup_{k\in\mathbb{Z}}[(2k-1)\pi,2k\pi]\right)$$

and lower semicontinuous on

$$\left(\mathbb{Q} \cap \bigcup_{k\in\mathbb{Z}}((2k-1)\pi,2k\pi)\right) \cup \left((\mathbb{R}\setminus\mathbb{Q}) \cap \bigcup_{k\in\mathbb{Z}}[2k\pi,(2k+1)\pi]\right).$$

1.4.13. We have

$$\varlimsup_{x\to x_0} f(x) = \begin{cases} x_0^2 - 1, & \text{if } x_0 < -1 \text{ or } x_0 > 1, \\ 0 & \text{if } x_0 \in [-1,1], \end{cases}$$

and

$$\varliminf_{x\to x_0} f(x) = \begin{cases} 0 & \text{if } x_0 < -1 \text{ or } x_0 > 1, \\ x_0^2 - 1 & \text{if } x_0 \in [-1,1]. \end{cases}$$

Thus f is upper semicontinuous at each irrational in $(-\infty, -1) \cup (1, \infty)$ and at each rational in the interval $[-1, 1]$; f is lower semicontinuous at each rational in $(-\infty, -1] \cup [1, \infty)$ and at each irrational in the interval $(-1, 1)$.

1.4.14. The function f is continuous at zero and at each irrational (see, e.g., 1.2.3). Assume that $0 \neq x_0 = \frac{p}{q}$, where $p \in \mathbb{Z}$ and $q \in \mathbb{N}$ are co-prime. Then $f(x_0) = \frac{1}{q}$ and $\overline{\lim}_{x \to x_0} f(x) = 0 < \frac{1}{q}$. Hence f is upper semicontinuous on $\mathbb{R}$.

1.4.15.

(a) The function f is continuous at zero and at each positive irrational (see, e.g., 1.2.3). Assume that x_0 is a negative irrational. Then $\overline{\lim}_{x \to x_0} f(x) = |x_0| = f(x_0)$. Therefore f is upper semicontinuous at zero and at each irrational. If $x_0 = \frac{p}{q} > 0$, then $\lim_{x \to x_0} f(x) = \frac{p}{q} > \frac{p}{q+1} = f(x_0)$. This means that f is lower semicontinuous at each positive rational. If $x_0 = \frac{p}{q} < 0$, then

$$\overline{\lim}_{x \to x_0} f(x) = -\frac{p}{q} > \frac{p}{q+1} = f(x_0)$$

and

$$\underline{\lim}_{x \to x_0} f(x) = \frac{p}{q} < \frac{p}{q+1} = f(x_0).$$

So f is neither upper nor lower semicontinuous at negative rationals.

(b) Note that for $x \in (0, 1]$,

$$\underline{\lim}_{t \to x} f(t) = -x < f(x) < x = \overline{\lim}_{t \to x} f(t).$$

Thus f is neither upper nor lower semicontinuous on $(0, 1]$.

1.4.16.

(a) If $x_0 \in \mathbf{A}$ is isolated in $\mathbf{A}$, then the assertion is obviously true. If $x_0 \in \mathbf{A}$ is a limit point of $\mathbf{A}$, then the assertion follows from the fact that

$$\overline{\lim}_{x \to x_0} af(x) = \begin{cases} a \overline{\lim}_{x \to x_0} f(x) & \text{if } a > 0, \\ a \underline{\lim}_{x \to x_0} f(x) & \text{if } a < 0. \end{cases}$$

1.4. Semicontinuous Functions

(b) Assume that x_0 is a limit point of $\mathbf{A}$ and, for example, f and g are lower semicontinuous at x_0. Then, by 1.4.8,

$$\varliminf_{x \to x_0} (f(x) + g(x)) \geq \varliminf_{x \to x_0} f(x) + \varliminf_{x \to x_0} g(x) \geq f(x_0) + g(x_0).$$

1.4.17. Assume, for example, that the f_n are lower semicontinuous at x_0. Since $\sup\limits_{n \in \mathbb{N}} f_n \geq f_n$ for $n \in \mathbb{N}$, we get

$$\varliminf_{x \to x_0} \sup_{n \in \mathbb{N}} f_n(x) \geq \varliminf_{x \to x_0} f_n(x) \geq f_n(x_0) \quad \text{for} \quad n \in \mathbb{N}.$$

Consequently,

$$\varliminf_{x \to x_0} \sup_{n \in \mathbb{N}} f_n(x) \geq \sup_{n \in \mathbb{N}} f_n(x_0).$$

1.4.18. It is enough to observe that if $\{f_n\}$ is an increasing (resp. decreasing) sequence, then $\lim\limits_{n \to \infty} f_n(x) = \sup\limits_{n \in \mathbb{N}} f_n(x)$ (resp. $\lim\limits_{n \to \infty} f_n(x) = \inf\limits_{n \in \mathbb{N}} f_n(x)$) (see, e.g., I, 2.1.1) and use the result in the foregoing problem.

1.4.19. By 1.4.1 we have

$$\begin{aligned} f_1(x) &= \max\{f(x), \varlimsup_{z \to x} f(z)\} \\ &= \inf_{\delta > 0} \sup\{f(z) : z \in \mathbf{A}, \ |z - x| < \delta\} \\ &= \lim_{\delta \to 0^+} \sup\{f(z) : z \in \mathbf{A}, \ |z - x| < \delta\}. \end{aligned}$$

Similarly,

$$f_2(x) = \lim_{\delta \to 0^+} \inf\{f(z) : z \in \mathbf{A}, \ |z - x| < \delta\}.$$

Hence

$$\begin{aligned} f_1(x) - f_2(x) &= \lim_{\delta \to 0^+} \sup\{f(z) : z \in \mathbf{A}, |z - x| < \delta\} \\ &\quad - \lim_{\delta \to 0^+} \inf\{f(u) : u \in \mathbf{A}, |u - x| < \delta\} \\ &= \lim_{\delta \to 0^+} \sup\{f(z) - f(u) : z, u \in \mathbf{A}, |z - x| < \delta, |u - x| < \delta\} \\ &= \lim_{\delta \to 0^+} \sup\{|f(z) - f(u)| : z, u \in \mathbf{A}, |z - x| < \delta, |u - x| < \delta\} \\ &= o_f(x). \end{aligned}$$

1.4.20. Let x be a limit point of $\mathbf{A}$, and let $\{x_n\}$ be a sequence of points in $\mathbf{A}$ converging to x. Set $\delta_n = |x_n - x| + \frac{1}{n}$. Then, $|z - x_n| < \delta_n$ implies $|z - x| < 2\delta_n$. Consequently, see the solution of the preceding problem,

$$f_2(x_k) = \lim_{n \to \infty} \inf\{f(z) : z \in \mathbf{A}, \ |z - x_k| < \delta_n\}$$
$$\geq \inf\{f(z) : z \in \mathbf{A}, \ |z - x_k| < \delta_k\}$$
$$\geq \inf\{f(z) : z \in \mathbf{A}, \ |z - x| < 2\delta_k\}.$$

Passage to the limit as $k \to \infty$ gives $\lim_{k \to \infty} f_2(x_k) \geq f_2(x)$. It then follows that $\lim_{z \to x} f_2(z) \geq f_2(x)$, and therefore the lower semicontinuity of f_2 is proved. In an entirely similar manner one can show that f_1 is upper semicontinuous. Now, by the result in the foregoing problem, $o_f(x) = f_1(x) - f_2(x)$, which together with 1.4.16 proves the upper semicontinuity of o_f.

1.4.21. We will prove our statement for lower semicontinuous functions. Assume first that the given condition is satisfied. Then for $a < f(x_0)$ there is $\delta > 0$ such that $f(x) > a$ whenever $|x - x_0| < \delta$. If $\{x_n\}$ is a sequence of points in $\mathbf{A}$ converging to x_0, then $|x_n - x_0| < \delta$ for sufficiently large n. Hence $f(x_n) > a$, which implies $\lim_{n \to \infty} f(x_n) \geq a$. Because of the arbitrariness of a we get $\lim_{x \to x_0} f(x) \geq f(x_0)$. Now assume that f is lower semicontinuous at x_0 and, contrary to our claim, the given condition is not satisfied. Then there is $a < f(x_0)$ such that for any $n \in \mathbb{N}$ there exists $x_n \in \mathbf{A}$ for which $|x_n - x_0| < \delta_n = \frac{1}{n}$ and $f(x_n) \leq a$. Thus the sequence $\{x_n\}$ converges to x_0 and $\lim_{n \to \infty} f(x_n) \leq a < f(x_0)$, a contradiction.

1.4.22. Suppose that for every $a \in \mathbb{R}$ the set $\{x \in \mathbf{A} : f(x) > a\}$ is open. Let x_0 be an element of $\mathbf{A}$ and take $a < f(x_0)$. Then there is $\delta > 0$ such that $(x_0 - \delta, x_0 + \delta) \subset \{x \in \mathbf{A} : f(x) > a\}$. It then follows by the result in the foregoing problem that f is lower semicontinuous.

Suppose now that f is lower semicontinuous on $\mathbf{A}$. We shall show that the set $\{x \in \mathbf{A} : f(x) \leq a\}$ is closed in $\mathbf{A}$. Let $\{x_n\}$ be a sequence of points in this set converging to x. Then $f(x_n) \leq a$, and

1.4. Semicontinuous Functions

consequently, $f(x) \leq \varliminf\limits_{n\to\infty} f(x_n) \leq a$, which implies that x is also an element of $\{x \in \mathbf{A} : f(x) \leq a\}$. So we have proved that this set is closed, or equivalently that its complement is open in $\mathbf{A}$.

1.4.23. Assume that f is lower semicontinuous on $\mathbb{R}$, and set $\mathbf{B} = \{(x,y) \in \mathbb{R}^2 : y \geq f(x)\}$. Our task is to show that $\mathbf{B}$ is closed in $\mathbb{R}^2$. Let $\{(x_n, y_n)\}$ be a sequence of points in $\mathbf{B}$ converging to (x_0, y_0). Then

$$y_0 = \lim_{n\to\infty} y_n \geq \varliminf_{n\to\infty} f(x_n) \geq \varliminf_{x\to x_0} f(x) \geq f(x_0).$$

Hence $(x_0, y_0) \in \mathbf{B}$.

Assume now that $\mathbf{B}$ is closed and f is not lower semicontinuous at an $x_0 \in \mathbb{R}$. Then the set $\mathbf{B}^c = \{(x,y) \in \mathbb{R}^2 : y < f(x)\}$ is open in $\mathbb{R}^2$ and there exists a sequence $\{x_n\}$, $x_n \neq x_0$, converging to x_0 and such that $y = \lim\limits_{n\to\infty} f(x_n) < f(x_0)$. Take g such that $y < g < f(x_0)$. Then (x_0, g) is in $\mathbf{B}^c$. Hence there is a ball centered at (x_0, g) contained in $\mathbf{B}^c$. This means that for sufficiently large n, (x_n, g) are in $\mathbf{B}^c$, or equivalently, $g < f(x_n)$. Therefore $g \leq y$, a contradiction.

Recall that f is upper semicontinuous on $\mathbb{R}$ if and only if $-f$ is lower semicontinuous on $\mathbb{R}$. So f is upper semicontinuous on $\mathbb{R}$ if and only the set $\{(x,y) \in \mathbb{R}^2 : y \leq f(x)\}$ is closed in $\mathbb{R}^2$.

1.4.24 [21]. We show first that f is lower semicontinuous if and only if the function $g(x) = \frac{2}{\pi} \arctan f(x)$ is lower semicontinuous. To this end we use the characterization given in 1.4.20. Suppose that f is lower semicontinuous. To prove that g is also lower semicontinuous it is enough to show that for every real a the set $\mathbf{B} = \{x \in \mathbf{A} : \frac{2}{\pi} \arctan f(x) > a\}$ is open in $\mathbf{A}$. Clearly, if $a \leq -1$, then $\mathbf{B} = \mathbf{A}$, and if $a \geq 1$, then $\mathbf{B} = \emptyset$. If $|a| < 1$, then $\mathbf{B} = \{x \in \mathbf{A} : f(x) > \tan(\frac{\pi}{2}a)\}$; so it is open by assumption. Suppose now that g is lower semicontinuous. Then $\{x \in \mathbf{A} : g(x) > \frac{2}{\pi} \arctan a\}$ is open for every real a. Consequently, the set $\{x \in \mathbf{A} : f(x) > a\}$ is open.

For $n \in \mathbb{N}$, $a \in \mathbf{A}$, define $\varphi_{a,n}$ by setting

$$\varphi_{a,n}(x) = g(a) + n|x - a|, \quad x \in \mathbb{R},$$

and put
$$g_n(x) = \inf_{a \in \mathbf{A}} \varphi_{a,n}(x).$$
Obviously,
$$g_n(x) \leq g_{n+1}(x) \quad \text{for} \quad x \in \mathbb{R}$$
and
$$g_n(x) \leq \varphi_{x,n}(x) = g(x) \quad \text{for} \quad x \in \mathbf{A}.$$
Hence for each $x \in \mathbf{A}$ the sequence $\{g_n(x)\}$ is convergent. Now we show that the functions g_n are continuous on $\mathbb{R}$. Indeed, for $x, x' \in \mathbb{R}$,
$$|\varphi_{a,n}(x) - \varphi_{a,n}(x')| \leq n|x - x'|.$$
It then follows that
$$\varphi_{a,n}(x') - n|x - x'| \leq \varphi_{a,n}(x) \leq \varphi_{a,n}(x') + n|x - x'|.$$
Consequently,
$$g_n(x') - n|x - x'| \leq g_n(x) \leq g_n(x') + n|x - x'|,$$
and therefore continuity of g_n is proved. It follows from the above that for $x \in \mathbf{A}$, $\lim_{n \to \infty} g_n(x) \leq g(x)$. Our task is to show that $\lim_{n \to \infty} g_n(x) \geq g(x)$. Let $x \in \mathbf{A}$ and let $\alpha < g(x)$. Since g is lower semicontinuous at x, there is $\delta > 0$ such that $g(a) > \alpha$ if $|x - a| < \delta$. Hence

(1) $$\varphi_{a,n}(x) \geq g(a) > \alpha \quad \text{for} \quad |x - a| < \delta.$$

On the other hand,

(2) $$\varphi_{a,n}(x) > -1 + n\delta \quad \text{for} \quad |x - a| \geq \delta,$$

which combined with (1) gives
$$g_n(x) = \inf_{a \in \mathbf{A}} \varphi_{a,n}(x) \geq \min\{\alpha, -1 + n\delta\}.$$

Therefore $g_n(x) \geq \alpha$ for sufficiently large n, and consequently we get $\lim_{n \to \infty} g_n(x) \geq \alpha$. Finally, upon passage to the limit as $\alpha \to g(x)$, we obtain $\lim_{n \to \infty} g_n(x) \geq g(x)$.

1.5. Uniform Continuity

1.4.25. It follows from the theorem of Baire (see the foregoing problem) that there are a decreasing sequence $\{f_n\}$ and an increasing sequence $\{g_n\}$ of continuous functions converging on $\mathbf{A}$ to f and g, respectively. Set

$$\varphi_1(x) = f_1(x), \qquad \psi_1(x) = \min\{\varphi_1(x), g_1(x)\},$$
$$\ldots \qquad \ldots$$
$$\varphi_n(x) = \max\{\psi_{n-1}(x), f_n(x)\}, \quad \psi_n(x) = \min\{\varphi_n(x), g_n(x)\}.$$

Then $\{\varphi_n\}$ is decreasing, because the inequalities $\psi_n \leq \varphi_n$ and $f_n \leq \varphi_n$ imply

$$\varphi_{n+1} = \max\{\psi_n, f_{n+1}\} \leq \max\{\psi_n, f_n\} \leq \max\{\varphi_n, f_n\} = \varphi_n.$$

Similarly, one can show that $\{\psi_n\}$ is increasing. Observe now that the sequences of continuous functions $\{\varphi_n\}$ and $\{\psi_n\}$ both converge, say to φ and ψ, respectively. One can show that $\varphi(x) = \max\{\psi(x), f(x)\}$ and $\psi(x) = \min\{\varphi(x), g(x)\}$ (see, e.g., I, 2.4.28). So if $\varphi(x) \neq \psi(x)$ for some x, then $\varphi(x) = f(x)$; and since $f(x) \leq g(x)$, we have also $\psi(x) = f(x)$, a contradiction. Consequently, the sequences $\{\varphi_n\}$ and $\{\psi_n\}$ have a common limit, say h, such that $f(x) \leq h(x) \leq g(x)$. By 1.4.18, h is lower and upper semicontinuous, hence continuous.

1.5. Uniform Continuity

1.5.1.

(a) The function can be continuously extended on $[0,1]$. Therefore f is uniformly continuous on $(0,1)$.

(b) Note that for $n \in \mathbb{N}$,

$$\left| f\left(\frac{1}{2n\pi}\right) - f\left(\frac{1}{2n\pi + \frac{\pi}{2}}\right) \right| = 1$$

although $\left|\frac{1}{2n\pi} - \frac{1}{2n\pi + \frac{\pi}{2}}\right|$ can be arbitrarily small. Consequently, the function is not uniformly continuous on $(0,1)$.

(c) Since there exists a continuous extension of f on $[0,1]$, the function f is uniformly continuous on $(0,1)$.

(d) We have
$$\left|f\left(\frac{1}{\ln n}\right) - f\left(\frac{1}{\ln(n+1)}\right)\right| = |n - (n+1)| = 1$$
and $\left|\frac{1}{\ln n} - \frac{1}{\ln(n+1)}\right| \underset{n\to\infty}{\longrightarrow} 0$. Hence f is not uniformly continuous on $(0,1)$.

(e) Since $\lim\limits_{x\to 0^+} e^{-\frac{1}{x}} = 0$, the function can be continuously extended on $[0,1]$. Thus f is uniformly continuous on $(0,1)$.

(f) The function is not uniformly continuous on $(0,1)$ because
$$\left|f\left(\frac{1}{2n\pi}\right) - f\left(\frac{1}{2n\pi + \pi}\right)\right| = e^{\frac{1}{2n\pi}} + e^{\frac{1}{2n\pi+\pi}} > 2, \quad n \in \mathbb{N}.$$

(g) To see that the function is not uniformly continuous on $(0,1)$, note that
$$\left|f\left(\frac{1}{e^n}\right) - f\left(\frac{1}{e^{n+1}}\right)\right| = 1.$$

(h) Observe that
$$\left|f\left(\frac{1}{2n}\right) - f\left(\frac{1}{2n+1}\right)\right| = \cos\frac{1}{2n} + \cos\frac{1}{2n+1} \underset{n\to\infty}{\longrightarrow} 2.$$
So, the function is not uniformly continuous on $(0,1)$.

(i) As above, one can show that the function is not uniformly continuous on $(0,1)$.

1.5.2.

(a) We will show that f is uniformly continuous on $[0,\infty)$. Indeed, in view of the inequality
$$|\sqrt{x_1} - \sqrt{x_2}| \leq \sqrt{|x_1 - x_2|} \quad \text{for} \quad x_1, x_2 \in [0,\infty)$$
we have
$$|x_1 - x_2| < \delta = \varepsilon^2 \quad \text{implies} \quad |\sqrt{x_1} - \sqrt{x_2}| < \varepsilon.$$

(b) Note that
$$\left|f(2n\pi) - f\left(2n\pi + \frac{1}{n}\right)\right| \underset{n\to\infty}{\longrightarrow} 2\pi.$$
So, f is not uniformly continuous on $[0,\infty)$.

1.5. Uniform Continuity

(c) Since
$$|\sin^2 x_1 - \sin^2 x_2| = |\sin x_1 - \sin x_2| \cdot |\sin x_1 + \sin x_2| \le 2|x_1 - x_2|,$$
the function is uniformly continuous on $[0, \infty)$.

(d) The function is not uniformly continuous on $[0, \infty)$ because
$$\left| f(\sqrt{2n\pi}) - f\left(\sqrt{2n\pi + \frac{\pi}{2}}\right) \right| = 1$$
although $\left|\sqrt{2n\pi} - \sqrt{2n\pi + \frac{\pi}{2}}\right| \underset{n\to\infty}{\longrightarrow} 0$.

(e) The function is not uniformly continuous on $[0, \infty)$. Indeed, it follows from the continuity of the logarithm function that
$$|\ln n - \ln(n+1)| = \ln\left(1 + \frac{1}{n}\right) \underset{n\to\infty}{\longrightarrow} 0.$$
Moreover,
$$|f(\ln n) - f(\ln(n+1))| = 1.$$

(f) One can show, as in (d), that the function is not uniformly continuous on $[0, \infty)$.

(g) Since
$$|\sin(\sin x_1) - \sin(\sin x_2)| \le 2\left|\sin \frac{\sin x_1 - \sin x_2}{2}\right| \le |x_1 - x_2|,$$
f is uniformly continuous on $[0, \infty)$.

(h) Note that
$$\left| f\left(2n\pi + \frac{1}{2n\pi}\right) - f(2n\pi) \right|$$
$$= \left| \sin\left(2n\pi \sin \frac{1}{2n\pi} + \frac{1}{2n\pi} \sin \frac{1}{2n\pi}\right) \right| \underset{n\to\infty}{\longrightarrow} \sin 1.$$
Consequently, the function is not uniformly continuous on $[0, \infty)$.

(i) Observe that
$$|\sin \sqrt{x_1} - \sin \sqrt{x_2}|$$
$$= \left| 2 \sin \frac{\sqrt{x_1} - \sqrt{x_2}}{2} \cos \frac{\sqrt{x_1} + \sqrt{x_2}}{2} \right| \le |\sqrt{x_1} - \sqrt{x_2}|.$$
Now reasoning as in (a) proves the uniform continuity of f.

1.5.3. We will show that $\lim_{x \to a^+} f(x)$ exists. By uniform continuity, given $\varepsilon > 0$ there exists $\delta > 0$ such that $|f(x_1) - f(x_2)| < \varepsilon$ whenever $|x_1 - x_2| < \delta$. Clearly, if $a < x_1 < a + \delta$ and $a < x_2 < a + \delta$, then $|x_1 - x_2| < \delta$. It follows from the Cauchy theorem (see, e.g., 1.1.37) that the left-hand limit of f at a exists. In an entirely similar manner one can show also that the right-hand limit of f at b exists.

1.5.4.

(a) It follows directly from the definition of uniform continuity that the sum of two uniformly continuous functions is also uniformly continuous.

(b) If f and g are uniformly continuous on a finite interval (a, b), then by the result in the foregoing problem the functions can be continuously extended on $[a, b]$. Thus f and g are bounded on (a, b). Consequently, the uniform continuity of fg on (a, b) follows from the inequality

$$|f(x_1)g(x_1) - f(x_2)g(x_2)|$$
$$\leq |f(x_1)||g(x_1) - g(x_2)| + |g(x_2)||f(x_1) - f(x_2)|.$$

On the other hand, the functions $f(x) = g(x) = x$ are uniformly continuous on $[a, \infty)$ but $f(x)g(x) = x^2$ is not uniformly continuous on this infinite interval.

(c) By (b), $x \to f(x)\sin x$ is uniformly continuous on (a, b). The function need not be uniformly continuous on $[a, \infty)$, as the example in 1.5.2(b) shows.

1.5.5.

(a) Given $\varepsilon > 0$, there are $\delta_1 > 0$ and $\delta_2 > 0$ such that $|f(x_1) - f(b)| < \frac{\varepsilon}{2}$ if $0 \leq b - x_1 < \frac{\delta_1}{2}$ and $|f(x_2) - f(b)| < \frac{\varepsilon}{2}$ if $0 \leq x_2 - b < \frac{\delta_2}{2}$. Setting $\delta = \min\{\delta_1, \delta_2\}$, we get

(1) $\qquad |f(x_1) - f(x_2)| < \varepsilon \quad \text{if} \quad |x_1 - x_2| < \delta.$

For $x_1, x_2 \in (a, b]$ or $x_1, x_2 \in [b, c)$, (1) is clearly satisfied with some positive $\delta > 0$.

1.5. Uniform Continuity

(b) No. Let $\mathbf{A} = \mathbb{N}$ and $\mathbf{B} = \{n + \frac{1}{n} : n \in \mathbb{N}\}$, and consider the function f defined by

$$f(x) = \begin{cases} 1 & \text{if } x \in \mathbf{A}, \\ 2 & \text{if } x \in \mathbf{B}. \end{cases}$$

1.5.6. If f is constant, then it is uniformly continuous on $\mathbb{R}$. If f is a nonconstant periodic function, then its fundamental period T exists (see 1.2.23). Clearly, f is uniformly continuous on each interval $[kT, (k+1)T]$, $k \in \mathbb{Z}$. So, as in the solution of 1.5.5(a), one can show that f is uniformly continuous on $\mathbb{R}$.

1.5.7.

(a) Set $\lim\limits_{x \to +\infty} f(x) = L$ and $\lim\limits_{x \to -\infty} f(x) = l$. Then, given $\varepsilon > 0$, there is $A > 0$ such that $|f(x) - L| < \frac{\varepsilon}{2}$ for $x \geq A$ and $|f(x) - l| < \frac{\varepsilon}{2}$ for $x \leq -A$. This implies that if $x_1, x_2 \in [A, \infty)$ or $x_1, x_2 \in (-\infty, -A]$, then $|f(x_1) - f(x_2)| < \varepsilon$. Obviously, f is uniformly continuous on $[-A, A]$. Finally, as in the solution of 1.5.5(a) one can show that f is uniformly continuous on $\mathbb{R}$.

(b) The proof runs as in (a).

1.5.8. It is enough to apply the result in the foregoing problem.

1.5.9. $\lim\limits_{x \to \infty} f(x)$ need not exist. To see this consider the function in 1.5.2(c). The limit $\lim\limits_{x \to 0^+} f(x)$ exists (see 1.5.3).

1.5.10. Assume that $\mathbf{I} = (a, b)$ is a bounded interval and, e.g., f is monotonically increasing. Then, as in 1.1.35, one can show that $\lim\limits_{x \to a^+} f(x) = \inf\limits_{x \in (a,b)} f(x)$ and $\lim\limits_{x \to b^-} f(x) = \sup\limits_{x \in (a,b)} f(x)$. Consequently, f can be continuously extended on $[a, b]$. So it is uniformly continuous on (a, b). If the interval $\mathbf{I}$ is unbounded, then the limits $\lim\limits_{x \to \infty} f(x)$ and/or $\lim\limits_{x \to -\infty} f(x)$ exist and are finite. By 1.5.7 f is uniformly continuous on $\mathbf{I}$.

1.5.11. No. The following function is uniformly continuous on $[0, \infty)$ but the limit $\lim\limits_{x \to \infty} f(x)$ does not exist:

$$f(x) = \begin{cases} x & \text{for } x \in [0,1], \\ -x+2 & \text{for } x \in [1,2], \\ \ldots \\ x - n(n+1) & \text{for } x \in [n(n+1), (n+1)^2], \\ -x + (n+1)(n+2) & \text{for } x \in [(n+1)^2, (n+1)(n+2)], \\ \ldots \end{cases}$$

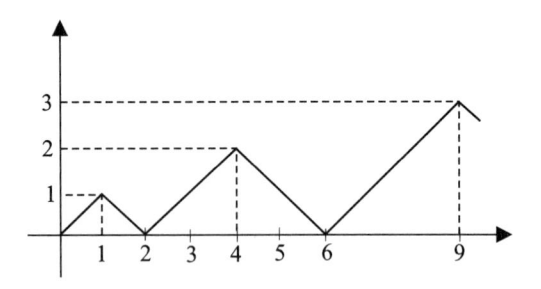

1.5.12. Let $\varepsilon > 0$ be arbitrarily fixed. Choose $\delta > 0$ so that for $x, x' \geq 0$
$$|x - x'| < \delta \quad \text{implies} \quad |f(x) - f(x')| < \frac{\varepsilon}{2}.$$
Let $x_1, x_2, \ldots, x_k$ be points in the interval $[0,1]$ such that for any $x \in [0,1]$ there is x_i for which $|x - x_i| < \delta$. Since $\lim\limits_{n \to \infty} f(x_i + n) = 0$ for $i = 1, 2, \ldots, k$, there is n_0 such that $|f(x_i + n)| < \frac{\varepsilon}{2}$ for $n > n_0$ and for $i = 1, 2, \ldots, k$. Suppose $x \geq n_0 + 1$ and set $n = [x]$. Then there is x_i such that $|x - (n + x_i)| < \delta$. It then follows that
$$|f(x)| \leq |f(x) - f(x_i + n)| + |f(x_i + n)| < \varepsilon.$$

1.5.13. By uniform continuity of f on $[1, \infty)$ there exists $\delta > 0$ such that $|f(x) - f(x')| < 1$ if $|x - x'| \leq \delta$. Any $x \geq 1$ can be written in the form $x = 1 + n\delta + r$, where $n \in \mathbb{N} \cup \{0\}$ and $0 \leq r < \delta$. Hence
$$|f(x)| \leq |f(1)| + |f(x) - f(1)| \leq |f(1)| + (n+1).$$

1.5. Uniform Continuity

Dividing by x gives

$$\frac{|f(x)|}{x} \leq \frac{|f(1)| + n + 1}{1 + n\delta + r} \leq \frac{|f(1)| + 2}{\delta} = M.$$

1.5.14. As in the solution of the foregoing problem, we find $\delta > 0$ such that if $x = n\delta + r$, then for any $u \geq 0$

$$|f(x + u) - f(u)| \leq n + 1.$$

Therefore

$$\frac{|f(x + u) - f(u)|}{x + 1} \leq \frac{n + 1}{1 + n\delta + r} \leq \frac{2}{\delta} = M.$$

1.5.15. Assume $\{x_n\}$ is a Cauchy sequence of elements in $\mathbf{A}$; that is, given $\delta > 0$ there is $n_0 \in \mathbb{N}$ such that $|x_n - x_m| < \delta$ for $n, m \geq n_0$. By uniform continuity of f, given $\varepsilon > 0$ there is $\delta_\varepsilon > 0$ such that $|f(x_n) - f(x_m)| < \varepsilon$ if $|x_n - x_m| < \delta_\varepsilon$. Thus $\{f(x_n)\}$ is a Cauchy sequence.

1.5.16. Assume, contrary to our claim, that f is not uniformly continuous on $\mathbf{A}$. Thus there is $\varepsilon > 0$ such that for any positive integer n there exist x_n and x'_n in $\mathbf{A}$ such that $|x_n - x'_n| < \frac{1}{n}$ and $|f(x_n) - f(x'_n)| \geq \varepsilon$. Since $\mathbf{A}$ is bounded, there is a convergent subsequence $\{x_{n_k}\}$ of $\{x_n\}$. It follows from the above that the sequence $\{x'_{n_k}\}$ is convergent to the same limit. Thus the sequence $\{z_k\}$ with terms $x_{n_1}, x'_{n_1}, x_{n_2}, x'_{n_2}, \ldots, x_{n_k}, x'_{n_k}, \ldots$ is convergent, and therefore it is a Cauchy sequence. But $|f(x_{n_k}) - f(x'_{n_k})| \geq \varepsilon$, and so $\{f(z_k)\}$ is not Cauchy. A contradiction.

The boundedness of $\mathbf{A}$ is essential. To see this, consider the function $f(x) = x^2$ on $(0, \infty)$.

1.5.17. The necessity of the condition follows immediately from the definition of uniform continuity. Now assume that the given condition is satisfied and f is not uniformly continuous on $\mathbf{A}$. Then there is $\varepsilon > 0$ such that for any positive integer n there exist x_n and y_n in $\mathbf{A}$ such that $|x_n - y_n| < \frac{1}{n}$ and $|f(x_n) - f(y_n)| \geq \varepsilon$, a contradiction.

1.5.18. No. Define f by setting

$$f(x) = \begin{cases} \frac{1}{2} & \text{for } x \in (0,2], \\ \frac{1}{n} & \text{for } x = n,\, n \geq 2, \\ \frac{2}{n} & \text{for } x = n+\frac{1}{n},\, n \geq 2, \\ x-n+\frac{1}{n} & \text{for } x \in (n, n+\frac{1}{n}),\, n \geq 2, \\ -\frac{n+2}{(n+1)(n-1)}\left(x-n-\frac{1}{n}\right)+\frac{2}{n} & \text{for } x \in (n+\frac{1}{n}, n+1),\, n \geq 2. \end{cases}$$

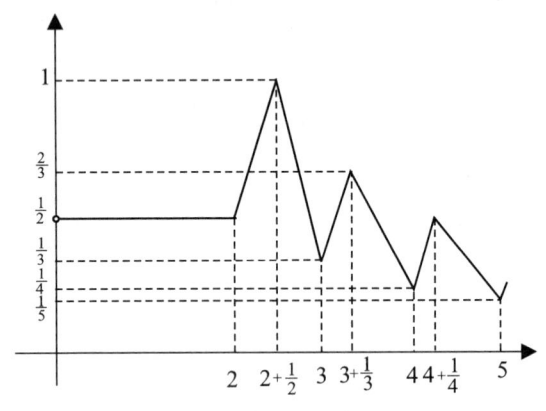

The function f is continuous on $(0,\infty)$, $\lim\limits_{x\to\infty} f(x) = 0$ and $\lim\limits_{x\to 0^+} f(x) = \frac{1}{2}$. It then follows from 1.5.7 that f is uniformly continuous on $(0,\infty)$. But

$$\lim_{n\to\infty} \frac{f\left(n+\frac{1}{n}\right)}{f(n)} = 2.$$

1.5.19. By the continuity of f at zero, given $\varepsilon > 0$ there is $\delta > 0$ such that $|f(x)| < \varepsilon$ for $|x| < \delta$. Hence the subadditivity of f implies that, for $x \in \mathbb{R}$ and $|t| < \delta$,

$$f(x+t) - f(x) \leq f(t) < \varepsilon \quad \text{and} \quad f(x) - f(x+t) \leq f(-t) < \varepsilon.$$

Consequently, $|f(x+t) - f(x)| < \varepsilon$, which proves the uniform continuity of f on $\mathbb{R}$.

1.5.20. Observe that ω_f is monotonically increasing on $(0,\infty)$. Thus (see 1.1.35)

$$\lim_{\delta \to 0^+} \omega_f(\delta) = \inf_{\delta > 0} \omega_f(\delta) \geq 0.$$

1.5. Uniform Continuity

If $\lim\limits_{\delta \to 0^+} \omega_f(\delta) = 0$, then given $\varepsilon > 0$ there is $\delta > 0$ such that $\omega_f(\delta) < \varepsilon$. Consequently, if $|x_1 - x_2| < \delta$, then $|f(x_1) - f(x_2)| \leq \omega_f(\delta) < \varepsilon$. This means that f is uniformly continuous on $\mathbf{A}$.

Assume now that f is uniformly continuous on $\mathbf{A}$. Then given $\varepsilon > 0$ there is $\delta_0 > 0$ such that $|f(x_1) - f(x_2)| < \varepsilon$ for $|x_1 - x_2| < \delta_0$. Hence $\lim\limits_{\delta \to 0^+} \omega_f(\delta) \leq \omega_f(\delta_0) \leq \varepsilon$. The arbitrariness of $\varepsilon > 0$ yields $\lim\limits_{\delta \to 0^+} \omega_f(\delta) = 0$.

1.5.21. Clearly, it is enough to prove that (b) implies (a). Let $\varepsilon > 0$ be arbitrarily fixed. Since fg is continuous at zero, there is $\delta_1 > 0$ such that

$$|x| < \delta_1 \quad \text{implies} \quad |f(x)g(x) - f(0)g(0)| < \frac{\varepsilon}{2}.$$

Thus if $|x_1| < \delta_1$ and $|x_2| < \delta_1$, then $|f(x_1)g(x_1) - f(x_2)g(x_2)| < \varepsilon$. For $|x_1| \geq \delta_1$ we have

$$|f(x_1)g(x_1) - f(x_2)g(x_2)|$$
$$\leq \frac{|g(x_1)|}{|x_1|}|x_1||f(x_1) - f(x_2)| + |f(x_2)||g(x_1) - g(x_2)|.$$

Consequently,

$$|f(x_1)g(x_1) - f(x_2)g(x_2)|$$
$$\leq \frac{|g(x_1)|}{|x_1|}(||x_1|f(x_1) - |x_2|f(x_2)| + |f(x_2)||x_2 - x_1|)$$
$$+ |f(x_2)||g(x_1) - g(x_2)|.$$

This combined with the result in 1.5.13 gives

$$|f(x_1)g(x_1) - f(x_2)g(x_2)| \leq M||x_1|f(x_1) - |x_2|f(x_2)|$$
$$+ ML|x_1 - x_2| + L|g(x_1) - g(x_2)|,$$

where

$$M = \sup\left\{\frac{|g(x)|}{|x|} : |x| \geq \delta_1\right\},$$
$$L = \max\left\{\sup\{|f(x)| : |x| \leq \delta_1\}, \sup\left\{\frac{|x||f(x)|}{|x|} : |x| \geq \delta_1\right\}\right\}.$$

Thus the desired result follows from the uniform continuity of $g(x)$ and $|x|f(x)$ on $\mathbb{R}$.

1.5.22. Suppose that f is uniformly continuous on **I**. Then, given $\varepsilon > 0$, there is $\delta > 0$ such that

(i) $\qquad |x_1 - x_2| < \delta \quad \text{implies} \quad |f(x_1) - f(x_2)| < \varepsilon.$

We will prove that, given $\varepsilon > 0$, there is $N > 0$ such that for every $x_1, x_2 \in \mathbf{I}$, $x_1 \neq x_2$,

(ii) $\qquad \left|\dfrac{f(x_1) - f(x_2)}{x_1 - x_2}\right| > N \quad \text{implies} \quad |f(x_1) - f(x_2)| < \varepsilon.$

Clearly, this implication is equivalent to

$$|f(x_1) - f(x_2)| \geq \varepsilon \quad \text{implies} \quad \left|\dfrac{f(x_1) - f(x_2)}{x_1 - x_2}\right| \leq N.$$

By (i), if $|f(x_1) - f(x_2)| \geq \varepsilon$, then $|x_1 - x_2| \geq \delta$. Without loss of generality we may assume that $x_1 < x_2$ and $f(x_1) < f(x_2)$. Since $f(x_2) - f(x_1) \geq \varepsilon$, there are $\eta \in [\varepsilon, 2\varepsilon]$ and a positive integer k such that $f(x_2) = f(x_1) + k\eta$. Now the intermediate value property of f on the interval $[x_1, x_2]$ implies that there are $x_1 = z_0 < z_1 < \cdots < z_k = x_2$ for which $f(z_i) = f(x_1) + i\eta$, $i = 1, 2, \ldots, k$. We have $|f(z_i) - f(z_{i-1})| = \eta \geq \varepsilon$, so $|z_i - z_{i-1}| \geq \delta$. Thus $|x_1 - x_2| \geq k\delta$. Setting $N = \frac{2\varepsilon}{\delta}$, we obtain

$$\left|\dfrac{f(x_1) - f(x_2)}{x_1 - x_2}\right| \leq \dfrac{k\eta}{k\delta} = \dfrac{\eta}{\delta} \leq \dfrac{2\varepsilon}{\delta} = N.$$

Assume now that (ii) is satisfied. Then, given $\varepsilon > 0$, there is $N > 0$ such that

$$|f(x_1) - f(x_2)| \geq \varepsilon \quad \text{implies} \quad \left|\dfrac{f(x_1) - f(x_2)}{x_1 - x_2}\right| \leq N.$$

Consequently,

$$|f(x_1) - f(x_2)| \geq \varepsilon \quad \text{implies} \quad |x_1 - x_2| \geq \dfrac{\varepsilon}{N}.$$

This means that (i) is satisfied with $\delta = \frac{\varepsilon}{N}$.

1.6. Functional Equations

1.6.1. Clearly, the functions $f(x) = ax$ are continuous and satisfy the Cauchy functional equation. We show that there are no other continuous solutions of this equation. Observe first that if f satisfies

(1) $\qquad f(x+y) = f(x) + f(y) \quad \text{for} \quad x, y \in \mathbb{R},$

then $f(2x) = 2f(x)$ for $x \in \mathbb{R}$. One can show by induction that for $n \in \mathbb{N}$,

(2) $\qquad\qquad\qquad f(nx) = nf(x).$

If in (2) we replace x by $\frac{x}{n}$, we get

(3) $\qquad\qquad\qquad f\left(\frac{x}{n}\right) = \frac{1}{n}f(x).$

If $r = \frac{p}{q}$, where $p, q \in \mathbb{N}$, then (2) and (3) imply

(4) $\qquad f(rx) = f\left(\frac{p}{q}x\right) = pf\left(\frac{1}{q}x\right) = \frac{p}{q}f(x) = rf(x).$

It follows from (2) that $f(0) = 0$. Combined with (1), this gives $0 = f(0) = f(x) + f(-x)$, or in other words, $f(-x) = -f(x)$. Thus, by (4), we get $-rf(x) = f(-rx) = -f(rx)$ for any negative rational r. Since for any real α there is a sequence $\{r_n\}$ of rationals converging to α, and since f is continuous, by (4) we get

$$f(\alpha x) = f(\lim_{n\to\infty} r_n x) = \lim_{n\to\infty} f(r_n x) = \lim_{n\to\infty} r_n f(x) = \alpha f(x).$$

Setting $x = 1$, gives $f(\alpha) = \alpha f(1)$. Consequently, $f(x) = ax$, where $a = f(1)$.

1.6.2.

(a) We will show that if f is continuous at at least one point and satisfies the Cauchy functional equation, then it is continuous on $\mathbb{R}$. So, the assertion follows from the preceding problem. Clearly, if f satisfies the Cauchy functional equation, then equalities (2)-(4) in the solution of 1.6.1 hold. We first show that the continuity of f at an x_0 implies the continuity at zero. Indeed, if $\{z_n\}$ is a sequence converging to zero, then $\{z_n + x_0\}$ converges to x_0. Moreover, it follows from the equality

$$f(z_n + x_0) = f(z_n) + f(x_0)$$

and from the continuity of f at x_0 that $\lim_{n\to\infty} f(z_n) = 0 = f(0)$. Now if x is any real number and $\{x_n\}$ converges to x, then $\{x_n-x\}$ converges to zero. The equality $f(x_n - x) = f(x_n) - f(x)$ and continuity of f at zero imply $\lim_{n\to\infty} f(x_n) = f(x)$.

(b) We first show that if f satisfies the Cauchy functional equation and is bounded above on the interval (a,b), then it is bounded on every interval $(-\varepsilon,\varepsilon)$, $\varepsilon > 0$. To this end, consider the function

$$g(x) = f(x) - f(1)x, \quad x \in \mathbb{R}.$$

Clearly, g satisfies the Cauchy functional equation, and it follows from the solution of 1.6.1 that $g(r) = 0$ for $r \in \mathbb{Q}$. For $x \in (-\varepsilon,\varepsilon)$, one can find a rational r such that $x + r \in (a,b)$. Then $g(x) = g(x) + g(r) = g(x+r) = f(x+r) - f(1)(x+r)$, which implies that g is bounded above on $(-\varepsilon,\varepsilon)$, and consequently, so is f. Since $f(-x) = -f(x)$, f is also bounded below on $(-\varepsilon,\varepsilon)$. Now our task is to show that f is continuous at zero. Let $\{x_n\}$ be a sequence converging to zero, and choose a sequence $\{r_n\}$ of rationals diverging to $+\infty$ so that $\lim_{n\to\infty} x_n r_n = 0$. Then the sequence $\{|f(r_n x_n)|\}$ is bounded above, say by M, and

$$|f(x_n)| = \left|f\left(\frac{1}{r_n}r_n x_n\right)\right| = \frac{1}{r_n}|f(r_n x_n)| \leq \frac{M}{r_n}.$$

Hence $\lim_{n\to\infty} f(x_n) = 0 = f(0)$. So our assertion follows from (a).

(c) Assume, for example, that f is monotonically increasing. It follows from (2)-(4) in the solution of 1.6.1 that for $-\frac{1}{n} < x < \frac{1}{n}$,

$$-\frac{1}{n}f(1) \leq f(x) \leq \frac{1}{n}f(1).$$

Thus f is continuous at zero, and our claim follows from (a).

1.6. Functional Equations

1.6.3. Observe that $f(x) = f^2\left(\frac{x}{2}\right) \geq 0$. If f attained zero at an x_0, then in view of $f(x+y) = f(x)f(y)$, f would be identically zero, which would contradict $f(1) > 0$. Thus f is positive on $\mathbb{R}$ and the function $g(x) = \ln f(x)$ is continuous and satisfies the Cauchy functional equation. It follows from 1.6.1 that $g(x) = ax$, where $a = g(1) = \ln f(1)$. Hence $f(x) = b^x$, $x \in \mathbb{R}$, with $b = f(1)$.

1.6.4. For $x, y \in (0, \infty)$, choose $t, s \in \mathbb{R}$ so that $x = e^t$ and $y = e^s$. Define g by the formula $g(t) = f(e^t)$. Then $g(t+s) = g(t) + g(s)$ for $t, s \in \mathbb{R}$, and, by 1.6.1, $g(t) = at$. Thus $f(x) = a \ln x = \log_b x$, where $b = e^{\frac{1}{a}}$.

1.6.5. As in the solution of the foregoing problem, for $x, y \in (0, \infty)$ we choose $t, s \in \mathbb{R}$ so that $x = e^t$, $y = e^s$. Next we define g by setting $g(t) = f(e^t)$. Then f satisfies the given equation if and only if $g(t+s) = g(t)g(s)$ for $t, s \in \mathbb{R}$. It follows from 1.6.3 that $g(t) = a^t$. Hence $f(x) = a^{\ln x} = x^b$, where $b = \ln a$.

1.6.6. If f is continuous on $\mathbb{R}$ and $f(x) - f(y)$ is rational for rational $x - y$, then $g(x) = f(x+1) - f(x)$ is continuous and assumes only rational values. It follows from the intermediate value property that g is constant. Let $f(x+1) - f(x) = q$, $q \in \mathbb{Q}$. If $f(0) = r$, then $f(1) = r+q$, and by induction, $f(n) = nq+r$, $n \in \mathbb{N}$. Since $f(x) = f(x+1)-q$, we get $f(-1) = -q+r$, and by induction, $f(-n) = -nq+r$, $n \in \mathbb{N}$. For a rational $p = \frac{n}{m}$, the function $f(x+p) - f(x)$ is also constant. Let $f(x+p) = f(x) + \tilde{q}$. As above, one can show that $f(kp) = k\tilde{q}+r$ for $k \in \mathbb{N}$. In particular, $f(n) = f(mp) = m\tilde{q}+r$. On the other hand, $f(n) = nq + r$. Hence $\tilde{q} = \frac{n}{m}q$ and $f\left(\frac{n}{m}\right) = \frac{n}{m}q + r$. Since p can be arbitrarily chosen, $f(x) = qx + r$ for $x \in \mathbb{Q}$. The continuity of f implies that f is defined by this formula for all $x \in \mathbb{R}$.

1.6.7. Observe that $f(0) = 0$. Moreover, for $x \in \mathbb{R}$ we get

$$f(x) = -f(qx) = f(q^2 x) = -f(q^3 x).$$

One can show by induction that $f(x) = (-1)^n f(q^n x)$. Letting $n \to \infty$ and using the continuity of f at zero, we see that $f(x) = 0$. Thus only the identically zero function satisfies the given equation.

1.6.8. We have $f(0) = 0$ and
$$f(x) = -f\left(\frac{2}{3}x\right) + x = f\left(\left(\frac{2}{3}\right)^2 x\right) - \frac{2}{3}x + x.$$
One can prove by induction that for $n \in \mathbb{N}$,
$$f(x) = (-1)^n f\left(\left(\frac{2}{3}\right)^n x\right) + (-1)^{n-1}\left(\frac{2}{3}\right)^{n-1} x + \cdots - \frac{2}{3}x + x.$$
We now pass to the limit as $n \to \infty$ and use the continuity of f at zero, and get $f(x) = \frac{3}{5}x$.

1.6.9. If in the equation we put $y = 2x$, we get
$$f(y) = \frac{1}{2}f\left(\frac{1}{2}y\right) + \frac{1}{2^2}y = \frac{1}{2^2}f\left(\frac{1}{2^2}y\right) + \frac{1}{2^4}y + \frac{1}{2^2}y.$$
One can prove by induction that
$$f(y) = \frac{1}{2^n}f\left(\frac{1}{2^n}y\right) + \frac{1}{2^{2n}}y + \frac{1}{2^{2(n-1)}}y + \cdots + \frac{1}{2^2}y.$$
Letting $n \to \infty$ and using the fact that $f(0) = 0$ and f is continuous at zero, we conclude that $f(y) = \frac{1}{3}y$.

1.6.10. Set $f(0) = c$. Putting $y = 0$ in the Jensen equation, we get
$$f\left(\frac{x}{2}\right) = \frac{f(x) + f(0)}{2} = \frac{f(x) + c}{2}.$$
Hence
$$\frac{f(x) + f(y)}{2} = f\left(\frac{x+y}{2}\right) = \frac{f(x+y) + c}{2},$$
which gives, $f(x) + f(y) = f(x+y) + c$. Now set $g(x) = f(x) - c$. Then g satisfies the Cauchy equation (see 1.6.1). Therefore $g(x) = ax$, or in other words, $f(x) = ax + c$.

1.6.11. We will first show that f is linear on every closed subinterval $[\alpha, \beta]$ of (a, b). By the Jensen equation,
$$f\left(\alpha + \frac{1}{2}(\beta - \alpha)\right) = f(\alpha) + \frac{1}{2}(f(\beta) - f(\alpha)).$$

1.6. Functional Equations

Furthermore,

$$f\left(\alpha + \frac{1}{4}(\beta - \alpha)\right) = f\left(\frac{\alpha + \frac{\alpha+\beta}{2}}{2}\right)$$
$$= \frac{1}{2}f(\alpha) + \frac{1}{2}f\left(\frac{\alpha + \beta}{2}\right)$$
$$= f(\alpha) + \frac{1}{4}(f(\beta) - f(\alpha))$$

and

$$f\left(\alpha + \frac{3}{4}(\beta - \alpha)\right) = f\left(\frac{1}{2}\beta + \frac{1}{2}\left(\alpha + \frac{1}{2}(\beta - \alpha)\right)\right)$$
$$= \frac{1}{2}f(\beta) + \frac{1}{2}f\left(\alpha + \frac{1}{2}(\beta - \alpha)\right)$$
$$= f(\alpha) + \frac{3}{4}(f(\beta) - f(\alpha)).$$

Now we prove by induction that

$$f\left(\alpha + \frac{k}{2^n}(\beta - \alpha)\right) = f(\alpha) + \frac{k}{2^n}(f(\beta) - f(\alpha))$$

for $k = 0, 1, 2, 3, \ldots, 2^n$ and $n \in \mathbb{N}$. Assuming the equality to hold for $m \leq n$, we will prove it for $n+1$. Indeed, if $k = 2l, l = 0, 1, \ldots, 2^n$, then, by the induction hypothesis,

$$f\left(\alpha + \frac{k}{2^{n+1}}(\beta - \alpha)\right) = f\left(\alpha + \frac{l}{2^n}(\beta - \alpha)\right)$$
$$= f(\alpha) + \frac{l}{2^n}(f(\beta) - f(\alpha))$$
$$= f(\alpha) + \frac{k}{2^{n+1}}(f(\beta) - f(\alpha)).$$

Similarly, if $k = 2l + 1, l = 0, 1, \ldots, 2^n - 1$, then

$$f\left(\alpha + \frac{k}{2^{n+1}}(\beta - \alpha)\right) = f\left(\frac{1}{2}\left(\alpha + \frac{l}{2^{n-1}}(\beta - \alpha)\right) + \frac{1}{2}\left(\alpha + \frac{1}{2^n}(\beta - \alpha)\right)\right)$$
$$= \frac{1}{2}f\left(\alpha + \frac{l}{2^{n-1}}(\beta - \alpha)\right) + \frac{1}{2}f\left(\alpha + \frac{1}{2^n}(\beta - \alpha)\right)$$
$$= f(\alpha) + \frac{k}{2^{n+1}}(f(\beta) - f(\alpha)).$$

Since the numbers $\frac{k}{2^n}$ form a dense set in $[0,1]$, the continuity of f implies that

$$f(\alpha + t(\beta - \alpha)) = f(\alpha) + t(f(\beta) - f(\alpha)) \quad \text{for} \quad t \in [0,1].$$

Setting $x = \alpha + t(\beta - \alpha)$ gives

$$f(x) = f(\alpha) + \frac{f(\beta) - f(\alpha)}{\beta - \alpha}(x - \alpha).$$

Now observe that under our hypothesis one-sided limits of f at a and b exist. Indeed, for example, we have

$$\lim_{y \to b^-} \frac{f(y)}{2} = f\left(\frac{x+b}{2}\right) - \frac{f(x)}{2} \quad \text{with} \quad x \in (a,b).$$

Clearly,

$$(a,b) = \bigcup_{n=1}^{\infty} [\alpha_n, \beta_n],$$

where $\{\alpha_n\}$ is a decreasing sequence of points in (a,b) converging to a, and $\{\beta_n\}$ is an increasing sequence of points in this interval converging to b. Thus for $x \in (a,b)$ there is $n_0 \in \mathbb{N}$ such that $x \in [\alpha_n, \beta_n]$ for all $n \geq n_0$. It then follows that

$$f(x) = f(\alpha_n) + \frac{f(\beta_n) - f(\alpha_n)}{\beta_n - \alpha_n}(x - \alpha_n).$$

If we let $n \to \infty$, we get

$$f(x) = f(a^+) + \frac{f(b^-) - f(a^+)}{b - a}(x - a).$$

1.6.12. For $x \in \mathbb{R}$, set

$$x_1 = x \quad \text{and} \quad x_{n+1} = \frac{x_n - 1}{2}, \quad n = 1, 2, 3, \ldots.$$

Then $\lim\limits_{n \to \infty} x_n = -1$ and $f(x_n) = f(2x_{n+1} + 1) = f(x_{n+1})$, $n \in \mathbb{N}$. Hence $f(x) = f(x_n)$. Letting $n \to \infty$, we see that $f(x) = f(-1)$. Thus only constant functions fulfill our assumptions.

1.6. Functional Equations

1.6.13. Note that $g(x) = f(x) - \frac{a}{2}x^2$ is continuous on $\mathbb{R}$ and satisfies the Cauchy functional equation (see 1.6.1). Thus $g(x) = g(1)x$, which gives
$$f(x) - \frac{a}{2}x^2 = \left(f(1) - \frac{a}{2}\right)x \quad \text{for} \quad x \in \mathbb{R}.$$

1.6.14. By assumption,
$$f(-1) = f\left(-\frac{1}{2}\right) = f\left(-\frac{1}{3}\right) = \cdots = f(0).$$
Moreover, for $t \neq 0, -1, -\frac{1}{2}, -\frac{1}{3}, \ldots$ we have
$$f(t) = f\left(\frac{t}{t+1}\right) = f\left(\frac{t}{2t+1}\right) = f\left(\frac{t}{3t+1}\right) = \cdots.$$
Since $\lim_{n\to\infty} \frac{t}{nt+1} = 0$, the continuity of f at zero implies that $f(t) = f(0)$. So the only solutions of the equation are constant functions.

1.6.15. No. In fact, there are infinitely many such functions. For $a \in (0, 1)$ let g be a strictly decreasing and continuous transformation of $[0, a]$ onto $[a, 1]$. Then f, defined as
$$f(x) = \begin{cases} g(x) & \text{for } x \in [0, a], \\ g^{-1}(x) & \text{for } x \in (a, 1], \end{cases}$$
where g^{-1} is the inverse of g, enjoys the desired property.

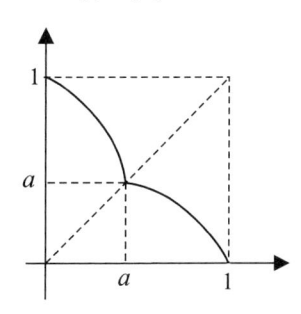

1.6.16. Suppose, contrary to our claim, that there is $y_0 \in \mathbb{R}$ such that $|g(y_0)| = a > 1$. Set $M = \sup\{|f(x)| : x \in \mathbb{R}\}$. By the definition of the supremum, there exists $x_0 \in \mathbb{R}$ for which $|f(x_0)| > \frac{M}{a}$. By assumption,

$$|f(x_0+y_0)| + |f(x_0-y_0)| \geq |f(x_0+y_0) + f(x_0-y_0)|$$
$$= 2|f(x_0)||g(y_0)| > 2\frac{M}{a}a = 2M.$$

Hence $|f(x_0+y_0)| > M$ or $|f(x_0-y_0)| > M$, a contradiction.

1.6.17. Note that $g(x) = f(x)e^{-x}$ satisfies the Cauchy functional equation. It follows from 1.6.1 that $f(x) = axe^x$.

1.6.18. By assumption, $f(0) = 0$ and $f(2x) = (f(x))^2$. By induction,

$$f(x) = \left(f\left(\frac{x}{2}\right)\right)^2 = \left(f\left(\frac{x}{2^2}\right)\right)^{2^2} = \cdots = \left(f\left(\frac{x}{2^n}\right)\right)^{2^n}.$$

Hence

$$f\left(\frac{x}{2^n}\right) = \sqrt[2^n]{f(x)}.$$

If $f(x) > 0$, then upon passage to the limit as $n \to \infty$ we get $0 = 1$, a contradiction. Thus only the identically zero function satisfies the given equation.

1.6.19. Replacing x by $\frac{x-1}{x}$ in

(i) $$f(x) + f\left(\frac{x-1}{x}\right) = 1 + x$$

gives

(ii) $$f\left(\frac{x-1}{x}\right) + f\left(\frac{-1}{x-1}\right) = \frac{2x-1}{x}.$$

Next replacing x by $\frac{-1}{x-1}$ in (i), we get

(iii) $$f\left(\frac{-1}{x-1}\right) + f(x) = \frac{x-2}{x-1}.$$

Adding (i) to (iii) and subtracting (ii) from the sum yield

$$2f(x) = 1 + x + \frac{x-2}{x-1} - \frac{2x-1}{x}.$$

Hence

$$f(x) = \frac{x^3 - x^2 - 1}{2x(x-1)}.$$

1.6. Functional Equations

One can easily check that this function satisfies the given functional equation.

1.6.20. For real x and y, define $\{x_n\}$ as follows: $x_{2k-1} = x$ and $x_{2k} = y$, $k = 1, 2, \ldots$. Then the equality $f(C\text{-}\lim_{n\to\infty} x_n) = C\text{-}\lim_{n\to\infty} f(x_n)$ implies

$$f\left(\lim_{n\to\infty} \frac{nx + ny}{2n}\right) = \lim_{n\to\infty} \frac{nf(x) + nf(y)}{2n},$$

which means that f satisfies the Jensen equation $f\left(\frac{x+y}{2}\right) = \frac{f(x)+f(y)}{2}$. As in the solution of 1.6.11, one can show that

(i) $\qquad f\left(x + \frac{k}{2^n}(y-x)\right) = f(x) + \frac{k}{2^n}(f(y) - f(x))$

for $k = 0, 1, 2, 3, \ldots, 2^n$ and $n \in \mathbb{N}$. For $t \in [0,1]$, one can find a sequence $\left\{\frac{k_n}{2^n}\right\}$ convergent to t. Since every convergent sequence is also Cesàro convergent (to the same limit), the sequence with the terms $x_n = x + \frac{k_n}{2^n}(y - x)$ converges in the Cesàro sense. By (i) the sequence $\{f(x_n)\}$ converges to $f(x) + t(f(y) - f(x))$. Consequently,

$$f(x + t(y - x)) = f(x) + t(f(y) - f(x)).$$

It follows from 1.2.33 that f is continuous on $\mathbb{R}$. Combined with 1.6.10, this shows that $f(x) = ax + c$.

1.6.21. Since $f(2x - f(x)) = x$ and f is an injection, we get $f^{-1}(x) = 2x - f(x)$. Thus

(i) $\qquad f(x) - x = x - f^{-1}(x).$

For $x_0 \in [0, 1]$, define the sequence $\{x_n\}$ recursively by $x_n = f(x_{n-1})$. It follows from (i) that $x_n - x_{n-1} = x_{n-1} - x_{n-2}$. Therefore $x_n = x_0 + n(x_1 - x_0)$. Since $|x_n - x_0| \leq 1$, we have $|x_1 - x_0| \leq \frac{1}{n}$ for $n \in \mathbb{N}$. Consequently, $f(x_0) = x_1 = x_0$.

1.6.22. We will show that the only continuous solutions of the given equation are the functions $f(x) = m(x - c)$. If $g(x) = 2x - \frac{f(x)}{m}$, then g is continuous and

(i) $\qquad g(g(x)) = 2g(x) - x \quad \text{for} \quad x \in \mathbb{R}.$

Thus g is a one-one function. Indeed, if $g(x_1) = g(x_2)$, then we get $g(g(x_1)) = g(g(x_2))$, which gives $x_1 = x_2$. By the result in 1.3.16, g is either strictly increasing or strictly decreasing on $\mathbb{R}$. We will show that the former case holds. By (i),

(ii) $\qquad g(g(x)) - g(x) = g(x) - x \quad \text{for} \quad x \in \mathbb{R}.$

If g were strictly decreasing, then for $x_1 < x_2$ we would get $g(x_1) > g(x_2)$, and consequently $g(g(x_1)) < g(g(x_2))$. On the other hand, (ii) gives

$$g(g(x_1)) - g(x_1) = g(x_1) - x_1, \quad g(g(x_2)) - g(x_2) = g(x_2) - x_2,$$

a contradiction.

It follows from (i), by induction, that

$$g^n(x) = ng(x) - (n-1)x \quad \text{for} \quad n \geq 1,$$

where g^n denotes the nth iteration of g. Hence $\lim\limits_{n \to \infty} \frac{g^n(x)}{n} = g(x) - x$. Moreover,

(iii) $\qquad g^n(x) - g^n(0) = n(g(x) - x - g(0)) + x.$

Thus, letting $n \to \infty$ and using the monotonicity of g, we get

(1) $\qquad \begin{aligned} g(x) &\leq x + g(0) \quad \text{for} \quad x < 0, \\ g(x) &\geq x + g(0) \quad \text{for} \quad x > 0, \end{aligned}$

which in turn gives $g(\mathbb{R}) = \mathbb{R}$. So the inverse function g^{-1} is defined on $\mathbb{R}$. Replacing in (i) x by $g^{-1}(g^{-1}(y))$ we see that $g^{-1}(g^{-1}(y)) = 2g^{-1}(y) - y$. Since g^{-1} satisfies (i), one can show by the same method that

$$g^{-n}(y) - g^{-n}(0) = n(g^{-1}(y) - y - g^{-1}(0)) + y.$$

Next, upon passage to the limit as $n \to \infty$, we get (as above)

(2) $\qquad \begin{aligned} g^{-1}(y) &\leq y + g^{-1}(0) \quad \text{for} \quad y < 0, \\ g^{-1}(y) &\geq y + g^{-1}(0) \quad \text{for} \quad y > 0. \end{aligned}$

We now show that $g^{-1}(0) = -g(0)$. Replacing x by $g^{-1}(y)$ in (ii), we obtain

$$g(y) - y = y - g^{-1}(y),$$

which gives $g^{-1}(0) = -g(0)$.

1.6. Functional Equations

Assume, for example, that $g(0) \geq 0$. Then $g(x) > 0$ for $x > 0$. By (2) with $y = g(x) > 0$, we see that $x \geq g(x) + g^{-1}(0) = g(x) - g(0)$. Thus by (1), for $x > 0$ we get $g(x) = x + g(0)$. Since $g^{-1}(0) \leq 0$, we have $g^{-1}(y) < 0$ for $y < 0$, and as above one can show that $g^{-1}(y) = y + g^{-1}(0)$, which means that $g(x) = x + g(0)$ for $x < 0$. Thus $g(x) = x + g(0)$, or equivalently, $f(x) = m(x - g(0))$ for $x \in \mathbb{R}$.

1.6.23. It is easy to verify that the given functions satisfy the desired conditions. Now we will show that there are no other solutions. If in the equation

(i) $$f(x+y) + f(y-x) = 2f(x)f(y)$$

we put $x = 0$ and a y such that $f(y) \neq 0$, we get $f(0) = 1$. Taking $y = 0$ in (i), we see that $f(x) = f(-x)$, which means that f is even. Since f is continuous and $f(0) = 1$, there exists an interval $[0, c]$ on which the function is positive. We consider two cases: $f(c) \leq 1$, and $f(c) > 1$. In the first case there exists θ, $0 \leq \theta < \frac{\pi}{2}$, such that $f(c) = \cos \theta$. Now rewrite (i) in the form

$$f(y+x) = 2f(x)f(y) - f(y-x).$$

Application of this equation with $x = c$, $y = c$, and $x = c$, $y = 2c$ gives $f(2c) = 2\cos^2 \theta - 1 = \cos 2\theta$ and $f(3c) = 2\cos\theta\cos 2\theta - \cos\theta = \cos 3\theta$, respectively. One can show by induction that $f(nc) = \cos n\theta$. Now applying (i) with $x = y = \frac{c}{2}$ gives

$$\left(f\left(\frac{c}{2}\right)\right)^2 = \frac{f(0) + f(c)}{2} = \frac{1 + \cos\theta}{2} = \cos^2\left(\frac{\theta}{2}\right).$$

Since $f\left(\frac{c}{2}\right)$ and $\cos\left(\frac{\theta}{2}\right)$ are positive, the last equation implies that $f\left(\frac{c}{2}\right) = \cos\left(\frac{\theta}{2}\right)$, and recursively $f\left(\frac{c}{2^n}\right) = \cos\left(\frac{\theta}{2^n}\right)$ for $n \in \mathbb{N}$. If we start with the equation $f(nc) = \cos n\theta$ and repeat the above procedure, we obtain

$$f\left(\frac{mc}{2^n}\right) = \cos\left(\frac{m\theta}{2^n}\right) \quad \text{for} \quad m, n \in \mathbb{N}.$$

Thus $f(cx) = \cos\theta x$ for $x = \frac{m}{2^n}$. Since the set of numbers of the form $\frac{m}{2^n}$, $m, n \in \mathbb{N}$, is a dense subset of $\mathbb{R}^+$, the continuity of f implies that

$f(cx) = \cos\theta x$ for $x > 0$. Since f is even, the equality holds also for negative x. Finally, $f(x) = \cos ax$ with $a = \frac{\theta}{c}$.

In the case where $f(c) > 1$, there is θ such that $f(c) = \cosh\theta$. To show that $f(x) = \cosh(ax)$, reasoning similar to the above can be used.

1.6.24. If we put $x = \tanh u$, $y = \tanh v$, then

$$\frac{x+y}{1+xy} = \frac{\tanh u + \tanh v}{1 + \tanh u \tanh v} = \tanh(u+v).$$

Therefore the function $g(u) = f(\tanh u)$ satisfies the Cauchy functional equation (see 1.6.1) and is continuous on $\mathbb{R}$. Consequently, $g(u) = au$. Hence $f(x) = \frac{1}{2}a\ln\frac{1+x}{1-x}$ for $|x| < 1$.

1.6.25. Assume that P is not identically zero and satisfies the equation. Set $Q(x) = P(1-x)$. Then $Q(1-x) = P(x)$, and the given equation can be rewritten as $Q((1-x)^2) = (Q(1-x))^2$ or

(i) $$Q(x^2) = (Q(x))^2 \quad \text{for} \quad x \in \mathbb{R}.$$

If Q is not a monomial, then it is of the form $Q(x) = ax^k + x^m R(x)$, where $a \neq 0$, $m > k \geq 0$, and R is a polynomial such that $R(0) \neq 0$. For such a Q, by (i),

$$ax^{2k} + x^{2m}R(x^2) = a^2 x^{2k} + 2ax^{k+m}R(x) + x^{2m}R^2(x).$$

Equating coefficients of like powers, we conclude that $Q(x) = ax^k$, $a \neq 0$, and that $a = 1$. Consequently, $P(x) = (1-x)^k$ with $k \in \mathbb{N} \cup \{0\}$. Clearly, the identically zero function also satisfies the given equation.

1.6.26 [S. Kotz, Amer. Math. Monthly 72 (1965), 1072-1075]. For simplicity of notation, we will write $f^m(x_i)$ instead of $(f(x_i))^m$. If in the equation

(i) $$f\left(\frac{1}{n}\sum_{i=1}^{n} x_i^m\right) = \frac{1}{n}\sum_{i=1}^{n} f^m(x_i)$$

we put $x_i = c$, $i = 1, 2, \ldots, n$, we get

(ii) $$f(c^m) = f^m(c).$$

1.6. Functional Equations

In particular, $f(1) = f^m(1)$, which implies $f(1) = 0$ or $f(1) = 1$; or $f(1) = -1$ in the case where m is odd. Likewise, $f(0) = 0$ or $f(0) = 1$, $f(0) = -1$ if m odd. Putting $c = x^{\frac{1}{m}}$, $x \geq 0$, in (ii), we get

$$f\left(x^{\frac{1}{m}}\right) = f^{\frac{1}{m}}(x).$$

Replacing x_i by $x_i^{\frac{1}{m}}$ in (i) and using the last equality, we obtain

(iii) $$f\left(\frac{1}{n}\sum_{i=1}^n x_i\right) = \frac{1}{n}\sum_{i=1}^n f^m(x_i^{\frac{1}{m}}) = \frac{1}{n}\sum_{i=1}^n f(x_i).$$

In particular, for $x_3 = x_4 = \cdots = x_n = 0$,

$$f\left(\frac{x_1 + x_2}{n}\right) = \frac{1}{n}f(x_1) + \frac{1}{n}f(x_2) + \frac{n-2}{n}f(0).$$

If in (iii) we put $x_2 = x_3 = \cdots = x_n = 0$, and replace x_1 by $x_1 + x_2$, we get

$$f\left(\frac{x_1 + x_2}{n}\right) = \frac{1}{n}f(x_1 + x_2) + \frac{n-1}{n}f(0).$$

Consequently,

$$f(x_1 + x_2) = f(x_1) + f(x_2) - f(0).$$

So, the function $g(x) = f(x) - f(0)$ satisfies the Cauchy functional equation and is continuous at at least one point. By the result in 1.6.2, $g(x) = ax$ for $x \geq 0$. Thus

$$f(x) = ax + b, \quad \text{where} \quad a = f(1) - f(0), \; b = f(0).$$

It follows from the above that $b = 0$ or $b = 1$; or additionally, if m is odd, $b = -1$. So, the only possible values of a are $-2, -1, 0, 1$ or 2. One can easily verify that

$$f(x) = 0, \quad f(x) = 1, \quad f(x) = x,$$

and, for odd m,

$$f(x) = -1, \quad f(x) = -x$$

are the only solutions.

1.6.27. If f satisfies the given condition, then for any real a, b, $b \neq 0$,
$$f(a+b) = f((ab^{-1}z + z)(z^{-1}b)) = f(ab^{-1}z + z)f(z^{-1}b)$$
$$= (f(ab^{-1}z) + f(z))f(z^{-1}b) = f(a) + f(b).$$
Hence $f(0) = 0$ and $f(-x) = -f(x)$. Moreover, $f(n) = nf(1)$ for any integer n. If f is not identically zero, then there is c such that $f(c) \neq 0$. But $f(c) = f(1)f(c)$, so $f(1) = 1$. If $x \neq 0$, then $1 = f(x)f(x^{-1})$, and consequently, $0 \neq f(x) = (f(x^{-1}))^{-1}$. It follows from the above that for integers p and $q \neq 0$,
$$f(pq^{-1}) = f(p)f(q^{-1}) = f(p)(f(q))^{-1} = pq^{-1}.$$
Note that for $x > 0$ we have $f(x) = (f(\sqrt{x}))^2 > 0$. Thus if $y - x > 0$, then $f(y - x) = f(y) - f(x) > 0$. This means that f is strictly increasing, and $f(x) = x$ if $x \in \mathbb{Q}$. It then follows that $f(x) = x$ for $x \in \mathbb{R}$.

1.6.28. A function f of the form

(i) $$f(x) = g(x) - g\left(\frac{1}{x}\right),$$

where g is any real function on $\mathbb{R} \setminus \{0\}$, satisfies the given functional equation. On the other hand, if f satisfies the given equation, then
$$f(x) = \frac{f(x) - f\left(\frac{1}{x}\right)}{2},$$
which means that f is of the form (i).

1.6.29. Observe that if f satisfies the given functional equation and if we set
$$g(x) = \frac{1}{2}\left(f(x) + f\left(\frac{1}{x}\right)\right), \quad h(x) = \frac{1}{2}\left(f(x) - f\left(\frac{1}{x}\right)\right),$$
then the functions g and h have the following properties:

(i) $$g(x) = g\left(\frac{1}{x}\right)$$

and

(ii) $\quad h(x) = -h\left(\frac{1}{x}\right), \quad h(x) + h(x^2) = 0, \quad h(-x) = h(x).$

1.6. Functional Equations

Now note that if g and h satisfy (i) and (ii), then $f = g + h$ satisfies the given functional equation. So our aim is to find functions g and h. As in the solution of the foregoing problem, one can show that all functions satisfying (i) are of the form

$$g(x) = k(x) + k\left(\frac{1}{x}\right),$$

where k is any function defined on $\mathbb{R} \setminus \{0\}$. To find functions h, observe first that (ii) implies that $h(1) = 0$. Now for $x > 1$ set $h(x) = s(\ln \ln x)$. Then s satisfies the functional equation

$$s(\ln \ln x) + s(\ln(2\ln x)) = 0,$$

which can be rewritten in the form

$$s(t) + s(\ln 2 + t) = 0 \quad \text{for} \quad t \in \mathbb{R}.$$

This means that s can be any function such that $s(t) = -s(\ln 2 + t)$ (note that s is periodic with period $2\ln 2$). There are infinitely many such functions, e.g., one can take $s(t) = \cos\frac{\pi t}{\ln 2}$. Next we extend the function h onto $(0,1)$ by setting $h(x) = -h\left(\frac{1}{x}\right)$, and then onto $(-\infty, 0)$ by setting $h(-x) = h(x)$.

1.6.30 [S. Haruki, Amer. Math. Monthly 86 (1979), 577-578]. If in the given equation we replace x by $x+y$ and y by $x-y$, we get

(1) $$\frac{f(x+y) - g(x-y)}{2y} = \phi(x).$$

Now replacing y by $-y$ in (1) gives

$$\frac{f(x-y) - g(x+y)}{-2y} = \phi(x).$$

Consequently, for $u, v \in \mathbb{R}$ we get

$$\phi(u+v) + \phi(u-v) = \frac{1}{2y}(f(u+v+y) - g(u+v-y)$$
$$+ f(u-v+y) - g(u-v-y))$$
$$= \frac{1}{2y}(f(u+v+y) - g(u-v-y))$$
$$+ \frac{1}{2y}(f(u-(v-y)) - g(u+(v-y))).$$

Thus
$$\phi(u+v) + \phi(u-v) = \frac{1}{2y}(2(v+y)\phi(u) - 2(v-y)\phi(u)) = 2\phi(u).$$

If we set $s = u+v$ and $t = u-v$, then this can be rewritten in the form
$$\frac{\phi(s)+\phi(t)}{2} = \phi\left(\frac{s+t}{2}\right), \quad s,t \in \mathbb{R}.$$

Let $A: \mathbb{R} \to \mathbb{R}$ be given by $A(s) = \phi(s) - \phi(0)$. Then $A(0) = 0$ and

(2)
$$\begin{aligned} A(t) + A(s) &= \phi(s) + \phi(t) - 2\phi(0) \\ &= 2\phi\left(\frac{s+t}{2}\right) - 2\phi(0) \\ &= 2A\left(\frac{s+t}{2}\right). \end{aligned}$$

Putting $t = 0$ gives $A(s) = 2A\left(\frac{s}{2}\right)$. Next, replacing s by $s+t$, we get
$$A(s+t) = 2A\left(\frac{s+t}{2}\right).$$

This and (2) imply

(3) $$A(s+t) = A(s) + A(t).$$

Thus equation (1) can be written in the form

(4) $$\frac{f(x+y) - g(x-y)}{2y} = B + A(x),$$

where $B = \phi(0)$ and $x \mapsto A(x)$ is a function satisfying (3). If in (4) we put $y = x$ and $y = -x$, respectively, then we obtain

$$f(2x) = g(0) + 2Bx + 2xA(x) \quad \text{and} \quad g(2x) = f(0) + 2Bx + 2xA(x).$$

Replacing $2x$ by x and using that fact that $A(s) = 2A\left(\frac{s}{2}\right)$, we get

$$f(x) = g(0) + Bx + \frac{1}{2}xA(x), \quad g(x) = f(0) + Bx + \frac{1}{2}xA(x).$$

Substituting these equations into (1) and applying (3), we arrive at

$$\frac{g(0) - f(0) + 2By + xA(y) + yA(x)}{2y} = \phi(x).$$

1.6. Functional Equations

Setting $x = 1$, we find that
$$A(y) = dy + f(0) - g(0), \quad \text{where} \quad d = 2\phi(1) - A(1) - 2B.$$

Since $A(0) = 0$, we have $f(0) = g(0)$. Hence $A(x) = dx$ and $f(x) = g(x) = f(0) + Bx + \frac{1}{2}dx^2$.

It is easy to check that $f(x) = g(x) = ax^2 + bx + c$ and $\phi(x) = f'(x) = 2ax + b$ satisfy the given functional equation.

1.6.31. The set $\mathbb{R}$ can be regarded as a vector space over $\mathbb{Q}$. A *Hamel basis* for $\mathbb{R}$ over $\mathbb{Q}$ is a maximal linearly independent set. There exists a Hamel basis $\mathbf{H}$ that contains 1. Thus each $x \in \mathbb{R}$ can be represented in a unique way as
$$x = \sum_{h \in \mathbf{H}} w_h(x) h,$$
where only finitely many coefficients $w_h(x) \in \mathbb{Q}$ are different from zero. Consequently, for $x, y \in \mathbb{R}$,
$$x + y = \sum_{h \in \mathbf{H}} w_h(x+y) h = \sum_{h \in \mathbf{H}} (w_h(x) + w_h(y)) h,$$
which implies $w_h(x+y) = w_h(x) + w_h(y)$. So, in particular, $f = w_1$ satisfies (a). We will show that it has the other properties also.

Note that $w_1(1) = 1$, because $1 = 1 \cdot 1$ and $1 \in \mathbf{H}$. Now we show that $w_1(x) = x$ for $x \in \mathbb{Q}$. By the additivity of w_1,
$$1 = w_1(1) = w_1\left(\frac{1}{q} + \frac{1}{q} + \cdots + \frac{1}{q}\right) = qw_1\left(\frac{1}{q}\right).$$
Hence
$$w_1\left(\frac{1}{q}\right) = \frac{1}{q}.$$
It then follows, by additivity again, that
$$w_1\left(\frac{p}{q}\right) = \frac{p}{q} \quad \text{for} \quad p, q \in \mathbb{N}.$$
Moreover, $w_1(0) = 0$, because $0 = 0 \cdot 1$ and $1 \in \mathbf{H}$. Thus
$$0 = w_1(0) = w_1\left(\frac{p}{q} + \left(-\frac{p}{q}\right)\right) = w_1\left(\frac{p}{q}\right) + w_1\left(-\frac{p}{q}\right),$$

or, in other words,
$$w_1\left(-\frac{p}{q}\right) = -\frac{p}{q}.$$
So we have proved that $w_1(x) = x$ for $x \in \mathbb{Q}$. Finally, we show that w_1 is not continuous. If it were, we would get $w_1(x) = x$ for all $x \in \mathbb{R}$. This would contradict the fact that w_1 assumes only rational values.

1.7. Continuous Functions in Metric Spaces

1.7.1. We will show first that (a) $\implies$ (b). Let $\mathbf{F}$ be a closed set in $\mathbf{Y}$. Then, if a sequence $\{x_n\}$ of elements in $f^{-1}(\mathbf{F})$ is convergent to x, then $f(x_n) \in \mathbf{F}$, and by continuity of f, $f(x_n) \to f(x)$. Since $\mathbf{F}$ is closed, $f(x) \in \mathbf{F}$, or in other words, $x \in f^{-1}(\mathbf{F})$. So we have proved that $f^{-1}(\mathbf{F})$ is closed.

To prove that (b) $\implies$ (c) it is enough to note that every open subset $\mathbf{G}$ of $\mathbf{Y}$ is the complement of a closed subset $\mathbf{F}$, that is, $\mathbf{G} = \mathbf{Y} \setminus \mathbf{F}$. Then, we have $f^{-1}(\mathbf{G}) = \mathbf{X} \setminus f^{-1}(\mathbf{F})$.

We will now prove that (c) $\implies$ (a). Let $x_0 \in \mathbf{X}$ and $\varepsilon > 0$ be arbitrarily fixed. By assumption, the set $f^{-1}(\mathbf{B_Y}(f(x_0), \varepsilon))$ is open in $\mathbf{X}$. Since x_0 is an element of $f^{-1}(\mathbf{B_Y}(f(x_0), \varepsilon))$, there is $\delta > 0$ such that $\mathbf{B_X}(x_0, \delta) \subset f^{-1}(\mathbf{B_Y}(f(x_0), \varepsilon))$. Therefore we have $f(\mathbf{B_X}(x_0, \delta)) \subset \mathbf{B_Y}(f(x_0), \varepsilon)$, which means that f is continuous at x_0.

So we have proved that the first three conditions are equivalent.

Next, we show that (a) $\implies$ (d). To this end, take $y_0 \in f(\overline{\mathbf{A}})$. By the definition of the image of a set under f, there is $x_0 \in \overline{\mathbf{A}}$ such that $f(x_0) = y_0$. By continuity of f at x_0, given $\varepsilon > 0$ there is a ball $\mathbf{B_X}(x_0, \delta)$ such that
$$f(\mathbf{B_X}(x_0, \delta)) \subset \mathbf{B_Y}(y_0, \varepsilon).$$
Since $x_0 \in \overline{\mathbf{A}}$, we see that $\mathbf{B_X}(x_0, \delta) \cap \mathbf{A} \neq \emptyset$. Thus
$$\emptyset \neq f(\mathbf{B_X}(x_0, \delta) \cap \mathbf{A}) \subset \mathbf{B_Y}(y_0, \varepsilon) \cap f(\mathbf{A}),$$
which means $y_0 \in \overline{f(\mathbf{A})}$.

1.7. Continuous Functions in Metric Spaces

To show that (d) $\Longrightarrow$ (e), set $\mathbf{A} = f^{-1}(\mathbf{B})$. Then
$$f(\overline{f^{-1}(\mathbf{B})}) \subset \overline{f(f^{-1}(\mathbf{B}))} \subset \overline{\mathbf{B}}.$$
Hence $\overline{f^{-1}(\mathbf{B}))} \subset f^{-1}(\overline{\mathbf{B}})$.

To end our proof we show that (e) $\Longrightarrow$ (b). If $\mathbf{F}$ is closed, then $\mathbf{F} = \overline{\mathbf{F}}$. By (e),
$$\overline{f^{-1}(\mathbf{F})} \subset f^{-1}(\mathbf{F}),$$
which means that $f^{-1}(\mathbf{F})$ is closed.

1.7.2. Let $\mathcal{B}(\mathbf{X})$ denote the family of all Borel subsets of $\mathbf{X}$, that is, the smallest σ-algebra of subsets of $\mathbf{X}$ containing all open sets. Denote by $\tilde{\mathcal{B}}$ the family of sets $\mathbf{B} \subset \mathbf{Y}$ such that $f^{-1}(\mathbf{B}) \in \mathcal{B}(\mathbf{X})$. Then $\tilde{\mathcal{B}}$ is a σ-algebra of subsets of $\mathbf{Y}$. Since f is continuous, it follows from the foregoing problem that the inverse image of any open set is open. Consequently, $\tilde{\mathcal{B}}$ contains all open subsets of $\mathbf{Y}$. Hence $\mathcal{B}(\mathbf{Y}) \subset \tilde{\mathcal{B}}$, which implies that if $\mathbf{B} \in \mathcal{B}(\mathbf{Y})$, then $f^{-1}(\mathbf{B}) \in \mathcal{B}(\mathbf{X})$.

1.7.3. Let $\mathbf{X} = \mathbf{Y} = \mathbb{R}$ be endowed with the usual Euclidean metric $d(x,y) = |x - y|$. Define $f(x) = \sin \pi x$ and $\mathbf{F} = \{n + \frac{1}{n} : n \geq 2\}$. Then $\mathbf{F}$ is closed in the metric space $\mathbf{X}$, because it contains only isolated points. On the other hand,
$$f(\mathbf{F}) = \left\{\sin\frac{\pi}{2}, -\sin\frac{\pi}{3}, \sin\frac{\pi}{4}, \dots\right\}$$
is not closed in $\mathbf{Y}$ because it does not contain zero, which is its accumulation point.

Let $\mathbf{X}$ and $\mathbf{Y}$ be as above and define $f(x) = x(x-2)^2$ and $\mathbf{G} = (1,3)$. Then $f(\mathbf{G}) = [0,3)$.

1.7.4. If $y_n \in f(\mathbf{F})$, then $y_n = f(x_n)$, where $x_n \in \mathbf{F}$, $n = 1, 2, 3, \dots$. If $\mathbf{F}$ is compact in $\mathbf{X}$, then there exists a subsequence $\{x_{n_k}\}$ of $\{x_n\}$ converging to an $x \in \mathbf{F}$. By the continuity of f, $\{y_{n_k}\}$ defined by $y_{n_k} = f(x_{n_k})$ is a subsequence of $\{y_n\}$ converging to $f(x) \in f(\mathbf{F})$. So the compactness of $f(\mathbf{F})$ is proved.

1.7.5. Let $\{x_n\}$ be a sequence of elements in $\mathbf{F}_1 \cup \mathbf{F}_2 \cup \cdots \cup \mathbf{F}_m$ converging to x. Then there is at least one set $\mathbf{F}_i$ containing a subsequence $\{x_{n_k}\}$. Consequently, the sequence $\{x_n\}$ can be decomposed into finitely many subsequences in such a way that every subsequence is contained in one set $\mathbf{F}_i$. Since $\mathbf{F}_i$ is closed and f is continuous on $\mathbf{F}_i$, $f(x_{n_k}) = f_{|\mathbf{F}_i}(x_{n_k}) \to f_{|\mathbf{F}_i}(x) = f(x)$. It then follows that $\{f(x_n)\}$ is decomposed into finitely many subsequences converging to $f(x)$, which means that $\{f(x_n)\}$ converges to $f(x)$.

To see that the statement does not hold in the case of infinitely many sets, consider $\mathbf{F}_i$ defined as follows: $\mathbf{F}_0 = \{0\}$, $\mathbf{F}_i = \left\{\frac{1}{i}\right\}$, $i = 1, 2, 3, \ldots$. The function given by

$$f(x) = \begin{cases} 1 & \text{for } x \in \mathbf{F}_i, \ i = 1, 2, 3, \ldots, \\ 0 & \text{for } x \in \mathbf{F}_0, \end{cases}$$

is continuous on each $\mathbf{F}_i$, $i = 0, 1, 2, 3, \ldots$, but is not continuous on the set $\bigcup_{i=0}^{\infty} \mathbf{F}_i$.

1.7.6. Let $x_0 \in \bigcup_{t \in \mathbf{T}} \mathbf{G}_t$ be arbitrarily chosen. Then there is $t_0 \in \mathbf{T}$ such that $x_0 \in \mathbf{G}_{t_0}$. Since $\mathbf{G}_{t_0}$ is open and the restriction of f to $\mathbf{G}_{t_0}$ is continuous, given $\varepsilon > 0$ there is $\delta > 0$ such that if $x \in \mathbf{B}(x_0, \delta) \subset \mathbf{G}_{t_0}$, then $f(x) = f_{|\mathbf{G}_{t_0}}(x) \in \mathbf{B}\left(f_{|\mathbf{G}_{t_0}}(x_0), \varepsilon\right) = \mathbf{B}(f(x_0), \varepsilon)$, which means that f is continuous at x_0.

1.7.7. Assume that for every compact $\mathbf{A} \subset \mathbf{X}$, $f_{|\mathbf{A}}$ is continuous. If a sequence $\{x_n\}$ of elements in $\mathbf{X}$ converges to x, then the set $\mathbf{A} = \{x, x_1, x_2, x_3, \ldots\}$ is compact in $\mathbf{X}$. So, $f(x_n) = f_{|\mathbf{A}}(x_n) \to f_{|\mathbf{A}}(x) = f(x)$. Thus f is continuous on $\mathbf{X}$. The other implication is obvious.

1.7.8. The continuity of f^{-1} is equivalent to the condition that $f(\mathbf{G})$ is open in $\mathbf{Y}$ for each open $\mathbf{G}$ in $\mathbf{X}$. If $\mathbf{G}$ is open in $\mathbf{X}$, then $\mathbf{G}^c = \mathbf{X} \setminus \mathbf{G}$, as a closed subset of the compact space $\mathbf{X}$, is compact. By the result in 1.7.4, $f(\mathbf{G}^c) = \mathbf{Y} \setminus f(\mathbf{G})$ is also compact, and therefore closed. This means that $f(\mathbf{G})$ is open.

To show that compactness is an essential assumption, consider $f : (0,1) \cup \{2\} \to (0,1]$ given by $f(x) = x$ for $x \in (0,1)$ and $f(2) = 1$. Obviously, f is a continuous bijection of $(0,1) \cup \{2\}$ onto $(0,1]$. Since $f^{-1}(x) = x$ for $x \in (0,1)$ and $f^{-1}(1) = 2$, the inverse function is not continuous on $(0,1]$.

1.7. Continuous Functions in Metric Spaces

1.7.9. Let d_1 and d_2 be metrics for $\mathbf{X}$ and $\mathbf{Y}$, respectively. By continuity of f, given $\varepsilon > 0$ and $x \in \mathbf{X}$ there exists $\delta(x) > 0$ such that

(1) $\qquad d_1(y, x) < \delta(x) \quad \text{implies} \quad d_2(f(y), f(x)) < \dfrac{\varepsilon}{2}.$

Since the family of the balls $\{\mathbf{B}\left(x, \frac{1}{2}\delta(x)\right) : x \in \mathbf{X}\}$ is an open cover of a compact space $\mathbf{X}$, there is a finite subcover

(2) $\qquad \left\{\mathbf{B}\left(x_i, \dfrac{1}{2}\delta(x_i)\right) : i = 1, 2, \ldots, n\right\}.$

Set $\delta = \frac{1}{2}\min\{\delta(x_1), \delta(x_2), \ldots, \delta(x_n)\}$ and take x and y in $\mathbf{X}$ such that $d_1(x, y) < \delta$. Since family (2) is a cover of $\mathbf{X}$, there exists an $i \in \{1, 2, \ldots, n\}$ such that $d_1(x, x_i) < \frac{1}{2}\delta(x_i)$. Then

$$d_1(y, x_i) < d_1(y, x) + d_1(x, x_i) < \delta + \dfrac{1}{2}\delta(x_i) \leq \delta(x_i).$$

Consequently, by (1),

$$d_2(f(x), f(y)) \leq d_2(f(x), f(x_i)) + d_2(f(x_i), f(y)) < \varepsilon.$$

1.7.10. For $x_0, x \in \mathbf{X}$ and $y \in \mathbf{A}$,

$$\operatorname{dist}(x, \mathbf{A}) \leq d(x, y) \leq d(x, x_0) + d(x_0, y).$$

Thus $\operatorname{dist}(x, \mathbf{A}) \leq d(x, x_0) + \operatorname{dist}(x_0, \mathbf{A})$. Hence

$$\operatorname{dist}(x, \mathbf{A}) - \operatorname{dist}(x_0, \mathbf{A}) \leq d(x, x_0).$$

Likewise, $\operatorname{dist}(x_0, \mathbf{A}) - \operatorname{dist}(x, \mathbf{A}) \leq d(x, x_0)$. Consequently,

$$|\operatorname{dist}(x, \mathbf{A}) - \operatorname{dist}(x_0, \mathbf{A})| \leq d(x, x_0),$$

and therefore f is uniformly continuous on $\mathbf{X}$.

1.7.11. If the set $f(\mathbf{X})$ were not connected, then there would exist nonempty, open and disjoint sets $\mathbf{G}_1$ and $\mathbf{G}_2$ such that $\mathbf{G}_1 \cup \mathbf{G}_2 = f(\mathbf{X})$. Continuity of f implies that $f^{-1}(\mathbf{G}_i)$, $i = 1, 2$, are open. Clearly, they are nonempty and disjoint and their union is $\mathbf{X}$, a contradiction.

1.7.12. Let d_1 and d_2 be metrics for $\mathbf{X}$ and $\mathbf{Y}$, respectively. Assume that f is continuous at $x_0 \in \mathbf{A}$. Then, given $\varepsilon > 0$, one can find $\delta > 0$ such that $f(x) \in \mathbf{B}(f(x_0), \varepsilon/2)$ whenever $x \in \mathbf{B}(x_0, \delta) \cap \mathbf{A}$. Consequently, $d_2(f(x), f(y)) < \varepsilon$ for $x, y \in \mathbf{B}(x_0, \delta) \cap \mathbf{A}$. It then follows that $o_f(x_0) = 0$. Conversely, if $o_f(x_0) = 0$, then given $\varepsilon > 0$ there is $\delta_\varepsilon > 0$ such that

$$0 < \delta < \delta_\varepsilon \text{ implies } \operatorname{diam}(f(\mathbf{A} \cap \mathbf{B}(x_0, \delta))) < \varepsilon.$$

Hence $d_1(x, x_0) < \delta$ implies

$$d_2(f(x), f(x_0)) \leq \operatorname{diam}(f(\mathbf{A} \cap \mathbf{B}(x_0, \delta))) < \varepsilon.$$

1.7.13. Set $\mathbf{B} = \{x \in \overline{\mathbf{A}} : o_f(x) \geq \varepsilon\}$ and let $\{x_n\}$ be a sequence of points of $\mathbf{B}$ converging to x_0. Since $\mathbf{B} \subset \overline{\mathbf{A}}$, $x_0 \in \overline{\mathbf{A}}$. Therefore $o_f(x_0)$ is well defined. Moreover, for any $\delta > 0$ there is $n \in \mathbb{N}$ such that $\mathbf{B}(x_n, \delta/2) \subset \mathbf{B}(x_0, \delta)$. Hence

$$\operatorname{diam}(f(\mathbf{A} \cap \mathbf{B}(x_0, \delta))) \geq \operatorname{diam}(f(\mathbf{A} \cap \mathbf{B}(x_n, \delta/2))) \geq o_f(x_n) \geq \varepsilon.$$

It then follows that $o_f(x_0) \geq \varepsilon$, or in other words, $x_0 \in \mathbf{B}$.

1.7.14. By the result in 1.7.12 the set $\mathbf{C}$ of points of continuity of f is equal to the set on which the oscillation vanishes. Put

$$\mathbf{B}_n = \left\{ x \in \mathbf{X} : o_f(x) < \frac{1}{n} \right\}.$$

It follows from the foregoing problem that the $\mathbf{B}_n$ are open in $\mathbf{X}$. On the other hand,

$$\mathbf{C} = \bigcap_{n=1}^{\infty} \mathbf{B}_n,$$

that is, the set of points of continuity of f is of type $\mathcal{G}_\delta$. It then follows that the set $\mathbf{X} \setminus \mathbf{C}$ of points of discontinuity of f is of type $\mathcal{F}_\sigma$ in $\mathbf{X}$.

1.7.15. Consider the function defined by (compare with 1.2.3(a))

$$f(x) = \begin{cases} 0 & \text{if } x \text{ is irrational,} \\ 1 & \text{if } x = 0, \\ \frac{1}{q} & \text{if } x = \frac{p}{q},\, p \in \mathbb{Z},\, q \in \mathbb{N}, \text{ and } p, q \text{ co-prime.} \end{cases}$$

1.7. Continuous Functions in Metric Spaces

1.7.16 [S. S. Kim, Amer. Math. Monthly 106 (1999), 258-259]. Let $\mathbf{A}$ be of type $\mathcal{F}_\sigma$ in $\mathbb{R}$, that is,

$$\bigcup_{n=1}^{\infty} \mathbf{F}_n = \mathbf{A},$$

where the $\mathbf{F}_n$ are closed. Without loss of generality we can assume that $\mathbf{F}_n \subset \mathbf{F}_{n+1}$ for $n \in \mathbb{N}$. Indeed, it is enough to replace $\mathbf{F}_n$ by $\mathbf{F}_1 \cup \mathbf{F}_2 \cup \cdots \cup \mathbf{F}_n$. If $\mathbf{A} = \mathbb{R}$, then, for example, $f(x) = \chi_\mathbb{Q}(x)$ is discontinuous at each $x \in \mathbb{R}$. If $\mathbf{A} \neq \mathbb{R}$, then we define a function g by setting

$$g(x) = \begin{cases} \sum_{n \in \mathbf{K}} \frac{1}{2^n} & \text{if } x \in \mathbf{A}, \\ 0 & \text{if } x \in \mathbb{R} \setminus \mathbf{A}, \end{cases}$$

where $\mathbf{K} = \{n : x \in \mathbf{F}_n\}$, and we put

$$f(x) = g(x)\left(\chi_\mathbb{Q}(x) - \frac{1}{2}\right).$$

First we show that each point of $\mathbf{A}$ is a point of discontinuity of f. Indeed, if $x \in \mathbf{A}^\circ$, then every neighborhood of x contains a point at which the sign of f is different from the sign of $f(x)$. If $x \in \partial \mathbf{A} \cap \mathbf{A}$, then $f(x) \neq 0$ and every neighborhood of x contains a point at which f vanishes. Since $\mathbf{A} = \mathbf{A}^\circ \cup (\partial \mathbf{A} \cap \mathbf{A})$, the function f is discontinuous on $\mathbf{A}$. Our task is now to show that f is continuous on $\mathbb{R} \setminus \mathbf{A}$. We have $f(x) = 0$ if $x \notin \mathbf{A}$. If a sequence $\{x_k\}$ converges to x and $x_k \in \mathbf{A}$, then for each n there is a k_n such that $x_k \notin \mathbf{F}_n$ for $k \geq k_n$. (If there were infinitely many x_k in some $\mathbf{F}_n$, then x would be also in $\mathbf{F}_n$.) Consequently, for $k \geq k_n$,

$$g(x_k) \leq \frac{1}{2^{n+1}} + \frac{1}{2^{n+2}} + \cdots = \frac{1}{2^n},$$

which means that $\lim_{k \to \infty} g(x_k) = 0 = g(x)$.

1.7.17. No. Every function defined on a discrete metric space is continuous.

1.7.18. Assume first that $x \in \partial \mathbf{A} = \overline{\mathbf{A}} \cap \overline{\mathbf{X} \setminus \mathbf{A}}$. Since each ball $\mathbf{B}(x, \delta)$ contains points of $\mathbf{A}$ and points of $\mathbf{X} \setminus \mathbf{A}$, we get $o_{\chi_\mathbf{A}}(x) = 1$.

Assume now that $o_{\chi_\mathbf{A}}(x) > 0$. This means that for every $\delta > 0$,

$$\sup\{|\chi_\mathbf{A}(x) - \chi_\mathbf{A}(y)| : y \in \mathbf{B}(x,\delta)\} = o_{\chi_\mathbf{A}}(x,\delta) > 0.$$

Consequently, each ball $\mathbf{B}(x,\delta)$ must contain points of $\mathbf{A}$ and points of $\mathbf{X} \setminus \mathbf{A}$. Hence $x \in \partial\mathbf{A} = \overline{\mathbf{A}} \cap \overline{\mathbf{X} \setminus \mathbf{A}}$.

Clearly, if $\mathbf{A}$ is both open and closed, then $\partial\mathbf{A} = \emptyset$. Therefore, by 1.7.12, $\chi_\mathbf{A}$ is continuous on $\mathbf{X}$. Conversely, if $\chi_\mathbf{A}$ is continuous on $\mathbf{X}$, then $\partial\mathbf{A} = \emptyset$. Now we show that $\overline{\mathbf{A}} \subset \mathbf{A}$. If not, there is $x \in \overline{\mathbf{A}} \setminus \mathbf{A} \subset \mathbf{X} \setminus \mathbf{A} \subset \overline{\mathbf{X} \setminus \mathbf{A}}$, a contradiction. One can show in an entirely similar manner that $\mathbf{X} \setminus \mathbf{A}$ is also closed.

1.7.19. For $x \in \mathbf{A}$ and $\delta > 0$ we have

$$\begin{aligned} o_f(x,\delta) &= \sup\{d_2(f(x),f(y)) : y \in \mathbf{B}(x,\delta)\} \\ &\leq \sup\{d_2(f(x),f(y)) : y \in \mathbf{A} \cap \mathbf{B}(x,\delta)\} \\ &\quad + \sup\{d_2(f(x),f(y)) : y \in (\mathbf{X} \setminus \mathbf{A}) \cap \mathbf{B}(x,\delta)\}. \end{aligned}$$

Thus

$$\begin{aligned} o_f(x,\delta) &\leq \sup\{d_2(g_1(x),g_1(y)) : y \in \mathbf{A} \cap \mathbf{B}(x,\delta)\} \\ &\quad + \sup\{d_2(g_1(x),g_2(y)) : y \in (\mathbf{X} \setminus \mathbf{A}) \cap \mathbf{B}(x,\delta)\} \\ &\leq o_{g_1}(x,\delta) + \sup\{d_2(g_1(x),g_2(y)) : y \in (\mathbf{X} \setminus \mathbf{A}) \cap \mathbf{B}(x,\delta)\} \\ &\leq o_{g_1}(x,\delta) \\ &\quad + \sup\{d_2(g_1(x),g_2(x)) + d_2(g_2(x),g_2(y)) : y \in (\mathbf{X} \setminus \mathbf{A}) \cap \mathbf{B}(x,\delta)\} \\ &\leq o_{g_1}(x,\delta) + d_2(g_1(x),g_2(x)) + o_{g_2}(x,\delta). \end{aligned}$$

Since g_1 and g_2 are continuous, we get, by 1.7.12,

(1) $$o_f(x) \leq d_2(g_1(x),g_2(x)).$$

Now our task is to show that for $x \in \mathbf{A}$,

(2) $$o_f(x) \geq d_2(g_1(x),g_2(x)).$$

Let $\{\delta_n\}$ be a sequence of positive numbers converging to zero. Since $\mathbf{A}° = \emptyset$, the set $\mathbf{X} \setminus \mathbf{A}$ is dense in $\mathbf{X}$. Thus each ball $\mathbf{B}(x,\delta_n)$ contains

1.7. Continuous Functions in Metric Spaces 205

a point y_n of $\mathbf{X} \setminus \mathbf{A}$. Consequently,

$$\sup\{d_2(f(x), f(y)) : y \in \mathbf{B}(x, \delta_n)\}$$
$$\geq \sup\{d_2(g_1(x), g_2(y)) : y \in \mathbf{B}(x, \delta_n) \cap (\mathbf{X} \setminus \mathbf{A})\}$$
$$\geq d_2(g_1(x), g_2(y_n)).$$

This combined with the continuity of g_2 implies

$$\lim_{n \to \infty} \sup\{d_2(f(x), f(y)) : y \in \mathbf{B}(x, \delta_n)\} \geq d_2(g_1(x), g_2(x)),$$

which in turn gives (2). It follows from (1) and (2) that the desired equality holds for $x \in \mathbf{A}$. In an entirely similar manner (using the density of $\mathbf{A}$) one can show that this equality holds also for $x \in \mathbf{X} \setminus \mathbf{A}$.

1.7.20. Assume that $\{f_n\}$ is a sequence of functions continuous on $\mathbf{X}$ and such that $f(x) = \lim_{n \to \infty} f_n(x)$. For $\varepsilon > 0$, put

$$\mathbf{P}_m(\varepsilon) = \{x \in \mathbf{X} : |f(x) - f_m(x)| \leq \varepsilon\}$$

and $\mathbf{G}(\varepsilon) = \bigcup_{m=1}^{\infty} (\mathbf{P}_m(\varepsilon))^\circ$. We will prove that $\mathbf{C} = \bigcap_{n=1}^{\infty} \mathbf{G}(1/n)$ is the set of all points of continuity of f. We show first that if f is continuous at x_0, then $x_0 \in \mathbf{C}$. Since $f(x) = \lim_{n \to \infty} f_n(x)$, there is an m such that

$$|f(x_0) - f_m(x_0)| \leq \frac{\varepsilon}{3}.$$

It follows from the continuity of f and f_m at x_0 that there exists a ball $\mathbf{B}(x_0, \delta)$ such that for $x \in \mathbf{B}(x_0, \delta)$,

$$|f(x) - f(x_0)| \leq \frac{\varepsilon}{3} \quad \text{and} \quad |f_m(x) - f_m(x_0)| \leq \frac{\varepsilon}{3}.$$

Consequently, $|f(x) - f_m(x)| \leq \varepsilon$ if $x \in \mathbf{B}(x_0, \delta)$. This means that $x_0 \in (\mathbf{P}_m(\varepsilon))^\circ \subset \mathbf{G}(\varepsilon)$. Since $\varepsilon > 0$ can be arbitrarily chosen, we see that $x_0 \in \mathbf{C}$.

Now if

$$x_0 \in \mathbf{C} = \bigcap_{n=1}^{\infty} \mathbf{G}(1/n),$$

then, for any $\varepsilon > 0$, $x_0 \in \mathbf{G}(\varepsilon/3)$. Thus there is a positive integer m such that $x_0 \in (\mathbf{P}_m(\varepsilon/3))^\circ$. Consequently, there exists a ball $\mathbf{B}(x_0, \delta)$

such that if $x \in \mathbf{B}(x_0,\delta)$, then
$$|f(x) - f_m(x)| \leq \frac{\varepsilon}{3}.$$
Since f_m is continuous, this shows that f is continuous at x_0. Now our task is to prove that $\mathbf{X} \setminus \mathbf{C}$ is of the first category. To this end, define
$$\mathbf{F}_m(\varepsilon) = \{x \in \mathbf{X} : |f_m(x) - f_{m+k}(x)| \leq \varepsilon \quad \text{for all} \quad k \in \mathbb{N}\}.$$
The continuity of f_n, $n \in \mathbb{N}$, implies that $\mathbf{F}_m(\varepsilon)$ is closed. Since $f(x) = \lim_{n \to \infty} f_n(x)$, $x \in \mathbf{X}$, we see that $\mathbf{X} = \bigcup_{m=1}^{\infty} \mathbf{F}_m(\varepsilon)$ and $\mathbf{F}_m(\varepsilon) \subset \mathbf{P}_m(\varepsilon)$. Consequently,
$$\bigcup_{m=1}^{\infty} (\mathbf{F}_m(\varepsilon))^\circ \subset \mathbf{G}(\varepsilon).$$
Now note that for any $\mathbf{F} \subset \mathbf{X}$ the interior of $\mathbf{F} \setminus \mathbf{F}^\circ$ is empty, because $(\mathbf{F} \setminus \mathbf{F}^\circ)^\circ \subset \mathbf{F}^\circ \setminus (\mathbf{F}^\circ)^\circ = \emptyset$. Moreover, if $\mathbf{F}$ is closed, then $\mathbf{F} \setminus \mathbf{F}^\circ$ is closed and therefore $\mathbf{F} \setminus \mathbf{F}^\circ$ is nowhere dense. Since
$$\mathbf{X} \setminus \bigcup_{m=1}^{\infty} (\mathbf{F}_m(\varepsilon))^\circ \subset \bigcup_{m=1}^{\infty} (\mathbf{F}_m(\varepsilon) \setminus (\mathbf{F}_m(\varepsilon))^\circ),$$
the set $\mathbf{X} \setminus \bigcup_{m=1}^{\infty} (\mathbf{F}_m(\varepsilon))^\circ$ is of the first category. Moreover, since $\mathbf{X} \setminus \mathbf{G}(\varepsilon) \subset \mathbf{X} \setminus \bigcup_{m=1}^{\infty} (\mathbf{F}_m(\varepsilon))^\circ$, the set $\mathbf{X} \setminus \mathbf{G}(\varepsilon)$ is also of the first category. Finally, observe that
$$\mathbf{X} \setminus \mathbf{C} = \mathbf{X} \setminus \bigcap_{n=1}^{\infty} \mathbf{G}(1/n) = \bigcup_{n=1}^{\infty} (\mathbf{X} \setminus \mathbf{G}(1/n)).$$
Therefore the set $\mathbf{X} \setminus \mathbf{C}$ of points of discontinuity of f is of the first category.

1.7.21. We will use the notation from the solution of the preceding problem. We have
$$\mathbf{X} \setminus \mathbf{G}(1/k) \subset \mathbf{X} \setminus \bigcup_{m=1}^{\infty} (\mathbf{F}_m(1/k))^\circ \subset \bigcup_{m=1}^{\infty} (\mathbf{F}_m(1/k) \setminus (\mathbf{F}_m(1/k))^\circ).$$

1.7. Continuous Functions in Metric Spaces

Hence

$$\bigcup_{k=1}^{\infty} (\mathbf{X} \setminus \mathbf{G}(1/k)) \subset \bigcup_{k=1}^{\infty} \bigcup_{m=1}^{\infty} (\mathbf{F}_m(1/k) \setminus (\mathbf{F}_m(1/k))^\circ).$$

So, $\mathbf{X} \setminus \mathbf{C}$ is a subset of the union of countably many closed and nowhere dense sets (their complements are open and dense in $\mathbf{X}$). It then follows that $\mathbf{C}$ contains the intersection of countably many open and dense sets. By the theorem of Baire, $\mathbf{C}$ is dense in $\mathbf{X}$.

1.7.22. For $\varepsilon > 0$ put

$$\mathbf{F}_k = \{0\} \cup \bigcap_{n \geq k} \left\{ x > 0 : \left| f\left(\frac{x}{n}\right) \right| \leq \varepsilon \right\}, \quad k = 1, 2, 3, \ldots.$$

Since f is continuous, the sets are closed (see, e.g., 1.7.1). By hypothesis, $\bigcup_{k=1}^{\infty} \mathbf{F}_k = [0, \infty)$. According to the theorem of Baire, at least one of the sets $\mathbf{F}_k$ has a nonempty interior. Consequently, there exist $a > 0, \delta > 0$, and $k \in \mathbb{N}$ such that $(a - \delta, a + \delta) \subset \mathbf{F}_k$. Without loss of generality we can assume that $\delta \leq \frac{a}{k}$. If $0 < x \leq \delta$ and $n = \left[\frac{a}{x}\right]$, then $a - \delta \leq a - x < nx \leq a < a + \delta$, and $n \geq k$. Thus $nx \in \mathbf{F}_k$, and, by the definition of $\mathbf{F}_k$,

$$f(x) = \left| f\left(\frac{nx}{n}\right) \right| \leq \varepsilon,$$

which implies $\lim_{x \to 0^+} f(x) = 0$.

1.7.23. Define $\mathbf{F}_n$ as follows:

$$\mathbf{F}_n = \{x \in \mathbf{X} : |f(x)| \leq n \quad \text{for all } f \in \mathcal{F}\}.$$

It follows from the continuity of f that the $\mathbf{F}_n$ are closed. By hypotheses, for every $x \in \mathbf{X}$ there is a positive integer n_x such that $|f(x)| \leq n_x$ for all $f \in \mathcal{F}$. Thus $x \in \mathbf{F}_{n_x}$. Consequently, $\mathbf{X} = \bigcup_{n=1}^{\infty} \mathbf{F}_n$. Since $(\mathbf{X}, d_1)$ is of the second category, there is an $\mathbf{F}_{n_0}$ with nonempty interior. Let $\mathbf{G} = \mathbf{F}_{n_0}^\circ$. Therefore $|f(x)| \leq n_0$ for every $f \in \mathbf{F}$ and each $x \in \mathbf{G}$.

1.7.24. We know that
$$f\left(\bigcap_{n=1}^{\infty} \mathbf{F}_n\right) \subset \bigcap_{n=1}^{\infty} f(\mathbf{F}_n).$$
Now we show that if f is continuous, then
$$\bigcap_{n=1}^{\infty} f(\mathbf{F}_n) \subset f\left(\bigcap_{n=1}^{\infty} \mathbf{F}_n\right).$$
Let $y \in \bigcap_{n=1}^{\infty} f(\mathbf{F}_n)$. Then, for any positive integer n, $y \in f(\mathbf{F}_n)$, or in other words, $y = f(x_n)$ with an $x_n \in \mathbf{F}_n$. By Cantor's nested set theorem, $\bigcap_{n=1}^{\infty} \mathbf{F}_n = \{x_0\}$ for some $x_0 \in \mathbf{X}$. By the continuity of f, $y = \lim_{n \to \infty} f(x_n) = f(x_0)$. Thus $y \in f\left(\bigcap_{n=1}^{\infty} \mathbf{F}_n\right)$.

1.7.25. For $u, v \in \mathbf{X}$ we have
$$d(f_u, f_v) = \sup\{|d_1(u, x) - d_1(v, x)| : x \in \mathbf{X}\} \leq d_1(u, v).$$
Moreover,
$$d(f_u, f_v) = \sup\{|d_1(u, x) - d_1(v, x)| : x \in \mathbf{X}\}$$
$$\geq |d_1(u, u) - d_1(u, v)| = d_1(u, v).$$

1.7.26. Assume first that $\mathbf{X}$ is a compact metric space and that $f : \mathbf{X} \to \mathbb{R}$ is continuous. Then, given $\varepsilon > 0$ and $x \in \mathbf{X}$, there is $\delta_x > 0$ such that $|f(y) - f(x)| < \varepsilon$ for $|y - x| < \delta_x$. Since the family $\{\mathbf{B}(x, \delta_x), x \in \mathbf{X}\}$ is an open cover of $\mathbf{X}$, there is a finite subcover $\mathbf{B}(x_1, \delta_{x_1}), \mathbf{B}(x_2, \delta_{x_2}), \ldots, \mathbf{B}(x_n, \delta_{x_n})$. Therefore for $x \in \mathbf{X}$ there exists $i \in \{1, 2, \ldots, n\}$ such that $x \in \mathbf{B}(x_i, \delta_{x_i})$. It then follows that
$$|f(x)| \leq |f(x) - f(x_i)| + |f(x_i)| \leq \varepsilon + \max\{f(x_1), f(x_2), \ldots, f(x_n)\},$$
which proves the boundedness of f on $\mathbf{X}$.

Assume now that every real function continuous on $\mathbf{X}$ is bounded, and suppose, contrary to our claim, that $\mathbf{X}$ is not compact. Then one can find a sequence $\{x_n\}$ of elements in $\mathbf{X}$ that does not contain any

1.7. Continuous Functions in Metric Spaces   209

convergent subsequence. Then $\mathbf{F} = \{x_n : n \in \mathbb{N}\}$ is closed in $\mathbf{X}$. The function f given by $f(x_n) = n$ is continuous on $\mathbf{F}$. According to the Tietze extension theorem, there exists a continuous extension of f defined on all of $\mathbf{X}$. So, we have constructed a continuous and unbounded function, a contradiction.

1.7.27. First we prove that (a) implies (b). So assume that (a) holds, let $\lim_{n\to\infty} \rho(x_n) = 0$, and suppose, contrary to our claim, that $\{x_n\}$ does not contain a convergent subsequence. Then there is a sequence $\{y_n\}$ of elements in $\mathbf{X}$ such that $\lim_{n\to\infty} d_1(x_n, y_n) = 0$ and $y_n \neq x_n$ for $n \in \mathbb{N}$. If $\{y_n\}$ contains a convergent subsequence $\{y_{n_k}\}$, then by $\lim_{k\to\infty} d_1(x_{n_k}, y_{n_k}) = 0$ the sequence $\{x_{n_k}\}$ is also convergent. Thus $\{y_n\}$ does not contain any convergent subsequence. It then follows that no term of the sequences $\{x_n\}$ and $\{y_n\}$ is repeated infinitely many times. Therefore there is a strictly increasing sequence $\{n_k\}$ of positive integers such that the infinite sets $\mathbf{F}_1 = \{x_{n_k} : k \in \mathbb{N}\}$ and $\mathbf{F}_2 = \{y_{n_k} : k \in \mathbb{N}\}$ are closed and disjoint. According to the Urysohn lemma, there is a continuous function $f : \mathbf{X} \to \mathbb{R}$ such that f is one on $\mathbf{F}_1$ and zero on $\mathbf{F}_2$. Thus

$$|f(x_{n_k}) - f(y_{n_k})| = 1 \quad \text{and} \quad \lim_{k\to\infty} d_1(x_{n_k}, y_{n_k}) = 0.$$

Hence f is continuous but not uniformly continuous on $\mathbf{X}$, which contradicts (a).

To show that (b) implies (a), denote by $\mathbf{A}$ the set of limit points of $\mathbf{X}$. By (b) every sequence of elements in $\mathbf{A}$ has a subsequence converging to an element in $\mathbf{A}$. Therefore $\mathbf{A}$ is compact. If $\mathbf{X} \neq \mathbf{A}$, then for $\delta_1 > 0$ put $\delta_2 = \inf\{\rho(x) : x \in \mathbf{X}, \mathrm{dist}(x, \mathbf{A}) > \delta_1\}$. We will show that $\delta_2 > 0$. If $\delta_2 = 0$, then there is a sequence $\{x_n\}$ of elements in $\mathbf{X}$ such that $\lim_{n\to\infty} \rho(x_n) = 0$ and $\mathrm{dist}(x_n, \mathbf{A}) > \delta_1$. By (b), $\{x_n\}$ has a subsequence converging to an element in $\mathbf{A}$, a contradiction. Let $f : \mathbf{X} \to \mathbb{R}$ be continuous and let $\varepsilon > 0$ be arbitrarily fixed. Then for $x \in \mathbf{A}$ there is $\delta_x > 0$ such that if $d_1(x, y) < \delta_x$, then $|f(x) - f(y)| < \frac{1}{2}\varepsilon$. Since $\mathbf{A}$ is compact, there are $x_1, \ldots, x_n \in \mathbf{A}$ such that

$$\mathbf{A} \subset \bigcup_{k=1}^{n} \mathbf{B}\left(x_k, \frac{1}{3}\delta_{x_k}\right).$$

Let $\delta_1 = \frac{1}{3}\min\{\delta_{x_1},\ldots,\delta_{x_n}\}$ and $\delta_2 > 0$ be as above. Put $\delta = \min\{\delta_1,\delta_2\}$ and let $x,y \in \mathbf{X}$ be such that $d_1(x,y) < \delta$. If $\operatorname{dist}(x,\mathbf{A}) > \delta_1$, then $\rho(x) > \delta_2$, so $d_1(x,y) < \delta \leq \delta_2$ only if $x = y$. Then, obviously, $|f(x) - f(y)| < \varepsilon$. If $\operatorname{dist}(x,\mathbf{A}) \leq \delta_1$, then there is an $a \in \mathbf{A}$ such that $d_1(x,a) \leq \delta_1$. It follows from the above that there is $k \in \{1,2,\ldots,n\}$ for which $d_1(a,x_k) < \frac{1}{3}\delta_{x_k}$. Consequently,

$$d_1(y,x_k) \leq d_1(y,x) + d_1(x,a) + d_1(a,x_k) < \delta + \delta_1 + \frac{1}{3}\delta_{x_k} \leq \delta_{x_k}.$$

Hence

$$|f(x) - f(y)| \leq |f(x) - f(x_k)| + |f(x_k) - f(y)| < \frac{1}{2}\varepsilon + \frac{1}{2}\varepsilon = \varepsilon.$$

This proves the uniform continuity of f on $\mathbf{X}$.

1.7.28. It is well known, see, e.g., 1.7.9, that every function continuous on a compact metric space is uniformly continuous. We claim that if $\mathbf{X}$ is compact, then each set $\{x \in \mathbf{X} : \rho(x) > \varepsilon\}$, $\varepsilon > 0$, is finite. On the contrary, suppose that there is an $\varepsilon > 0$ for which the set $\{x \in \mathbf{X} : \rho(x) > \varepsilon\}$ is infinite. Since the family of balls $\{\mathbf{B}(x,\varepsilon) : x \in \mathbf{X}\}$ is an open cover of $\mathbf{X}$, it has a finite subcover, which contradicts the fact that $\rho(x) > \varepsilon$ for infinitely many x.

Assume now that every real continuous function on $\mathbf{X}$ is uniformly continuous and that every set $\{x \in \mathbf{X} : \rho(x) > \varepsilon\}$ is finite. We will show that $\mathbf{X}$ is compact. Let $\{x_n\}$ be a sequence of points in $\mathbf{X}$. If a term in the sequence is repeated infinitely many times, then obviously there is a convergent subsequence. If not, then $\lim\limits_{n\to\infty} \rho(x_n) = 0$, because the sets $\{x \in \mathbf{X} : \rho(x) > \varepsilon\}$ are finite. By the result in the foregoing problem, $\{x_n\}$ contains a convergent subsequence.

1.7.29. It is enough to consider $\mathbf{X} = [0,1] \cup \{2\} \cup \{3\} \cup \{4\} \cup \ldots$ equipped with the usual Euclidean metric $d_1(x,y) = |x - y|$.

Chapter 2

Differentiation

2.1. The Derivative of a Real Function

2.1.1.

(a) We have
$$f(x) = \begin{cases} x^2 & \text{if } x \geq 0, \\ -x^2 & \text{if } x < 0. \end{cases}$$
Hence
$$f'(x) = \begin{cases} 2x & \text{if } x \geq 0, \\ -2x & \text{if } x < 0, \end{cases}$$
because
$$f'_+(0) = \lim_{h \to 0^+} \frac{h^2 - 0}{h} = 0 = f'_-(0).$$

(b) We get
$$f'(x) = \begin{cases} \frac{1}{2\sqrt{x}} & \text{if } x > 0, \\ -\frac{1}{2\sqrt{-x}} & \text{if } x < 0. \end{cases}$$
Since
$$f'_+(0) = \lim_{h \to 0^+} \frac{\sqrt{h} - 0}{h} = +\infty$$
and
$$f'_-(0) = \lim_{h \to 0^-} \frac{\sqrt{-h} - 0}{h} = -\infty,$$
the derivative of f at zero does not exist.

(c) $f'(x) = n\pi \sin(2\pi x)$ for $x \in (n, n+1)$, $n \in \mathbb{Z}$. Moreover, for $n \in \mathbb{Z}$,

$$f'_+(n) = \lim_{x \to n^+} \frac{n \sin^2(\pi x) - 0}{x - n} = \lim_{x \to n^+} \frac{n \sin^2(\pi x - \pi n)}{x - n} = 0,$$

$$f'_-(n) = \lim_{x \to n^-} \frac{(n-1) \sin^2(\pi x) - 0}{x - n} = 0.$$

It then follows that $f'(x) = \pi[x] \sin(2\pi x)$.

(d) It follows from (c) that

$$f'(x) = (x \sin^2(\pi x))' - ([x] \sin^2(\pi x))'$$
$$= \sin^2(\pi x) + \pi(x - [x]) \sin(2\pi x).$$

(e) $f'(x) = \frac{1}{x}$ for $x \neq 0$.

(f) $f'(x) = \frac{1}{x\sqrt{x^2-1}}$ if $|x| > 1$.

2.1.2.

(a) Since $\log_x 2 = \frac{\ln 2}{\ln x}$, we get

$$f'(x) = -\frac{\ln 2}{x \ln^2 x} = -\frac{\log_x 2 \cdot \log_x e}{x}.$$

(b) As in (a) we show that

$$f'(x) = \frac{-\tan x \ln x - \frac{1}{x} \ln \cos x}{\ln^2 x}$$
$$= -\tan x \log_x e - \frac{1}{x} \log_x \cos x \cdot \log_x e.$$

2.1.3.

(a) Clearly,

$$f'(x) = \begin{cases} \frac{1}{1+x^2} & \text{if } |x| < 1, \\ \frac{1}{2} & \text{if } |x| > 1. \end{cases}$$

We will now check whether the derivative exists at $x = 1$ and at $x = -1$. We have

$$f'_+(1) = \lim_{x \to 1^+} \frac{\frac{\pi}{4} + \frac{x-1}{2} - \frac{\pi}{4}}{x - 1} = \frac{1}{2},$$

$$f'_-(1) = \lim_{x \to 1^-} \frac{\arctan x - \frac{\pi}{4}}{x - 1} = \arctan'(1) = \frac{1}{2}.$$

2.1. The Derivative of a Real Function

So $f'(1) = \frac{1}{2}$. We have also

$$f'_+(-1) = \lim_{x \to -1^+} \frac{\arctan x + \frac{\pi}{4}}{x+1} = \arctan'(-1) = \frac{1}{2},$$

$$f'_-(-1) = \lim_{x \to -1^-} \frac{-\frac{\pi}{4} + \frac{x-1}{2} + \frac{\pi}{4}}{x+1} = +\infty.$$

Therefore $f'(-1)$ does not exist.

(b) We have

$$f'(x) = \begin{cases} 2xe^{-x^2}(1-x^2) & \text{if } |x| < 1, \\ 0 & \text{if } |x| > 1. \end{cases}$$

Moreover,

$$f'_+(1) = \lim_{x \to 1^+} \frac{\frac{1}{e} - \frac{1}{e}}{x-1} = 0,$$

$$f'_-(1) = \lim_{x \to 1^-} \frac{x^2 e^{-x^2} - \frac{1}{e}}{x-1} = (x^2 e^{-x^2})'\big|_{x=1} = 0.$$

Since f is even, $f'(-1) = 0$.

(c) Observe that f is continuous at zero. Moreover,

$$f'_+(0) = \lim_{x \to 0^+} \frac{\arctan \frac{1}{x} - \frac{\pi}{2}}{x} = \lim_{t \to \frac{\pi}{2}^-} \frac{t - \frac{\pi}{2}}{\frac{1}{\tan t}}$$

$$= \lim_{t \to \frac{\pi}{2}^-} \left(t - \frac{\pi}{2}\right)\tan t = -1$$

and

$$f'_-(0) = \lim_{x \to 0^-} \frac{\arctan\left(-\frac{1}{x}\right) - \frac{\pi}{2}}{x} = \lim_{t \to \frac{\pi}{2}^-} \frac{t - \frac{\pi}{2}}{-\frac{1}{\tan t}}$$

$$= -\lim_{t \to \frac{\pi}{2}^-} \left(t - \frac{\pi}{2}\right)\tan t = 1.$$

Thus the function is not differentiable only at zero.

2.1.4. Note first that

$$f'(0) = \lim_{x \to 0} \frac{x^2 \left|\cos \frac{\pi}{x}\right|}{x} = 0.$$

Clearly, for $x \neq \frac{2}{2n+1}$, $n \in \mathbb{Z}$, $f'(x)$ exists. For $x_n = \frac{2}{2n+1}$, $n = 0, 2, 4, \ldots$, we get

$$f'_+(x_n) = \lim_{x \to x_n^+} \frac{x^2 \cos \frac{\pi}{x}}{x - x_n} = \left(x^2 \cos \frac{\pi}{x}\right)'\Big|_{x = x_n} = \pi,$$

$$f'_-(x_n) = \lim_{x \to x_n^-} \frac{-x^2 \cos \frac{\pi}{x}}{x - x_n} = \left(-x^2 \cos \frac{\pi}{x}\right)'\Big|_{x = x_n} = -\pi.$$

Similarly, if $x_n = \frac{2}{2n+1}$, $n = 1, 3, 5, \ldots$, then $f'_+(x_n) = \pi$ and $f'_-(x_n) = -\pi$. Since f is even, f is not differentiable at x_n, $n \in \mathbb{Z}$.

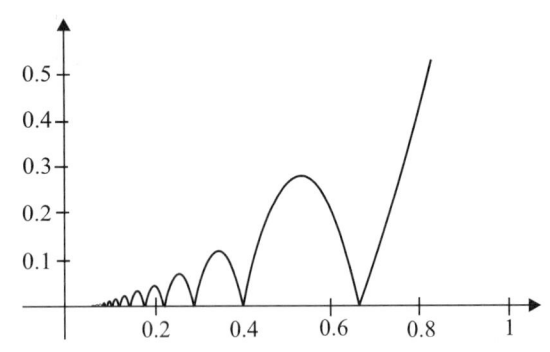

2.1.5.

(a) Since f must be continuous, we get $c = 0$ and $a + b = 1$. Since $f'_-(0) = 4$, $f'_+(0) = b$, we get $b = 4$ and $a = -3$. It is easy to verify that for such a, b and c the function f is differentiable on $\mathbb{R}$.

(b) $a = d = -1, b = 0, c = 1$.

(c) $b = c = 1, a = 0, d = \frac{1}{4}$.

2.1.6.

(a) For $x \neq 0$,

$$\sum_{k=0}^{n} e^{kx} = \frac{1 - e^{(n+1)x}}{1 - e^x}.$$

Differentiating both sides of this equality, we get

$$\sum_{k=0}^{n} k e^{kx} = \frac{n e^{(n+2)x} - (n+1) e^{(n+1)x} + e^x}{(1 - e^x)^2}.$$

2.1. The Derivative of a Real Function

(b) Differentiating both sides of the equality

$$\sum_{k=0}^{2n}(-1)^k \binom{2n}{k} e^{kx} = (e^x - 1)^{2n}$$

n times, we obtain

$$\sum_{k=0}^{2n}(-1)^k k^n e^{kx} \binom{2n}{k} = \left((e^x - 1)^{2n}\right)^{(n)}.$$

To calculate $\left((e^x - 1)^{2n}\right)^{(n)}$ at zero we set $g(x) = e^x - 1$ and note that the nth derivative of $(g(x))^{2n}$ is a sum whose every term contains a power of $g(x)$ at least of order n (compare with 2.1.38). So, the nth derivative of $x \mapsto (e^x - 1)^{2n}$ at zero is 0. Consequently,

$$\sum_{k=0}^{2n}(-1)^k k^n \binom{2n}{k} = 0.$$

(c) Differentiating the equality

$$\sum_{k=1}^{n} \sin(kx) = \frac{\sin \frac{nx}{2} \sin \frac{(n+1)x}{2}}{\sin \frac{x}{2}}, \quad x \neq 2l\pi, \ l \in \mathbb{Z},$$

we get

$$\sum_{k=1}^{n} k \cos(kx) = \frac{n \sin \frac{x}{2} \sin \frac{(2n+1)x}{2} - \sin^2 \frac{nx}{2}}{2 \sin^2 \frac{x}{2}}, \quad x \neq 2l\pi, \ l \in \mathbb{Z}.$$

For $x = 2l\pi$,

$$\sum_{k=1}^{n} k \cos(kx) = \frac{1}{2}n(n + 1).$$

2.1.7. Put $f(x) = a_1 \sin x + a_2 \sin 2x + \cdots + a_n \sin nx$. Then

$$|a_1 + 2a_2 + \cdots + na_n| = |f'(0)| = \lim_{x \to 0} \left|\frac{f(x) - f(0)}{x}\right|$$

$$= \lim_{x \to 0} \left|\frac{f(x)}{\sin x}\right| \cdot \left|\frac{\sin x}{x}\right| = \lim_{x \to 0} \left|\frac{f(x)}{\sin x}\right| \leq 1.$$

2.1.8.

(a) We have

$$\lim_{x \to a} \frac{xf(a) - af(x)}{x - a} = \lim_{x \to a} \frac{(x-a)f(a) - a(f(x) - f(a))}{x - a}$$
$$= f(a) - af'(a).$$

(b) As in (a) we have

$$\lim_{x \to a} \frac{f(x)g(a) - f(a)g(x)}{x - a}$$
$$= \lim_{x \to a} \frac{(f(x) - f(a))g(a) - f(a)(g(x) - g(a))}{x - a}$$
$$= f'(a)g(a) - f(a)g'(a).$$

2.1.9.

(a) Since f is continuous at a and $f(a) > 0$, we see that $f\left(a + \frac{1}{n}\right) > 0$ for sufficiently large n. Moreover, since f is differentiable at a, the function $x \mapsto \ln(f(x))$ is also. Consequently,

$$\lim_{n \to \infty} \ln \left(\frac{f\left(a + \frac{1}{n}\right)}{f(a)} \right)^{\frac{1}{n}} = \lim_{n \to \infty} \frac{1}{n^2} \frac{\ln f\left(a + \frac{1}{n}\right) - \ln f(a)}{\frac{1}{n}}$$
$$= 0 \cdot (\ln f(x))'\big|_{x=a} = 0.$$

Hence

$$\lim_{n \to \infty} \left(\frac{f\left(a + \frac{1}{n}\right)}{f(a)} \right)^{\frac{1}{n}} = 1.$$

(b) As in (a) we get

$$\lim_{x \to a} \ln \left(\frac{f(x)}{f(a)} \right)^{\frac{1}{\ln x - \ln a}} = \lim_{x \to a} \frac{\ln f(x) - \ln f(a)}{x - a} \cdot \frac{x - a}{\ln x - \ln a} = \frac{f'(a)}{f(a)} a.$$

2.1.10.

(a) By 2.1.8(b) with $g(x) = x^n$,

$$\lim_{x \to a} \frac{a^n f(x) - x^n f(a)}{x - a} = -na^{n-1} f(a) + a^n f'(a).$$

2.1. The Derivative of a Real Function 217

(b)

$$\lim_{x \to 0} \frac{f(x)e^x - f(0)}{f(x)\cos x - f(0)} = \lim_{x \to 0} \frac{f(x)e^x - f(0)}{x} \cdot \frac{x}{f(x)\cos x - f(0)}$$

$$= (f(x)e^x)'|_{x=0} \cdot \frac{1}{(f(x)\cos x)'|_{x=0}}$$

$$= \frac{f'(0) + f(0)}{f'(0)}.$$

(c)

$$\lim_{n \to \infty} n \left(f\left(a + \frac{1}{n}\right) + f\left(a + \frac{2}{n}\right) + \cdots + f\left(a + \frac{k}{n}\right) - kf(a) \right)$$

$$= \lim_{n \to \infty} \left(\frac{f\left(a + \frac{1}{n}\right) - f(a)}{\frac{1}{n}} + 2 \frac{f\left(a + \frac{2}{n}\right) - f(a)}{\frac{2}{n}} \right.$$

$$\left. + \cdots + k \frac{f\left(a + \frac{k}{n}\right) - f(a)}{\frac{k}{n}} \right)$$

$$= (1 + 2 + \cdots + k)f'(a) = \frac{k(k+1)}{2} f'(a).$$

(d) For $k \in \mathbb{N}$,

$$\lim_{n \to \infty} \frac{f\left(a + \frac{k}{n^2}\right) - f(a)}{\frac{k}{n^2}} = f'(a).$$

This implies that given $\varepsilon > 0$ there is n_0 such that if $n \geq n_0$, then

$$\frac{k}{n^2} f'(a) - \frac{k}{n^2} \varepsilon < f\left(a + \frac{k}{n^2}\right) - f(a) < \frac{k}{n^2} f'(a) + \frac{k}{n^2} \varepsilon$$

for $k \in \{1, 2, \ldots, n\}$. Summing over k, we obtain

$$\frac{n(n+1)}{2n^2} f'(a) - \frac{n(n+1)}{2n^2} \varepsilon < \sum_{k=1}^{n} \left(f\left(a + \frac{k}{n^2}\right) - f(a) \right)$$

$$< \frac{n(n+1)}{2n^2} f'(a) + \frac{n(n+1)}{2n^2} \varepsilon.$$

It then follows that the limit is $\frac{1}{2} f'(a)$.

2.1.11.

(a) We have

$$\lim_{n\to\infty} \left(\frac{(n+1)^m + (n+2)^m + \cdots + (n+k)^m}{n^{m-1}} - kn \right)$$
$$= \lim_{n\to\infty} \frac{(n+1)^m - n^m + (n+2)^m - n^m + \cdots + (n+k)^m - n^m}{n^{m-1}}$$
$$= \lim_{n\to\infty} \left(\frac{\left(1+\frac{1}{n}\right)^m - 1}{\frac{1}{n}} + 2\frac{\left(1+\frac{2}{n}\right)^m - 1}{\frac{2}{n}} + \cdots + k\frac{\left(1+\frac{k}{n}\right)^m - 1}{\frac{k}{n}} \right)$$
$$= \frac{k(k+1)}{2} m.$$

Compare with 2.1.10(c).

(b) By 2.1.10(c),

$$\lim_{n\to\infty} \ln \left(\frac{\left(a+\frac{1}{n}\right)^n \left(a+\frac{2}{n}\right)^n \cdots \left(a+\frac{k}{n}\right)^n}{a^{nk}} \right) = \frac{k(k+1)}{2} \frac{1}{a}.$$

Thus

$$\lim_{n\to\infty} \frac{\left(a+\frac{1}{n}\right)^n \left(a+\frac{2}{n}\right)^n \cdots \left(a+\frac{k}{n}\right)^n}{a^{nk}} = e^{\frac{k(k+1)}{2a}}.$$

(c) Note that

$$\lim_{n\to\infty} \ln \left(\left(1+\frac{a}{n^2}\right)\left(1+\frac{2a}{n^2}\right) \cdots \left(1+\frac{na}{n^2}\right) \right)$$
$$= \lim_{n\to\infty} \left(\ln\left(1+\frac{a}{n^2}\right) + \ln\left(1+\frac{2a}{n^2}\right) + \cdots + \ln\left(1+\frac{na}{n^2}\right) \right)$$
$$= \lim_{n\to\infty} \left(\ln\left(\frac{1}{a}+\frac{1}{n^2}\right) + \cdots + \ln\left(\frac{1}{a}+\frac{n}{n^2}\right) - n\ln\frac{1}{a} \right).$$

It then follows by 2.1.10(d) that

$$\lim_{n\to\infty} \left(\left(1+\frac{a}{n^2}\right)\left(1+\frac{2a}{n^2}\right) \cdots \left(1+\frac{na}{n^2}\right) \right) = e^{\frac{a}{2}}.$$

2.1. The Derivative of a Real Function 219

2.1.12. We have

$$\lim_{x \to 0} \frac{1}{x}\left(f(x) + f\left(\frac{x}{2}\right) + f\left(\frac{x}{3}\right) + \cdots + f\left(\frac{x}{k}\right)\right)$$

$$= \lim_{x \to 0}\left(\frac{f(x) - f(0)}{x} + \frac{f\left(\frac{x}{2}\right) - f(0)}{x} + \cdots + \frac{f\left(\frac{x}{k}\right) - f(0)}{x}\right)$$

$$= \left(1 + \frac{1}{2} + \frac{1}{3} + \cdots + \frac{1}{k}\right) f'(0).$$

2.1.13.

(a) If $f(x) = x^m$, $m \in \mathbb{N}$, then

$$\lim_{n \to \infty} \frac{f(x_n) - f(z_n)}{x_n - z_n} = \lim_{n \to \infty} \frac{x_n^m - z_n^m}{x_n - z_n} = ma^{m-1} = f'(a).$$

(b) Consider the function given by

$$f(x) = \begin{cases} x^2 \sin \frac{1}{x} & \text{if } x \neq 0, \\ 0 & \text{if } x = 0. \end{cases}$$

For

$$x_n = \frac{2}{\pi(4n+1)} \quad \text{and} \quad z_n = \frac{1}{2n\pi}$$

we have

$$\lim_{n \to \infty} \frac{f(x_n) - f(z_n)}{x_n - z_n} = -\frac{2}{\pi} \neq 0 = f'(0).$$

On the other hand, if

$$g(x) = \begin{cases} x^{\frac{3}{2}} \sin \frac{1}{x} & \text{if } x \neq 0, \\ 0 & \text{if } x = 0, \end{cases}$$

and $\{x_n\}$, $\{z_n\}$ are as above, then

$$\lim_{n \to \infty} \frac{g(x_n) - g(z_n)}{x_n - z_n} = -\infty.$$

2.1.14. By hypotheses,

$$\frac{f(x_n) - f(z_n)}{x_n - z_n} = \frac{f(x_n) - f(a)}{x_n - a} \cdot \frac{x_n - a}{x_n - z_n} + \frac{f(z_n) - f(a)}{z_n - a} \cdot \frac{a - z_n}{x_n - z_n},$$

where

$$0 < \frac{a - z_n}{x_n - z_n} < 1, \quad 0 < \frac{x_n - a}{x_n - z_n} < 1$$

and
$$\frac{a-z_n}{x_n-z_n}+\frac{x_n-a}{x_n-z_n}=1.$$
It then follows that
$$\frac{f(x_n)-f(z_n)}{x_n-z_n}$$
is between
$$\frac{f(x_n)-f(a)}{x_n-a} \quad \text{and} \quad \frac{f(z_n)-f(a)}{z_n-a}.$$
By the squeeze law for sequences,
$$\lim_{n\to\infty}\frac{f(x_n)-f(z_n)}{x_n-z_n}=f'(a).$$

2.1.15 [W. R. Jones, M. D. Landau, Amer. Math. Monthly 76 (1969), 816-817].

(a) Note first that f is continuous only at 1. If $\{x_n\}$ is a sequence of rationals different from 1 converging to 1, then
$$\lim_{n\to\infty}\frac{f(x_n)-1}{x_n-1}=\lim_{n\to\infty}(x_n+1)=2.$$
If $\{x_n\}$ is a sequence of irrationals converging to 1, then
$$\lim_{n\to\infty}\frac{f(x_n)-1}{x_n-1}=\lim_{n\to\infty}2=2.$$
Thus $f'(1)=2$. Clearly, f is one-to-one on $(0,2)$. The inverse function f^{-1} is defined on $(0,3)$ except for the rationals with irrational square roots. This means that there are no interior points in the domain of f^{-1}. So, $(f^{-1})'(1)$ cannot be defined.

(b) Note first that f is defined on $(0,2)\cup \mathbf{B}$, where $\mathbf{B}\subset (2,7/2)$. Note also that the restriction of f to $(0,2)$ is a function defined in (a). Thus $f'(1)=2$. Since $f(\mathbf{B})=\mathbf{A}$, the range of f contains $(0,3)$. However, $(f^{-1})'(1)$ does not exist, because each neighborhood of $1=f(1)$ contains images under f of points in $(0,2)$ and images of points in $\mathbf{B}$. Consequently, even the limit of f^{-1} at 1 does not exist.

2.1. The Derivative of a Real Function

2.1.16. By a theorem of Liouville (see, e.g., J.C. Oxtoby, Measure and Category, Springer-Verlag, 1980, p. 7), any algebraic surd x of degree k is badly approximable by rationals, in the sense that there exists $M > 0$ such that $\left|x - \frac{p}{q}\right| > \frac{1}{Mq^k}$ for all rationals $\frac{p}{q}$. Consequently,

$$\left|\frac{f\left(\frac{p}{q}\right) - f(x)}{\frac{p}{q} - x}\right| \leq Mq^k |a_q|.$$

It then follows by assumption that $f'(x) = 0$.

It is worth noting here that if, e.g., $a_q = 2^{-q}$, then f is differentiable at each algebraic surd.

2.1.17. Let $P(x) = a(x - x_1)(x - x_2) \cdots (x - x_n)$. Then

$$P'(x_k) = a \prod_{\substack{j=1 \\ j \neq k}}^{n} (x_k - x_j), \quad k = 1, 2, \ldots, n.$$

The identity to be proved,

$$\frac{Q(x)}{P(x)} = \sum_{k=1}^{n} \frac{Q(x_k)}{P'(x_k)(x - x_k)},$$

is equivalent to

$$Q(x) = \sum_{k=1}^{n} \frac{Q(x_k) P(x)}{P'(x_k)(x - x_k)},$$

which, in turn, can be written as

$$Q(x) = \sum_{k=1}^{n} Q(x_k) \frac{\prod_{\substack{j=1 \\ j \neq k}}^{n} (x - x_j)}{\prod_{\substack{j=1 \\ j \neq k}}^{n} (x_k - x_j)}.$$

Since Q is a polynomial of degree at most $n - 1$, it is enough to prove that this equality holds at n different points. Clearly, the equality holds at $x = x_k$, $k = 1, 2, \ldots, n$.

In particular, if $Q(x) \equiv 1$, then

$$1 = \sum_{k=1}^{n} \frac{1}{P'(x_k)} \prod_{\substack{j=1 \\ j \neq k}}^{n} (x - x_j).$$

Equating the coefficients of x^{n-1}, we get

$$\sum_{k=1}^{n} \frac{1}{P'(x_k)} = 0 \quad \text{for} \quad n \geq 2.$$

2.1.18. Apply the result in the foregoing problem with
(a) $P(x) = x(x+1)(x+2)\cdots(x+n)$ and $Q(x) \equiv n!$.
(b) $P(x) = x(x+2)(x+4)\cdots(x+2n)$ and $Q(x) \equiv n!2^n$.

2.1.19. Clearly, the derivative of $|f|$ exists at each x such that $f(x) \neq 0$. Moreover, if $f(x) = 0$ and $f'(x) = 0$, then $|f|'(x) = 0$.

2.1.20. There is a neighborhood of x where each of the functions f_k does not change its sign. Consequently, $|f_k|$ is differentiable at x and we have

$$\frac{\left(\prod_{k=1}^{n}|f_k|\right)'}{\prod_{k=1}^{n}|f_k|}(x) = \left(\ln \prod_{k=1}^{n}|f_k|\right)'(x) = \sum_{k=1}^{n} \frac{|f_k|'(x)}{|f_k(x)|}.$$

Our proof ends with the observation $|f_k|'(x) = \text{sgn}(f_k(x))f_k'(x)$.

2.1.21. Apply the result in the preceding problem with f_k replaced by $\frac{f_k}{g_k}$.

2.1.22.

(a) Clearly, f and $|f|$ are continuous only at $x = 0$. Moreover, $f'(0) = 1$ and $|f|'(0)$ does not exist (compare with 2.1.19).

(b) f and $|f|$ are continuous only at $x_k = \frac{3}{2^k}$, $k = 2, 3, \ldots$. It is easily verifiable that $f'(x_k) = 1$, and that $|f|'(x_k)$ does not exist.

2.1. The Derivative of a Real Function

2.1.23. Let $\varepsilon > 0$ be chosen arbitrarily. By the definition of $f'_+(x_0)$,

(1) $(f'_+(x_0) - \varepsilon)(x - x_0) \leq f(x) - f(x_0) \leq (f'_+(x_0) + \varepsilon)(x - x_0)$

for $x > x_0$ sufficiently close to x_0. Likewise,

(2) $(f'_-(x_0) - \varepsilon)(x - x_0) \geq f(x) - f(x_0) \geq (f'_-(x_0) + \varepsilon)(x - x_0)$

for $x < x_0$ sufficiently close to x_0. Continuity of f at x_0 is an immediate consequence of (1) and (2).

2.1.24. Since $f(c) = \max\{f(x) : x \in (a,b)\}$, we have $f(x) - f(c) \leq 0$ for $x \in (a,b)$. Therefore

$$f'_-(c) = \lim_{x \to c^-} \frac{f(x) - f(c)}{x - c} \geq 0.$$

Similarly, $f'_+(c) \leq 0$.

If $f(c_0) = \min\{f(x) : x \in (a,b)\}$, then we get $f'_+(c_0) \geq 0$ and $f'_-(c_0) \leq 0$.

2.1.25. Clearly, the assertion is true if f is constant. Suppose, then, that f is not constant. Without loss of generality we can assume that $f(a) = f(b) = 0$. Then, for example, there exists $x_1 \in (a,b)$ for which $f(x_1) > 0$. Let k be a real number such that $0 = f(b) < k < f(x_1)$. Set $c = \sup\{x \in (x_1, b) : f(x) > k\}$. Then $f(x) \leq k$ for $x \in [c, b]$. Moreover, there exists a negative-valued sequence $\{h_n\}$ convergent to zero and such that $f(c + h_n) > k$. Since f'_- exists,

$$f'_-(c) = \lim_{n \to \infty} \frac{f(c + h_n) - f(c)}{h_n} \leq 0.$$

So, we have proved that $\inf\{f'_-(x) : x \in (a,b)\} \leq 0$. In an entirely similar manner one can show that $\sup\{f'_-(x) : x \in (a,b)\} \geq 0$.

It is worth noting here that an analogous result can be obtained for f'_+. Namely,

$$\inf\{f'_+(x) : x \in (a,b)\} \leq 0 \leq \sup\{f'_+(x) : x \in (a,b)\}.$$

2.1.26. To prove the assertion we apply the above result to the auxiliary function
$$x \mapsto f(x) - \frac{f(b) - f(a)}{b - a}(x - a).$$
A similar assertion can be proved for f'_+, that is,
$$\inf\{f'_+(x) : x \in (a,b)\} \leq \frac{f(b) - f(a)}{b - a} \leq \sup\{f'_+(x) : x \in (a,b)\}.$$

2.1.27. By the result in the foregoing problem,
$$\inf\{f'_-(z) : z \in (x, x+h)\} \leq \frac{f(x+h) - f(x)}{h}$$
$$\leq \sup\{f'_-(z) : z \in (x, x+h)\}$$
for $x \in (a,b)$ and $0 < h$ so small that $x + h$ is in (a,b). Since f'_- is continuous on (a,b), upon passage to the limit as $h \to 0^+$ we get $f'_+(x) = f'_-(x)$.

2.1.28. It follows from the result in the preceding problem that such a function does not exist.

2.1.29. By assumption, f vanishes at at least one point of an open interval (a,b). Set
$$c = \inf\{x \in (a,b) : f(x) = 0\}.$$
Then $f(c) = 0$. Since $f'(a) > 0$, we have $f(x) > 0$ for $x \in (a,c)$. Moreover, since $f'(c)$ exists,
$$f'(c) = \lim_{h \to 0^-} \frac{f(c+h) - f(c)}{h} = \lim_{h \to 0^-} \frac{f(c+h)}{h} \leq 0.$$

2.1.30. Clearly, $(1 + x^2)f'(x) = 1$, which implies $(1 + x^2)f''(x) + 2xf'(x) = 0$. Using induction one can show that
$$(1 + x^2)f^{(n)}(x) + 2(n-1)xf^{(n-1)}(x) + (n-2)(n-1)f^{(n-2)}(x) = 0.$$
If we take $x = 0$, then by induction again, we get $f^{(2m)}(0) = 0$ and $f^{(2m+1)}(0) = (-1)^m (2m)!$.

2.1.31. The identities can be established easily by induction.

2.1. The Derivative of a Real Function 225

2.1.32.

(a) Apply the Leibniz formula

$$(f(x)g(x))^{(n)} = \sum_{k=0}^{n} \binom{n}{k} f^{(n-k)}(x) g^{(k)}(x)$$

and the identity (a) in the foregoing problem.

(b) Apply the Leibniz formula and the identity (b) in the foregoing problem.

2.1.33. It is easy to see that if $x > 1$, then $f(x) > 0$, $f'(x) > 0$ and $f''(x) < 0$. Now differentiating

$$(f(x))^2 = x^2 - 1$$

n times, $n \geq 3$, and using the Leibniz formula, we get

$$2f(x)f^{(n)}(x) + \sum_{k=1}^{n-1} \binom{n}{k} f^{(k)}(x) f^{(n-k)}(x) \equiv 0.$$

The desired result can be obtained by induction.

2.1.34. We have

$$f_{2n}(x) = \ln(1 + x^{2n}) = \sum_{k=1}^{2n} \ln(x - \omega_k),$$

where $\omega_k = \cos \frac{(2k-1)\pi}{2n} + i \sin \frac{(2k-1)\pi}{2n}$. Hence

$$f_{2n}^{(2n)}(x) = -(2n-1)! \sum_{k=1}^{2n} \frac{1}{(x - \omega_k)^{2n}}.$$

Putting $x = -1$, we get

$$f_{2n}^{(2n)}(-1) = -(2n-1)! \sum_{k=1}^{2n} \frac{1}{(1 + \omega_k)^{2n}}.$$

An easy calculation shows that

$$f_{2n}^{(2n)}(-1) = i \frac{(2n-1)!}{2^{2n}} \sum_{k=1}^{2n} \frac{(-1)^k}{\cos^{2n} \frac{(2k-1)\pi}{4n}}.$$

Since $f_{2n}^{(2n)}(-1)$ is real, we see that $f_{2n}^{(2n)}(-1) = 0$.

2.1.35. Denote by $L(x)$ and $R(x)$ the left and right side of the identity to be proved. Clearly, L and R are polynomials of degree $n+1$ and $L(0) = R(0) = 0$. So, it is enough to show that $L'(x) = R'(x)$, $x \in \mathbb{R}$. We have

$$L'(x) = \sum_{k=0}^{n} \frac{P^{(k)}(0)}{k!} x^k = P(x),$$

$$R'(x) = \sum_{k=0}^{n} (-1)^k \frac{P^{(k)}(x)}{k!} x^k + \sum_{k=0}^{n} (-1)^k \frac{P^{(k+1)}(x)}{(k+1)!} x^{k+1}$$

$$= P(x) + (-1)^n \frac{P^{(n+1)}(x)}{(n+1)!} x^{n+1} = P(x).$$

2.1.36. There is a neighborhood of zero where f is positive. Thus

$$(\ln f(x))' = \frac{f'(x)}{f(x)} = \frac{\lambda_1}{1 - \lambda_1 x} + \cdots + \frac{\lambda_n}{1 - \lambda_n x} = g(x).$$

Hence $f'(x) = f(x)g(x)$ and $f'(0) = \lambda_1 + \lambda_2 + \cdots + \lambda_n > 0$. Moreover,

$$(1) \qquad g^{(i)}(x) = i! \left(\frac{\lambda_1^{i+1}}{(1 - \lambda_1 x)^{i+1}} + \cdots + \frac{\lambda_n^{i+1}}{(1 - \lambda_n x)^{i+1}} \right).$$

By the Leibniz formula,

$$f^{(k)}(x) = \sum_{i=0}^{k-1} \binom{k-1}{i} g^{(i)}(x) f^{(k-1-i)}(x).$$

In view of (1), it then follows by induction that $f^{(k)}(0) > 0$, $k \in \mathbb{N}$.

2.1.37. We will proceed by induction. For $n = 1$ the equality is obvious. Assuming the equality to hold for $k \leq n$, we will show it for

2.1. The Derivative of a Real Function

$n+1$. We have

$$(-1)^{n+1}\left(x^n f\left(\frac{1}{x}\right)\right)^{(n+1)} = (-1)^{n+1}\left(\left(x^n f\left(\frac{1}{x}\right)\right)'\right)^{(n)}$$

$$= (-1)^{n+1} n \left(x^{n-1} f\left(\frac{1}{x}\right)\right)^{(n)} - (-1)^{n+1}\left(x^{n-2} f'\left(\frac{1}{x}\right)\right)^{(n)}$$

$$= -\frac{n}{x^{n+1}} f^{(n)}\left(\frac{1}{x}\right) - (-1)^{n-1}\left(x^{n-2} f'\left(\frac{1}{x}\right)\right)^{(n)}.$$

Moreover,

$$(-1)^{n-1}\left(x^{n-2} f'\left(\frac{1}{x}\right)\right)^{(n)} = (-1)^{n-1}\left(\left(x^{n-2} f'\left(\frac{1}{x}\right)\right)^{(n-1)}\right)'.$$

The induction hypotheses applied to f' with $k = n-1$ gives

$$\frac{1}{x^n} f^{(n)}\left(\frac{1}{x}\right) = (-1)^{n-1}\left(x^{n-2} f'\left(\frac{1}{x}\right)\right)^{(n-1)}.$$

Consequently,

$$(-1)^{n+1}\left(x^n f\left(\frac{1}{x}\right)\right)^{(n+1)} = -\frac{n}{x^{n+1}} f^{(n)}\left(\frac{1}{x}\right) - \left(\frac{1}{x^n} f^{(n)}\left(\frac{1}{x}\right)\right)'$$

$$= \frac{1}{x^{n+2}} f^{(n+1)}\left(\frac{1}{x}\right).$$

2.1.38. The proof of this well known formula presented here is based on S. Roman's paper [Amer. Math. Monthly 87 (1980), 805-809]. Although methods of functional analysis are applied, the proof is elementary. Linear functionals $L : \mathcal{P} \to \mathbb{R}$ defined on the set $\mathcal{P}$ of all polynomials with real coefficients will be considered. Let $\langle L, P(x) \rangle$ denote the value of L at the polynomial $P(x)$. Let A^k be a linear functional such that

$$\langle A^k, x^n \rangle = n! \delta_{n,k},$$

where

$$\delta_{n,k} = \begin{cases} 1 & \text{if } n = k, \\ 0 & \text{if } n \neq k. \end{cases}$$

It is worth noting here that the value of A^k at x^n is $(x^n)^{(k)}_{|x=0}$. Let $\sum_{k=0}^{\infty} a_k A^k, a_k \in \mathbb{R}$, denote the linear functional defined by

$$\left\langle \sum_{k=0}^{\infty} a_k A^k, P(x) \right\rangle = \sum_{k=0}^{\infty} a_k \langle A^k, P(x) \rangle.$$

Since $\langle A^k, P(x) \rangle = 0$ for almost all k, there are only finitely many nonzero terms in the sum on the right side of this equality. The task is now to show that if L is a linear functional on $\mathcal{P}$, then

(1)
$$L = \sum_{k=0}^{\infty} \frac{\langle L, x^k \rangle}{k!} A^k.$$

Indeed, for $n \geq 0$,

$$\left\langle \sum_{k=0}^{\infty} \frac{\langle L, x^k \rangle}{k!} A^k, x^n \right\rangle = \sum_{k=0}^{\infty} \frac{\langle L, x^k \rangle}{k!} \langle A^k, x^n \rangle = \langle L, x^n \rangle.$$

Since L and A^k are linear, we get

$$\langle L, P(x) \rangle = \left\langle \sum_{k=0}^{\infty} \frac{\langle L, x^k \rangle}{k!} A^k, P(x) \right\rangle$$

for any polynomial P, which proves (1). According to the fact mentioned above that A^k at x^n is $(x^n)^{(k)}_{|x=0}$, it seems natural to define the operation on A^k by setting

$$A^k A^j = A^{k+j}.$$

In view of (1) this operation can be extended to the operation defined for any $L, M : \mathcal{P} \to \mathbb{R}$ as follows

$$LM = \sum_{k=0}^{\infty} \frac{\langle L, x^k \rangle}{k!} A^k \sum_{j=0}^{\infty} \frac{\langle M, x^j \rangle}{j!} A^j = \sum_{n=0}^{\infty} c_n A^n,$$

where

$$c_n = \sum_{k=0}^{n} \frac{\langle L, x^k \rangle}{k!} \frac{\langle M, x^{n-k} \rangle}{(n-k)!} = \frac{1}{n!} \sum_{k=0}^{n} \binom{n}{k} \langle L, x^k \rangle \langle M, x^{n-k} \rangle.$$

2.1. The Derivative of a Real Function

Hence by (1),

(2) $$\langle LM, x^n \rangle = \sum_{k=0}^{n} \binom{n}{k} \langle L, x^k \rangle \langle M, x^{n-k} \rangle.$$

Using induction, one can show that

(3) $$\langle L_1 \cdots L_j, x^n \rangle = \sum_{\substack{k_1,\ldots,k_j=0 \\ k_1+\cdots+k_j=n}}^{n} \frac{n!}{k_1! \cdots k_j!} \langle L_1, x^{k_1} \rangle \langle L_2, x^{k_2} \rangle \cdots \langle L_j, x^{k_j} \rangle.$$

Now define the formal derivative L' of L by

$$\left(A^0\right)' = 0, \quad \left(A^k\right)' = kA^{k-1} \quad \text{for} \quad k \in \mathbb{N}$$

and

$$L' = \sum_{k=1}^{\infty} \frac{\langle L, x^k \rangle}{k!} kA^{k-1}.$$

Now we show that for any $P \in \mathcal{P}$,

(4) $$\langle L', P(x) \rangle = \langle L, xP(x) \rangle.$$

Clearly, it is enough to show that $\langle (A^k)', x^n \rangle = \langle A^k, x^{n+1} \rangle$. We have

$$\langle (A^k)', x^n \rangle = \langle kA^{k-1}, x^n \rangle = kn!\delta_{n,k-1} = (n+1)!\delta_{n+1,k} = \langle A^k, x^{n+1} \rangle.$$

To prove the Faà di Bruno formula, set

$$h_n = h^{(n)}(t), \quad g_n = g^{(n)}(t), \quad f_n = f^{(n)}(u)\big|_{u = g(t)}.$$

Clearly,

$$h_1 = f_1 g_1, \quad h_2 = f_1 g_2 + f_2 g_1^2, \quad h_3 = f_1 g_3 + f_2 3 g_1 g_2 + f_3 g_1^3.$$

It can be shown by induction that

(5) $$h_n = \sum_{k=1}^{n} f_k l_{n,k}(g_1, g_2, \ldots, g_n),$$

where $l_{n,k}(g_1, g_2, \ldots, g_n)$ is independent of $f_j, j = 0, 1, 2, \ldots, n$. To determine $l_{n,k}(g_1, g_2, \ldots, g_n)$, choose $f(t) = e^{at}$, $a \in \mathbb{R}$. Then $f_k = a^k e^{ag(t)}$ and $h_n = \left(e^{ag(t)}\right)^{(n)}$. It follows from (5) that

$$(6) \qquad e^{-ag(t)} \left(e^{ag(t)}\right)^{(n)} = \sum_{k=1}^{n} a^k l_{n,k}(g_1, g_2, \ldots, g_n).$$

Put $B_n(t) = e^{-ag(t)} \left(e^{ag(t)}\right)^{(n)}$, $n \geq 0$. It then follows by the Leibniz formula that

$$B_n(t) = e^{-ag(t)} \left(ag_1(t) e^{ag(t)}\right)^{(n-1)}$$

$$(7) \qquad = a \cdot e^{-ag(t)} \sum_{k=0}^{n-1} \binom{n-1}{k} g_{k+1}(t) \left(e^{ag(t)}\right)^{(n-k-1)}$$

$$= a \sum_{k=0}^{n-1} \binom{n-1}{k} g_{k+1}(t) B_{n-k-1}(t).$$

For an arbitrarily fixed $t \in \mathbf{I}$, set $B_n = B_n(t)$ and define functionals L and M on $\mathcal{P}$ by $\langle L, x^n \rangle = B_n$, $\langle M, x^n \rangle = g_n$. Then $\langle L, 1 \rangle = B_0 = 1$ and $\langle M, 1 \rangle = g_0 = g(t)$. Moreover, by (1),

$$L = \sum_{k=0}^{\infty} \frac{B_k}{k!} A^k \quad \text{and} \quad M = \sum_{k=0}^{\infty} \frac{g_k}{k!} A^k.$$

Now (7) combined with (2) and (4) yields

$$\langle L, x^n \rangle = a \sum_{k=0}^{n-1} \binom{n-1}{k} \langle M, x^{k+1} \rangle \langle L, x^{n-1-k} \rangle$$

$$= a \sum_{k=0}^{n-1} \binom{n-1}{k} \langle M', x^k \rangle \langle L, x^{n-1-k} \rangle$$

$$= a \langle M'L, x^{n-1} \rangle.$$

So, $\langle L', x^{n-1} \rangle = a \langle M'L, x^{n-1} \rangle$, or in other words,

$$L' = aM'L.$$

This formal differential equation has solutions of the form $L = ce^{a(M-g_0)}$, where c is a real constant. By the initial value condition, $1 = B_0 = \langle L, 1 \rangle = \langle ce^{a(M-g_0)}, 1 \rangle = c$. Hence $L = e^{a(M-g_0)}$. It

2.1. The Derivative of a Real Function

then follows that

$$B_n = \langle L, x^n \rangle = \langle e^{a(M-g_0)}, x^n \rangle = \sum_{k=0}^{\infty} \frac{a^k}{k!} \langle (M-g_0)^k, x^n \rangle$$

$$= \sum_{k=0}^{\infty} \frac{a^k}{k!} \sum_{\substack{j_1,\ldots,j_k=0 \\ j_1+\cdots+j_k=n}}^{n} \frac{n!}{j_1!\cdots j_k!} \langle M-g_0, x^{j_1}\rangle \langle M-g_0, x^{j_2}\rangle \cdots \langle M-g_0, x^{j_k}\rangle$$

$$= \sum_{k=0}^{\infty} \frac{a^k}{k!} \sum_{\substack{j_1,\ldots,k=1 \\ j_1+\cdots+j_k=n}}^{n} \frac{n!}{j_1! j_2!\cdots j_k!} g_{j_1} g_{j_2} \cdots g_{j_k}.$$

Equating the coefficients of a^k in (6) gives

$$l_{n,k}(g_1, g_2, \ldots, g_n) = \frac{n!}{k!} \sum_{\substack{j_1,\ldots,j_k=1 \\ j_1+\cdots+j_k=n}}^{n} \frac{g_{j_1}}{j_1!} \cdot \frac{g_{j_2}}{j_2!} \cdots \frac{g_{j_k}}{j_k!}$$

$$= \frac{n!}{k!} \sum_{\substack{k_1,\ldots,k_n=0 \\ k_1+\cdots+k_n=k \\ k_1+2k_2+\cdots+nk_n=n}}^{n} \frac{k!}{k_1!\cdots k_n!} \left(\frac{g_1}{1!}\right)^{k_1} \left(\frac{g_2}{2!}\right)^{k_2} \cdots \left(\frac{g_n}{n!}\right)^{k_n}.$$

Finally,

$$\sum_{k=1}^{n} f_k l_{n,k}(g_1,\ldots,g_n) = \sum_{k=1}^{n} f_k \sum_{\substack{k_1,\ldots,k_n=0 \\ k_1+\cdots+k_n=k \\ k_1+2k_2+\cdots+nk_n=n}}^{n} \frac{n!}{k_1!\cdots k_n!} \left(\frac{g_1}{1!}\right)^{k_1} \cdots \left(\frac{g_n}{n!}\right)^{k_n},$$

which ends the proof.

2.1.39.

(a) We have

$$f'(x) = \begin{cases} \frac{2}{x^3} e^{-\frac{1}{x^2}} & \text{if } x \neq 0, \\ 0 & \text{if } x = 0, \end{cases}$$

because (see, e.g., 1.1.12)

$$\lim_{x \to 0} \frac{e^{-\frac{1}{x^2}}}{x} = 0.$$

It then follows that f' is continuous on $\mathbb{R}$. Moreover, for $x \neq 0$,
$$f''(x) = e^{-\frac{1}{x^2}} \left(\frac{4}{x^6} - \frac{2 \cdot 3}{x^4} \right).$$
Again by the result in 1.1.12, it can be shown that $f''(0) = 0$. Consequently, f'' is also continuous on $\mathbb{R}$. Finally, observe that
$$f^{(n)}(x) = \begin{cases} e^{-\frac{1}{x^2}} P\left(\frac{1}{x}\right) & \text{if } x \neq 0, \\ 0 & \text{if } x = 0, \end{cases}$$
where P is a polynomial. Therefore for every $n \in \mathbb{N}$, $f^{(n)}$ is continuous on $\mathbb{R}$.

(b) As in (a) one can show that $g^{(n)}(0) = 0$ for $n \in \mathbb{N}$, and that g is in $C^\infty(\mathbb{R})$.

(c) The function is a product of two functions $f_1, f_2 \in C^\infty(\mathbb{R})$. Indeed, $f_1(x) = g(x-a)$ and $f_2(x) = g(b-x)$, where g is defined in (b).

2.1.40. We have
$$f''(x) = g'(f(x))f'(x) = g'(f(x))g(f(x)),$$
$$f'''(x) = g''(f(x))(g(f(x)))^2 + (g'(f(x)))^2 g(f(x)).$$
Therefore f'' and f''' are continuous on (a,b). One can show by induction that $f^{(n)}$, $n \geq 3$, are sums of products of derivatives $g^{(k)}(f)$, $k = 0, 1, 2, \ldots, n-1$. Consequently, they are continuous on (a,b).

2.1.41. If $\alpha \neq 0$, then
$$f''(x) = \frac{-\beta f'(x) - \gamma f(x)}{\alpha}.$$
Consequently,
$$f'''(x) = \frac{-\beta f''(x) - \gamma f'(x)}{\alpha} = \frac{(\beta^2 - \gamma\alpha)f'(x) + \gamma\beta f(x)}{\alpha^2}.$$
One can show by induction that the nth derivative of f is a linear combination of f and f'.

If $\alpha = 0$, then $\beta \neq 0$ and $f'(x) = \frac{-\gamma}{\beta} f(x)$. By induction once again,
$$f^{(n)}(x) = (-1)^n \frac{\gamma^n}{\beta^n} f(x).$$

2.2. Mean Value Theorems

2.2.1. The auxiliary function $h(x) = e^{\alpha x}f(x)$, $x \in [a,b]$, satisfies the conditions of Rolle's theorem. Hence there is $x_0 \in (a,b)$ such that
$$0 = h'(x_0) = (\alpha f(x_0) + f'(x_0))e^{\alpha x_0}.$$
Consequently, $\alpha f(x_0) + f'(x_0) = 0$.

2.2.2. The function $h(x) = e^{g(x)}f(x)$, $x \in [a,b]$, satisfies the conditions of Rolle's theorem. Therefore there is $x_0 \in (a,b)$ such that
$$0 = h'(x_0) = (g'(x_0)f(x_0) + f'(x_0))e^{g(x_0)}.$$
Hence $g'(x_0)f(x_0) + f'(x_0) = 0$.

2.2.3. Apply Rolle's theorem to the function $h(x) = \frac{f(x)}{x}$, $x \in [a,b]$.

2.2.4. Take $h(x) = f^2(x) - x^2$, $x \in [a,b]$, in Rolle's theorem.

2.2.5. Apply Rolle's theorem to the function $h(x) = \frac{f(x)}{g(x)}$, $x \in [a,b]$.

2.2.6. Note that the polynomial
$$Q(x) = \frac{a_0}{n+1}x^{n+1} + \frac{a_1}{n}x^n + \cdots + a_n x$$
satisfies the conditions of Rolle's theorem on the interval $[0,1]$.

2.2.7. The function
$$h(x) = \frac{a_n}{n+1}\ln^{n+1}x + \cdots + \frac{a_2}{3}\ln^3 x + \frac{a_1}{2}\ln^2 x + \frac{a_0}{1}\ln x, \quad x \in [1, e^2],$$
satisfies the hypotheses of Rolle's theorem.

2.2.8. By Rolle's theorem, between two real zeros of the polynomial P there is at least one real zero of P'. Moreover, each zero of P of order k ($k \geq 2$) is a zero of P' of order $(k-1)$. Thus there are $n-1$ zeros of P', counted according to multiplicity.

2.2.9. By Rolle's theorem applied to f on $[a,b]$ there is $c \in (a,b)$ such that $f'(c) = 0$. Next, by Rolle's theorem again applied to f' on $[a,c]$, we see that there is $x_1 \in (a,c) \subset (a,b)$ such that $f''(x_1) = 0$.

2.2.10. Apply the reasoning similar to that used in the solution of the foregoing problem.

2.2.11.

(a) Set $P(x) = x^{13} + 7x^3 - 5$. Then $P(0) = -5$ and $\lim\limits_{x\to\infty} P(x) = +\infty$. By the intermediate value property there is at least one positive root of $P(x) = 0$. If there were two distinct positive roots, then Rolle's theorem would imply $P'(x_0) = 0$ for some positive x_0. This would contradict the fact that $P'(x) = 0$ if and only if $x = 0$. Finally, observe that $P(x) < 0$ for $x < 0$.

(b) Consider the function
$$f(x) = \left(\frac{3}{5}\right)^x + \left(\frac{4}{5}\right)^x - 1.$$
Then $f(2) = 0$. If f vanished at another point, then by Rolle's theorem its derivative would vanish at at least one point, which would contradict the fact that $f'(x) < 0$ for all $x \in \mathbb{R}$.

2.2.12. We will proceed by induction. For $n = 1$ the equation $a_1 x^{\alpha_1} = 0$ does not possess zeros in $(0, \infty)$. Assuming that for an arbitrarily chosen $n \in \mathbb{N}$ the equation
$$a_1 x^{\alpha_1} + a_2 x^{\alpha_2} + \cdots + a_n x^{\alpha_n} = 0$$
has at most $n - 1$ roots in $(0, \infty)$, we consider the equation
$$a_1 x^{\alpha_1} + a_2 x^{\alpha_2} + \cdots + a_n x^{\alpha_n} + a_{n+1} x^{\alpha_{n+1}} = 0,$$
which can be rewritten as
$$a_1 + a_2 x^{\alpha_2 - \alpha_1} + \cdots + a_{n+1} x^{\alpha_{n+1} - \alpha_1} = 0.$$
If the last equation had more than n positive roots, then by Rolle's theorem the derivative of the function on the left side of this equality would have at least n positive zeros. This would contradict the induction hypotheses.

2.2.13. Apply the last result, replacing x by e^x.

2.2. Mean Value Theorems 235

2.2.14. Clearly, $F(a) = F(b) = 0$, and F is continuous on $[a, b]$. Moreover, F is differentiable on (a, b) and

$$F'(x) = \det \begin{vmatrix} f'(x) & g'(x) & h'(x) \\ f(a) & g(a) & h(a) \\ f(b) & g(b) & h(b) \end{vmatrix}.$$

By Rolle's theorem there is $x_0 \in (a, b)$ such that $F'(x_0) = 0$.

Taking $g(x) = x$ and $h(x) = 1$ for $x \in [a, b]$, we get

$$F'(x_0) = \det \begin{vmatrix} f'(x_0) & 1 & 0 \\ f(a) & a & 1 \\ f(b) & b & 1 \end{vmatrix} = 0,$$

which gives $f(b) - f(a) = f'(x_0)(b - a)$. So, we have derived the mean value theorem. To get the generalized mean value theorem it is enough to take $h(x) \equiv 1$.

2.2.15. It follows from the mean value theorem that there exist x_1 in $(0, 1)$ and x_2 in $(1, 2)$ such that

$$f'(x_1) = f(1) - f(0) = 1 \quad \text{and} \quad f'(x_2) = f(2) - f(1) = 1.$$

Now the assertion follows from Rolle's theorem applied to f' on $[x_1, x_2]$.

2.2.16. Since f is not a linear function, there is $c \in (a, b)$ such that

$$f(c) < f(a) + \frac{f(b) - f(a)}{b - a}(c - a) \quad \text{or} \quad f(c) > f(a) + \frac{f(b) - f(a)}{b - a}(c - a).$$

Suppose, for example, that

$$f(c) < f(a) + \frac{f(b) - f(a)}{b - a}(c - a).$$

Then

$$\frac{f(c) - f(a)}{c - a} < \frac{f(b) - f(a)}{b - a} \quad \text{and} \quad \frac{f(c) - f(b)}{c - b} > \frac{f(b) - f(a)}{b - a}.$$

Consequently, the assertion follows from the mean value theorem. Analogous reasoning can be applied to the case where

$$f(c) > f(a) + \frac{f(b) - f(a)}{b - a}(c - a).$$

2.2.17. Suppose first that $x_0 \neq \frac{1}{2}$. Then either $[0, x_0]$ or $[x_0, 1]$ has length less than $\frac{1}{2}$. Suppose, for example, that this is $[x_0, 1]$. By the mean value theorem,

$$\frac{-1}{1-x_0} = \frac{f(1) - f(x_0)}{1 - x_0} = f'(c),$$

and consequently, $|f'(c)| > 2$. Suppose now that $x_0 = \frac{1}{2}$ and that f is linear on $\left[0, \frac{1}{2}\right]$. Then $f(x) = 2x$ for $x \in \left[0, \frac{1}{2}\right]$. Since $f'\left(\frac{1}{2}\right) = 2$, there is $x_1 > \frac{1}{2}$ such that $f(x_1) > 1$. In this case the assertion follows from the mean value theorem applied to f on $[x_1, 1]$. Finally, suppose that f is not linear on $\left[0, \frac{1}{2}\right]$. If there is $x_2 \in \left(0, \frac{1}{2}\right)$ such that $f(x_2) > 2x_2$, then to get the desired result it is enough to apply the mean value theorem on $[0, x_2]$. If $f(x_2) < 2x_2$, then one can apply the mean value theorem on $\left[x_2, \frac{1}{2}\right]$.

2.2.18. Applying the generalized mean value theorem to the functions $x \mapsto \frac{f(x)}{x}$ and $x \mapsto \frac{1}{x}$ on $[a, b]$, we see that

$$\frac{bf(a) - af(b)}{b - a} = \frac{\frac{f(b)}{b} - \frac{f(a)}{a}}{\frac{1}{b} - \frac{1}{a}} = \frac{\frac{x_1 f'(x_1) - f(x_1)}{x_1^2}}{-\frac{1}{x_1^2}} = f(x_1) - x_1 f'(x_1).$$

2.2.19. By the mean value theorem, for $x_1, x_2 \in [0, \infty)$,

$$|\ln(1 + x_1) - \ln(1 + x_2)| = \frac{1}{1 + x_0}|x_1 - x_2| \leq |x_1 - x_2|.$$

Likewise,

$$|\ln(1 + x_1^2) - \ln(1 + x_2^2)| = \frac{2x_0}{1 + x_0^2}|x_1 - x_2| \leq |x_1 - x_2|$$

and

$$|\arctan x_1 - \arctan x_2| = \frac{1}{1 + x_0^2}|x_1 - x_2| \leq |x_1 - x_2|.$$

2.2.20. Fix $x_0 \in (a, b)$. Then for every $x \in (a, b)$ (by the mean value theorem) there is c between x_0 and x such that $f'(x) - f'(x_0) = f''(c)(x - x_0)$. Hence

$$|f'(x)| \leq M|x - x_0| + |f'(x_0)| \leq M(b - a) + |f'(x_0)|,$$

2.2. Mean Value Theorems 237

which means that f' is bounded. It then follows (as in the solution of the foregoing problem) that f is uniformly continuous on (a, b).

2.2.21. Consider the function $x \mapsto \arctan f(x)$. By the mean value theorem, for $a < x_1 < x_2 < b$, $x_2 - x_1 > \pi$, we have

$$|\arctan f(x_2) - \arctan f(x_1)| = \frac{|f'(x_0)|}{1 + f^2(x_0)}(x_2 - x_1).$$

Hence,

$$\pi \geq \frac{|f'(x_0)|}{1 + f^2(x_0)}(x_2 - x_1),$$

and consequently,

$$\frac{|f'(x_0)|}{1 + f^2(x_0)} \leq \frac{\pi}{x_2 - x_1} < 1.$$

2.2.22. We have

$$\arctan f(x_2) - \arctan f(x_1) = \frac{f'(x_0)}{1 + f^2(x_0)}(x_2 - x_1)$$

for $a < x_1 < x_2 < b$. By (ii),

$$\arctan f(x_2) - \arctan f(x_1) \geq -(x_2 - x_1).$$

Letting $x_2 \to b^-$ and $x_1 \to a^+$ and using (i), we see that $-\pi \geq -(b-a)$.

2.2.23. By the mean value theorem,

$$f'_-(b) = \lim_{h \to 0^-} \frac{f(b+h) - f(b)}{h} = \lim_{h \to 0^-} f'(b + \theta h) = A.$$

2.2.24. Since $f'(x) = O(x)$, there are $M > 0$ and $x_0 \in (0, \infty)$ such that $|f'(x)| \leq Mx$ for $x \geq x_0$. By the mean value theorem,

$$|f(x) - f(x_0)| = |f'(x_0 + \theta(x - x_0))|(x - x_0)$$
$$\leq M(x_0 + \theta(x - x_0))(x - x_0) \leq Mx(x - x_0) \leq Mx^2$$

for $x \geq x_0$.

2.2.25. The result follows from Rolle's theorem applied to the auxiliary function
$$h(x) = \sum_{k=1}^{n} \left(f_k(x) - f_k(a) - (g_k(x) - g_k(a)) \frac{f_k(b) - f_k(a)}{g_k(b) - g_k(a)} \right).$$

2.2.26. Assume first that f is uniformly differentiable on $[a,b]$. Then for any sequence $\{h_n\}$ converging to zero such that $h_n \neq 0$ and $x + h_n \in \mathbf{I}$ for $x \in [a,b]$, the sequence of functions $\{\frac{f(x+h_n)-f(x)}{h_n}\}$ is uniformly convergent on $[a,b]$ to f'. By the result in 1.2.34, f' is continuous on $[a,b]$.

Assume now that f' is continuous on $[a,b]$. By the mean value theorem, for $x \in [a,b]$, $x + h \in \mathbf{I}$,
$$\frac{f(x+h) - f(x)}{h} - f'(x) = f'(x + \theta h) - f'(x)$$
with some $0 < \theta < 1$. It then follows by uniform continuity of f' on $[a,b]$ that f is uniformly differentiable.

2.2.27. Since f is continuous on $[a,b]$, it is bounded; that is, there is $A \geq 0$ such that $|f(x)| \leq A$ for $x \in [a,b]$. By assumption,
$$|g'(x)| \leq \frac{1+A}{|\lambda|} |g(x)|.$$
Now let $[c,d]$ be a subinterval of $[a,b]$ whose length is not greater than $\frac{1}{2} \frac{|\lambda|}{1+A} = \frac{B}{2}$ and such that $g(c) = 0$. For an $x_0 \in [c,d]$, we get
$$|g(x_0) - g(c)| = |g(x_0)| = (x_0 - c)|g'(x_1)| \leq \frac{B}{2} \frac{|g(x_1)|}{B}.$$
By repeating the process, one can find a decreasing sequence $\{x_n\}$ of points in $[c,d]$ such that
$$|g(x_0)| \leq \frac{1}{2}|g(x_1)| \leq \cdots \leq \frac{1}{2^n}|g(x_n)| \leq \cdots.$$
Consequently, $g(x_0) = 0$. To end the proof it is enough to decompose $[a,b]$ into a finite number of subintervals with lengths less than or equal to $\frac{B}{2}$.

It is worth noting here that the assumption of continuity of f on $[a,b]$ can be replaced by its boundedness on $[a,b]$.

2.2. Mean Value Theorems

2.2.28. By the generalized mean value theorem,

$$\frac{\frac{f(2x)}{2x} - \frac{f(x)}{x}}{\frac{1}{2x} - \frac{1}{x}} = f(\zeta) - \zeta f'(\zeta),$$

where $x < \zeta < 2x$. Hence

$$\frac{f(2x)}{2x} - \frac{f(x)}{x} = \frac{\zeta}{2x}\left(f'(\zeta) - \frac{f(\zeta)}{\zeta}\right).$$

This implies that

$$0 \leq |f'(\zeta)| \leq 2\left|\frac{f(2x)}{2x} - \frac{f(x)}{x}\right| + \left|\frac{f(\zeta)}{\zeta}\right|.$$

Upon passage to the limit as $x \to \infty$ we obtain the desired result.

2.2.29. The result is a direct consequence of 1.6.30.

2.2.30. By assumption,

(1) $$f'(px + qy) = f'(qx + py) \quad \text{for} \quad x \neq y.$$

If $p \neq q$, then f' is a constant function. Indeed, if $f'(x_1) \neq f'(x_2)$, then, taking

$$x = \frac{p}{2p-1}x_1 + \frac{p-1}{2p-1}x_2 \quad \text{and} \quad y = \frac{p-1}{2p-1}x_1 + \frac{p}{2p-1}x_2,$$

we have $x_1 = px + (1-p)y$ and $x_2 = py + (1-p)x$, which contradicts (1). So we have proved that if $p \neq q$, then f is a linear function. If $p = q = \frac{1}{2}$, then by the result in the foregoing problem, f is a polynomial of the second degree.

2.2.31. For $[a,b] \subset \mathbf{I}$, assume, for example, that $f'(a) < f'(b)$. Let λ be a number such that $f'(a) < \lambda < f'(b)$. Consider the function given by setting $g(x) = f(x) - \lambda x$. Then $g'(a) < 0$ and $g'(b) > 0$. Consequently, g attains its minimum on $[a,b]$ at an x_0 in the open interval (a,b). Therefore $g'(x_0) = 0$, or in other words, $f'(x_0) = \lambda$.

2.2.32.

(a) Given $\varepsilon > 0$, let $a > 0$ be such that $|f(x) + f'(x)| < \varepsilon$ for $x \geq a$. Then by the generalized mean value theorem there is $\xi \in (a, x)$ such that
$$\frac{e^x f(x) - e^a f(a)}{e^x - e^a} = f(\xi) + f'(\xi).$$
Thus
$$|f(x) - f(a)e^{a-x}| < \varepsilon|1 - e^{a-x}|,$$
which gives
$$|f(x)| < |f(a)|e^{a-x} + \varepsilon|1 - e^{a-x}|.$$
Consequently, $|f(x)| < 2\varepsilon$ for sufficiently large x.

(b) Apply the generalized mean value theorem to $x \mapsto e^{\sqrt{x}} f(x)$ and $x \mapsto e^{\sqrt{x}}$, and proceed as in (a).

2.2.33. It follows from the hypotheses that the function $x \mapsto e^{-x} f(x)$ has at least three distinct zeros in $[a, b]$. Hence by Rolle's theorem its derivative $x \mapsto e^{-x}(f'(x) - f(x))$ has at least two distinct zeros in this interval. This in turn implies that the second derivative has at least one zero, which means that the equation $e^{-x}(f(x) + f''(x) - 2f'(x)) = 0$ has at least one root in $[a, b]$.

2.2.34. Observe first that $Q(x) = F(x)G(x)$, where
$$F(x) = P'(x) + xP(x) = e^{-\frac{x^2}{2}} \left(e^{\frac{x^2}{2}} P(x) \right)',$$
$$G(x) = xP'(x) + P(x) = (xP(x))'.$$
Let $1 < a_1 < a_2 < \cdots < a_n$ be zeros of the polynomial P. By Rolle's theorem F has $n - 1$ zeros, say $b_i, i = 1, 2, \ldots, n - 1$, and G has n zeros, say $c_i, i = 1, 2, \ldots, n$. We can assume that
$$1 < a_1 < b_1 < a_2 < b_2 < \cdots < b_{n-1} < a_n,$$
$$0 < c_1 < a_1 < c_2 < a_2 < \cdots < c_n < a_n.$$
If $b_i \neq c_{i+1}$ for $i = 1, 2, \ldots, n - 1$, then the polynomial Q has at least $2n - 1$ zeros. Now suppose that there is i such that $b_i = c_{i+1} = r$. Then $P'(r) + rP(r) = 0 = rP'(r) + P(r)$. Hence $(r^2 - 1)P(r) = 0$. Since $r > 1$, this gives $P(r) = 0$, a contradiction.

2.2. Mean Value Theorems

2.2.35. Let $x_1 < x_2 < \cdots < x_m$ be zeros of P. It follows from the hypotheses that $P'(x_m) > 0$ and $P'(x_{m-1}) < 0$, and $P'(x_{m-2}) > 0, \ldots$. Moreover, we see that $Q(x_m) < 0$, $Q(x_{m-1}) > 0, \ldots$. If m is odd, then $Q(x_1) < 0$. If m is even, then $Q(x_1) > 0$. Consequently, by Rolle's theorem, Q has at least $m+1$ real zeros when m is odd, and at least m real zeros when m is even. Now we show that all real zeros of Q are distinct. Since all zeros of P are real and distinct, $(P'(x))^2 > P(x)P''(x)$ for $x \in \mathbb{R}$. Indeed, since

$$P(x) = a_m(x - x_1)(x - x_2) \cdots (x - x_m),$$

we see that for $x \neq x_j$, $j = 1, 2, \ldots, m$,

$$\frac{P'(x)}{P(x)} = \sum_{j=1}^{m} \frac{1}{x - x_j}.$$

Hence

$$P(x)P''(x) - (P'(x))^2 = -P^2(x) \sum_{j=1}^{m} \frac{1}{(x - x_j)^2} < 0.$$

Moreover, for $x = x_j$,

$$(P'(x_j))^2 > 0 = P(x_j)P''(x_j).$$

Thus the inequality $(P'(x))^2 > P(x)P''(x)$ is proved. Consequently,

$$P(x)Q'(x) = P(x)(2P(x)P'(x) - P''(x))$$
$$= 2P'(x)(P^2(x) - P'(x)) + 2(P'(x))^2 - P(x)P''(x)$$
$$> 2P'(x)P^2(x) - (P'(x))^2.$$

This means that

(1) $$P(x)Q'(x) > 2P'(x)Q(x),$$

which shows that all zeros of Q are of the first order. If y_1 and y_2 are two consecutive zeros of Q, then $Q'(y_1)$ and $Q'(y_2)$ have distinct signs. Then by (1) $P(y_1)$ and $P(y_2)$ also have distinct signs. Therefore between two consecutive zeros of Q there is at least one zero of P. Thus, when m is odd, if Q had more than $m+1$ real zeros, then P would have more than m real zeros, which would contradict the hypotheses. Similarly, when m is even, if Q had more than m real zeros, it would have at least $m+2$ real zeros, and consequently, P would have more than m real zeros. A contradiction.

2.2.36 [G. Peyser, Amer. Math. Monthly 74 (1967), 1102-1104]. Let us observe that if all zeros of a polynomial P of degree n are real, then by Rolle's theorem all zeros of P' are real and lie between the zeros of P. Thus P' is of the form given in the problem. We will prove only the first assertion, because the proof of the second one is analogous. Clearly, $P(x) = Q(x)(x - a_n)$. Hence

$$(*) \qquad P'(x) = Q'(x)(x - a_n) + Q(x).$$

It is enough to consider the case $a_i < a_{i+1}$. Suppose, for example, that $P(x) > 0$ for $x \in (a_i, a_{i+1})$. Then $Q(x) < 0$ for $x \in (a_i, a_{i+1})$. Moreover, it follows from $(*)$ that $Q'(x) < 0$ for $x \in (a_i, c_i)$, and $Q'(c_i) < 0$. Consequently, $d_i > c_i$, which ends the proof of the first assertion.

2.2.37 [G. Peyser, Amer. Math. Monthly 74 (1967), 1102-1104]. Assume that $a_{n-1} < a_n$ and $\varepsilon > 0$. Clearly,

$$S(x) = P(x) - \varepsilon R(x),$$

where $R(x) = (x - a_2) \cdots (x - a_n)$. Suppose that, e.g., $P(x) < 0$ for $x \in (a_{n-1}, a_n)$. Then also $S(x) < 0$ and $R(x) < 0$ for $x \in (a_{n-1}, a_n)$. Since

$$(1) \qquad S'(x) = P'(x) - \varepsilon R'(x),$$

we see that $S'(c_{n-1}) = -\varepsilon R'(c_{n-1})$. It follows from the foregoing problem that $R'(c_{n-1}) > 0$. By (1) we have $S'(c_{n-1}) < 0$. Since S' changes its sign from negative to positive at a point in the interval (a_{n-1}, a_n), we see that $f_{n-1} > c_{n-1}$. The other assertion can be proved in an entirely similar manner.

2.2.38 [G. Peyser, Amer. Math. Monthly 74 (1967), 1102-1104]. Set $W(x) = (x - a_i)^i (x - a_{i+1})$. If $i = 2, 3, \ldots, n-1$, then $W'(x) = 0$ for $x = a_i$ and for

$$x = c = \frac{i a_{i+1} + a_i}{i + 1} = a_{i+1} - \frac{a_{i+1} - a_i}{i + 1}.$$

2.2. Mean Value Theorems

If $i = 1$, then W' vanishes only at c. Applying the first result in 2.2.36 $(n - i - 1)$ times, and next the first result in the foregoing problem $(i-1)$ times with ε equal to $a_i - a_1, a_i - a_2, \ldots, a_i - a_{i-1}$, successively, we arrive at

$$c_i \leq c = a_{i+1} - \frac{a_{i+1} - a_i}{i+1}.$$

To get the left inequality one can apply the second parts of the last two problems.

2.2.39. Observe that, by the mean value theorem, for every x in $(0, 1/K) \cap [0, 1]$ we have

$$|f(x)| \leq Kx|f(x_1)| \leq K^2 x x_1 |f(x_2)| \leq \cdots \leq K^n x x_1 \cdots x_{n-1} |f(x_n)|,$$

where $0 < x_n < x_{n-1} < \cdots < x_1 < x$. Thus $|f(x)| \leq (Kx)^n |f(x_n)|$. Since f is bounded, it then follows that $f(x) \equiv 0$ on $[0, 1/K] \cap [0, 1]$. If $K \geq 1$, one can show in an entirely similar manner that $f(x) \equiv 0$ on $[1/K, 2/K]$. Repeating the process finitely many times gives $f(x) \equiv 0$ on $[0, 1]$.

2.2.40. For $x_1 \in \mathbf{J_1}$ and $x_3 \in \mathbf{J_3}$,

$$\frac{f^{(k-1)}(x_3) - f^{(k-1)}(x_1)}{x_3 - x_1} = f^{(k)}(\zeta)$$

with some $\zeta \in (x_1, x_3)$. Hence

$$m_k(\mathbf{J}) \leq \frac{1}{x_3 - x_1}(|f^{(k-1)}(x_3)| + |f^{(k-1)}(x_1)|)$$

$$\leq \frac{1}{\lambda_2}(|f^{(k-1)}(x_3)| + |f^{(k-1)}(x_1)|).$$

Taking the infimum over all $x_1 \in \mathbf{J_1}$ and $x_3 \in \mathbf{J_3}$ gives the desired inequality.

2.2.41. We will proceed by induction on k. For $k = 1$, the inequality can be concluded from the mean value theorem and from the fact that $|f(x)| \leq 1$. Assume that the inequality to be proved holds for an arbitrarily chosen positive integer k. Then by the result in the foregoing problem we get

$$m_{k+1}(\mathbf{J}) \leq \frac{1}{\lambda_2}\left(m_k(\mathbf{J_1}) + m_k(\mathbf{J_3})\right)$$

$$\leq \frac{1}{\lambda_2}\left(\frac{1}{\lambda_1^k}2^{\frac{k(k+1)}{2}}k^k + \frac{1}{\lambda_3^k}2^{\frac{k(k+1)}{2}}k^k\right)$$

$$= 2^{\frac{k(k+1)}{2}}k^k\left(\frac{1}{\lambda_1^k\lambda_2} + \frac{1}{\lambda_3^k\lambda_2}\right).$$

Putting $\lambda_1 = \lambda_3 = \frac{k\lambda}{2(k+1)}$ and $\lambda_2 = \frac{\lambda}{k+1}$, we obtain

$$m_{k+1}(\mathbf{J}) \leq \frac{2^{\frac{(k+1)(k+2)}{2}}(k+1)^{k+1}}{\lambda^{k+1}}.$$

2.2.42. We have

$$P^{(p-1)}(x) = (p-1)!a_{p-1} + \frac{(p+1)!}{2!}a_{p+1}x^2 + \cdots + \frac{n!}{(n-p+1)!}a_n x^{n-p+1}.$$

It follows from Rolle's theorem that between two consecutive real zeros of P there is exactly one real zero of P'. Consequently, the polynomial $P^{(p-1)}$ has $n-p+1$ distinct real zeros and $P^{(p)}$ has $n-p$ distinct real zeros. As we have already mentioned, between two consecutive real zeros of $P^{(p-1)}$ there is exactly one real zero of $P^{(p)}$.

Suppose, contrary to our claim, that a_{p-1} and a_{p+1} have the same sign. Without loss of generality, we can assume that both are positive. Then there is $\varepsilon > 0$ such that $P^{(p-1)}$ is decreasing in $(-\varepsilon, 0)$ and is increasing in $(0, \varepsilon)$. Clearly, $P^{(p)}(0) = 0$. If there were no other zeros of $P^{(p)}$, then we would get $P^{(p-1)}(x) > P^{(p-1)}(0) > 0$ for $x \neq 0$, a contradiction. If $P^{(p)}$ has nonzero zeros, then denote by $x_0 \neq 0$ the nearest to zero. So between 0 and x_0 there is a zero of $P^{(p-1)}$. On the other hand, $P^{(p-1)}(x) > 0$ on an open interval with ends 0 and x_0, a contradiction.

2.3. Taylor's Formula and L'Hospital's Rule

2.3.1. Note that for $n = 1$ the formula follows immediately from the definition of $f'(x_0)$. Now for $n > 1$ put

$$r_n(x) = f(x) - \left(f(x_0) + \frac{f'(x_0)}{1!}(x - x_0) + \cdots + \frac{f^{(n)}(x_0)}{n!}(x - x_0)^n \right).$$

Then $r_n(x_0) = r'_n(x_0) = \cdots = r_n^{(n)}(x_0) = 0$. By the definition of the nth derivative,

$$r_n^{(n-1)}(x) = r_n^{(n-1)}(x) - r_n^{(n-1)}(x_0) = r_n^{(n)}(x_0)(x - x_0) + o(x - x_0).$$

Consequently, $r_n^{(n-1)}(x) = o(x - x_0)$. Using the mean value theorem, we get

$$r_n^{(n-2)}(x) = r_n^{(n-2)}(x) - r_n^{(n-2)}(x_0) = r_n^{(n-1)}(c)(x - x_0),$$

where c is a point in an open interval with the endpoints x and x_0. Since $|c - x_0| < |x - x_0|$, we see that $r_n^{(n-2)}(x) = o((x - x_0)^2)$. By repeating the process n times we obtain $r_n(x) = o((x - x_0)^n)$.

2.3.2. For $x, x_0 \in [a, b]$, set

$$r_n(x) = f(x) - \left(f(x_0) + \frac{f'(x_0)}{1!}(x - x_0) + \cdots + \frac{f^{(n)}(x_0)}{n!}(x - x_0)^n \right).$$

Without loss of generality we may assume that $x > x_0$. On $[x_0, x]$ define the auxiliary function φ by

$$\varphi(z) = f(x) - \left(f(z) + \frac{f'(z)}{1!}(x - z) + \cdots + \frac{f^{(n)}(z)}{n!}(x - z)^n \right).$$

We have

(1) $\qquad\qquad \varphi(x_0) = r_n(x) \quad \text{and} \quad \varphi(x) = 0.$

Moreover, φ' exists in (x_0, x) and

(2) $\qquad\qquad \varphi'(z) = -\frac{f^{(n+1)}(z)}{n!}(x - z)^n.$

By the generalized mean value theorem,

$$\frac{\varphi(x) - \varphi(x_0)}{\psi(x) - \psi(x_0)} = \frac{\varphi'(c)}{\psi'(c)},$$

where ψ is continuous on $[x_0, x]$ with nonvanishing derivative on (x, x_0). This combined with (1) and (2) gives

$$r_n(x) = \frac{\psi(x) - \psi(x_0)}{\psi'(c)} \cdot \frac{f^{(n+1)}(c)}{n!}(x-c)^n.$$

Taking $\psi(z) = (x-z)^p$ and writing $c = x_0 + \theta(x-x_0)$, we arrive at

$$r_n(x) = \frac{f^{(n+1)}(x_0 + \theta(x-x_0))}{n!p}(1-\theta)^{n+1-p}(x-x_0)^{n+1}.$$

2.3.3. Note that these results are contained as special cases in the foregoing problem. Indeed, it is enough to take
(a) $p = n+1$,
(b) $p = 1$.

2.3.4. Integration by parts gives

$$f(x) - f(x_0) = \int_{x_0}^x f'(t)dt = \left[-(x-t)f'(t)\right]_{x_0}^x + \int_{x_0}^x (x-t)f''(t)dt.$$

Hence

$$f(x) = f(x_0) + \frac{f'(x_0)}{1!}(x-x_0) + \int_{x_0}^x (x-t)f''(t)dt.$$

To get the desired version of Taylor's formula it suffices to repeat the above reasoning n times.

2.3.5. For $n = 1$,

$$R_2(x) = \int_{x_0}^x \int_{x_0}^{t_2} f^{(2)}(t_1)dt_1 dt_2 = \int_{x_0}^x (f'(t_2) - f'(x_0))dt_2$$
$$= f(x) - f(x_0) - (x-x_0)f'(x_0).$$

To derive the formula, induction can be used.

2.3.6. By Taylor's formula, with the Lagrange form for the remainder (see, e.g., 2.3.3 (a)),

$$\sqrt{1+x} = 1 + \frac{1}{2}x - \frac{1}{8}x^2 + \frac{3}{3!8}(1+\theta x)^{-\frac{5}{2}}x^3.$$

2.3. Taylor's Formula and L'Hospital's Rule

with some $0 < \theta < 1$. Consequently,

$$|\sqrt{1+x} - (1 + \tfrac{1}{2}x - \tfrac{1}{8}x^2)| \leq \frac{3|x|^3}{48(\tfrac{1}{2})^{\frac{5}{2}}} = \frac{\sqrt{2}|x|^3}{4} < \frac{1}{2}|x|^3.$$

2.3.7. Applying Taylor's formula with the Lagrange form for the remainder to $f(x) = (1+x)^\alpha$, we get

$$(1+x)^\alpha = 1 + \alpha x + \frac{\alpha(\alpha-1)(1+\theta x)^{\alpha-2}}{2}x^2$$

with some $0 < \theta < 1$. Now it is enough to observe that

$$\frac{\alpha(\alpha-1)(1+\theta x)^{\alpha-2}}{2} > 0 \quad \text{for} \quad \alpha > 1 \quad \text{or} \quad \alpha < 0,$$

and

$$\frac{\alpha(\alpha-1)(1+\theta x)^{\alpha-2}}{2} < 0 \quad \text{for} \quad 0 < \alpha < 1.$$

2.3.8. By Taylor's formula,

$$\frac{f(x) - f(0)}{g(x) - g(0)} = \frac{f'(0)x + \tfrac{1}{2}f''(\theta_1(x))x^2}{g'(0)x + \tfrac{1}{2}g''(\theta_2(x))x^2}.$$

On the other hand, the mean value theorem yields

$$\frac{f'(\theta(x))}{g'(\theta(x))} = \frac{f'(0) + \theta(x)f''(\theta_3(x))}{g'(0) + \theta(x)g''(\theta_4(x))}.$$

Using the above equalities and continuity of f'' and g'' at zero, one easily checks that

$$\lim_{x \to 0^+} \frac{\theta(x)}{x} = \frac{1}{2}.$$

2.3.9.

(a) By Taylor's formula,

$$f(0) = f(x + (-x)) = f(x) + \frac{f'(x)}{1!}(-x) + \frac{f''(x)}{2!}(-x)^2$$
$$+ \cdots + \frac{f^{(n)}(x)}{n!}(-x)^n + \frac{f^{(n+1)}(x - \theta_1 x)}{(n+1)!}(-x)^{n+1}.$$

Taking $\theta = 1 - \theta_1$ gives the desired equality.

(b) Observe that $f\left(\frac{x}{1+x}\right) = f\left(x - \frac{x^2}{1+x}\right)$, and proceed as in (a).

2.3.10. We have

$$f(x) = f\left(\frac{x}{2} + \frac{x}{2}\right) = f\left(\frac{x}{2}\right) + \frac{f'\left(\frac{x}{2}\right)}{1!}\left(\frac{x}{2}\right) + \cdots + \frac{f^{(2n)}\left(\frac{x}{2}\right)}{(2n)!}\left(\frac{x}{2}\right)^{2n}$$
$$+ \frac{f^{(2n+1)}\left(\frac{x}{2} + \theta_1\frac{x}{2}\right)}{(2n+1)!}\left(\frac{x}{2}\right)^{2n+1}.$$

Similarly,

$$f(0) = f\left(\frac{x}{2} - \frac{x}{2}\right) = f\left(\frac{x}{2}\right) - \frac{f'\left(\frac{x}{2}\right)}{1!}\left(\frac{x}{2}\right) + \cdots + \frac{f^{(2n)}\left(\frac{x}{2}\right)}{(2n)!}\left(\frac{x}{2}\right)^{2n}$$
$$- \frac{f^{(2n+1)}\left(\frac{x}{2} - \theta_2\frac{x}{2}\right)}{(2n+1)!}\left(\frac{x}{2}\right)^{2n+1}.$$

Subtracting the above equalities yields

$$f(x) = f(0) + \frac{2}{1!}f'\left(\frac{x}{2}\right)\left(\frac{x}{2}\right) + \frac{2}{3!}f^{(3)}\left(\frac{x}{2}\right)\left(\frac{x}{2}\right)^3$$
$$+ \cdots + \frac{2}{(2n-1)!}f^{(2n-1)}\left(\frac{x}{2}\right)\left(\frac{x}{2}\right)^{2n-1}$$
$$+ \frac{f^{(2n+1)}\left(\frac{x}{2} + \theta_1\frac{x}{2}\right) + f^{(2n+1)}\left(\frac{x}{2} - \theta_2\frac{x}{2}\right)}{(2n+1)!}\left(\frac{x}{2}\right)^{2n+1}.$$

Since the derivative enjoys the intermediate value property (see, e.g., 2.2.31), the desired result follows.

2.3.11. Apply the foregoing problem with $f(x) = \ln(1+x)$, $x > 0$, and observe that the odd derivatives of f are positive for positive x.

2.3.12. By Taylor's formula with the Peano form for the remainder (see, e.g., 2.3.1),

(a)
$$\lim_{h \to 0} \frac{f(x+h) - 2f(x) + f(x-h)}{h^2}$$
$$= \lim_{h \to 0} \left\{ \frac{f(x) + hf'(x) + \frac{h^2}{2}f''(x) + o(h^2)}{h^2} \right.$$
$$\left. - \frac{2f(x) - f(x) + hf'(x) - \frac{h^2}{2}f''(x) + o(h^2)}{h^2} \right\} = f''(x).$$

2.3. Taylor's Formula and L'Hospital's Rule

(b)
$$\lim_{h \to 0} \frac{f(x+2h) - 2f(x+h) + f(x)}{h^2}$$
$$= \lim_{h \to 0} \frac{h^2 f''(x) + o(4h^2) - 2o(h^2)}{h^2} = f''(x).$$

2.3.13. As in the solution of the foregoing problem one can apply Taylor's formula with the Peano form for the remainder.

2.3.14.

(a) It follows from Taylor's formula that for $x > 0$,

$$e^x = \sum_{k=0}^{n} \frac{x^k}{k!} + \frac{x^{n+1}}{(n+1)!} e^{\theta x} > \sum_{k=0}^{n} \frac{x^k}{k!}.$$

(b) For $x > 0$ we have

$$\ln(1+x) = x - \frac{x^2}{2} + \frac{x^3}{3} - \frac{x^4}{4} + \frac{x^5}{5} \frac{1}{(1+\theta_1 x)^5} > x - \frac{x^2}{2} + \frac{x^3}{3} - \frac{x^4}{4}.$$

Similarly, for $x > -1$, $x \neq 0$,

$$\ln(1+x) = x - \frac{x^2}{2} + \frac{x^3}{3} - \frac{x^4}{4} \frac{1}{(1+\theta_2 x)^4} < x - \frac{x^2}{2} + \frac{x^3}{3}.$$

(c) Applying Taylor's formula to the function $x \mapsto \sqrt{1+x}$, we get

$$\sqrt{1+x} = 1 + \frac{1}{2}x - \frac{1}{8}x^2 + \frac{1}{16}x^3 - \frac{1}{128}(1+\theta_1 x)^{-\frac{7}{2}} x^4$$
$$< 1 + \frac{1}{2}x - \frac{1}{8}x^2 + \frac{1}{16}x^3$$

and

$$\sqrt{1+x} = 1 + \frac{1}{2}x - \frac{1}{8}x^2 + \frac{1}{16}(1+\theta_2 x)^{-\frac{5}{2}} x^3 > 1 + \frac{1}{2}x - \frac{1}{8}x^2.$$

2.3.15. By 2.3.1,

$$f(x+h) = f(x) + hf'(x) + \cdots + \frac{h^n}{n!} f^{(n)}(x) + \frac{h^{n+1}}{(n+1)!} f^{(n+1)}(x) + o(h^{n+1}).$$

On the other hand,

$$f(x+h) = f(x) + hf'(x) + \cdots + \frac{h^{n-1}}{(n-1)!} f^{(n-1)}(x) + \frac{h^n}{n!} f^{(n)}(x + \theta(h)h).$$

Subtracting the first equality from the latter, we get

$$\frac{f^{(n)}(x + \theta(h)h) - f^{(n)}(x)}{h} = \frac{f^{(n+1)}(x)}{n+1} + \frac{o(h)}{h}.$$

Consequently,

$$\theta(h) = \frac{\frac{f^{(n+1)}(x)}{n+1} + \frac{o(h)}{h}}{\frac{f^{(n)}(x+\theta(h)h) - f^{(n)}(x)}{\theta(h)h}}.$$

The desired result follows from the fact that $f^{(n+1)}(x)$ exists and is different from zero.

2.3.16. For $0 < x \leq 1$,

(1) $\quad f(0) = f(x - x) = f(x) - f'(x)x + f''(x - \theta_1 x)\dfrac{x^2}{2},$

and for $0 \leq x < 1$,

(2) $\quad \begin{aligned} f(1) &= f(x + (1-x)) \\ &= f(x) + f'(x)(1-x) + f''(x + \theta_2(1-x))\dfrac{(1-x)^2}{2}. \end{aligned}$

Note that (1) implies $|f'(1)| \leq \frac{A}{2}$, and (2) implies $|f'(0)| \leq \frac{A}{2}$. Moreover, subtracting (2) from (1) gives

$$f'(x) = \frac{1}{2}(f''(x - \theta_1 x)x^2 - f''(x + \theta_2(1-x))(1-x)^2) \quad \text{for} \quad 0 < x < 1.$$

Hence

$$|f'(x)| \leq \frac{A}{2}(2x^2 - 2x + 1) < \frac{A}{2}, \quad 0 < x < 1.$$

2.3. Taylor's Formula and L'Hospital's Rule

2.3.17.

(a) For $x \in [-c, c]$,

(1) $\quad f(c) - f(x) = f'(x)(c-x) + \dfrac{f''(x+\theta_1(c-x))}{2}(c-x)^2$

and

$$f(-c) - f(x) = -f'(x)(c+x) + \dfrac{f''(x - \theta_2(c+x))}{2}(c+x)^2.$$

Hence

$$f'(x) = \dfrac{f(c) - f(-c)}{2c}$$
$$- \dfrac{(c-x)^2 f''(x+\theta_1(c-x)) - (c+x)^2 f''(x-\theta_2(c+x))}{4c}.$$

Consequently,

$$|f'(x)| \leq \dfrac{M_0}{c} + (c^2 + x^2)\dfrac{M_2}{2c}.$$

(b) By (1) in the solution of (a), for $x \in [-c, c)$, we get

$$f'(x) = \dfrac{f(c) - f(x)}{h} - \dfrac{f''(x + \theta_1 h)}{2} h,$$

where $h = c - x > 0$. Thus $|f'(x)| \leq 2\dfrac{M_0}{h} + \dfrac{1}{2} M_2 h$. Now taking $h = 2\sqrt{\dfrac{M_0}{M_2}}$ gives $|f'(x)| \leq 2\sqrt{M_0 M_2}$, which in turn implies that $M_1 \leq 2\sqrt{M_0 M_2}$.

2.3.18. The inequality $M_1 \leq 2\sqrt{M_0 M_2}$ is proved in the solution of 2.3.17 (b). The equality is attained, for example, for f defined by

$$f(x) = \begin{cases} 2x^2 - 1 & \text{if } -1 < x < 0, \\ \dfrac{x^2 - 1}{x^2 + 1} & \text{if } 0 \leq x < \infty. \end{cases}$$

Indeed, we have $M_0 = 1$ and $M_1 = M_2 = 4$.

2.3.19. For $h > 0$ and $x \in \mathbb{R}$,

$$f(x+h) = f(x) + f'(x)h + f''(x+\theta h)\dfrac{h^2}{2}$$

and

$$f(x-h) = f(x) - f'(x)h + f''(x - \theta_1 h)\dfrac{h^2}{2}.$$

Consequently,
$$f'(x) = \frac{1}{2h}(f(x+h) - f(x-h)) - \frac{h}{4}(f''(x+\theta h) - f''(x-\theta_1 h)),$$
which implies
$$|f'(x)| \leq \frac{M_0}{h} + \frac{h}{2}M_2 \quad \text{for} \quad h > 0.$$
To get the desired inequality it suffices to take $h = \sqrt{2\frac{M_0}{M_2}}$.

2.3.20. For $p = 2$ the result is contained in the foregoing problem. Now we proceed by induction. Assuming the assertion to hold for $2, 3, \ldots, p$, we will prove it for $p + 1$. We have
$$f^{(p-1)}(x+h) = f^{(p-1)}(x) + f^{(p)}(x)h + f^{(p+1)}(x+\theta h)\frac{h^2}{2}$$
and
$$f^{(p-1)}(x-h) = f^{(p-1)}(x) - f^{(p)}(x)h + f^{(p+1)}(x-\theta_1 h)\frac{h^2}{2}.$$
Hence
$$f^{(p)}(x) = \frac{1}{2h}(f^{(p-1)}(x+h) - f^{(p-1)}(x-h))$$
$$- \frac{h}{4}(f^{(p+1)}(x+\theta h) - f^{(p+1)}(x-\theta_1 h)).$$
Consequently,
$$|f^{(p)}(x)| \leq \frac{M_{p-1}}{h} + \frac{h}{2}M_{p+1}, \quad h > 0.$$
Taking $h = \sqrt{2\frac{M_{p-1}}{M_{p+1}}}$ gives $M_p \leq \sqrt{2M_{p-1}M_{p+1}}$. By the induction hypothesis for $k = p - 1$, upon simple calculation, we get

(1) $$M_p \leq 2^{\frac{p}{2}} M_0^{\frac{1}{p+1}} M_{p+1}^{\frac{p}{p+1}}.$$

So the inequality is proved for $k = p$. Now we will prove it for $1 \leq k \leq p - 1$. By the induction hypothesis,
$$M_k \leq 2^{\frac{k(p-k)}{2}} M_0^{1-\frac{k}{p}} M_p^{\frac{k}{p}},$$

2.3. Taylor's Formula and L'Hospital's Rule   253

which combined with (1) gives

$$M_k \leq 2^{\frac{k(p+1-k)}{2}} M_0^{1-\frac{k}{p+1}} M_{p+1}^{\frac{k}{p+1}}.$$

2.3.21. Suppose $|f''(x)| \leq M$ ($M > 0$) for $x \in (0, \infty)$. By Taylor's formula for $x, h \in (0, \infty)$,

$$f(x+h) = f(x) + f'(x)h + f''(x+\theta h)\frac{h^2}{2}.$$

It then follows that

$$|f'(x)| \leq \frac{|f(x+h) - f(x)|}{h} + \frac{Mh}{2}.$$

Since $\lim\limits_{x \to \infty} f(x) = 0$, given $\varepsilon > 0$ there is an x_0 such that

$$|f'(x)| \leq \frac{\varepsilon}{h} + \frac{Mh}{2} \quad \text{for} \quad x > x_0, h > 0.$$

Taking $h = \sqrt{2\frac{\varepsilon}{M}}$ we get $|f'(x)| \leq \sqrt{2\varepsilon M}$, $x > x_0$, which means that $\lim\limits_{x \to \infty} f'(x) = 0$.

2.3.22. For $x > 0$,

$$f(x+1) = f(x) + f'(x) + \frac{1}{2}f''(\xi) \quad \text{with some} \quad \xi \in (x, x+1).$$

Hence

$$xf'(x) = \frac{x}{x+1}(x+1)f(x+1) - xf(x) - \frac{1}{2} \cdot \frac{x}{\xi} \cdot \xi f''(\xi).$$

Consequently, $\lim\limits_{x \to +\infty} xf'(x) = 0$.

2.3.23. For $u, x \in (0, 1)$, $u > x$, by Taylor's formula we get

$$f(u) = f(x) + f'(x)(u-x) + \frac{1}{2}f''(\xi)(u-x)^2$$

with some $\xi \in (x, u)$. Taking $u = x + \varepsilon(1-x)$, $0 < \varepsilon < \frac{1}{2}$, we obtain

$$f(u) - f(x) = \varepsilon(1-x)f'(x) + \frac{1}{2}\varepsilon^2 f''(x+\theta\varepsilon(1-x))(1-x)^2$$

with some $\theta \in (0,1)$. Upon letting $x \to 1^-$ we see that

(1) $\quad 0 = \lim_{x \to 1^-} \left((1-x)f'(x) + \frac{1}{2}\varepsilon f''(x + \theta\varepsilon(1-x))(1-x)^2 \right).$

By the definition of the limit, if $\varepsilon_1 > 0$, then

$$(1-x)|f'(x)| \leq \varepsilon_1 + \frac{1}{2}\varepsilon |f''(x+\theta\varepsilon(1-x))|(1-x)^2$$
$$\leq \varepsilon_1 + \frac{1}{2}\frac{M\varepsilon}{(\theta\varepsilon-1)^2}$$

for x sufficiently close to 1. Since $\varepsilon > 0$ can be chosen arbitrarily, $(1-x)|f'(x)| \leq \varepsilon_1$, which means that $\lim_{x \to 1^-}(1-x)f'(x) = 0$.

2.3.24. We have

$$f\left(\frac{a+b}{2}\right) = f(a) + \frac{f''(x_1)}{2!}\left(\frac{b-a}{2}\right)^2$$

and

$$f\left(\frac{a+b}{2}\right) = f(b) + \frac{f''(x_2)}{2!}\left(\frac{b-a}{2}\right)^2,$$

with some $x_1 \in \left(a, \frac{a+b}{2}\right)$ and $x_2 \in \left(\frac{a+b}{2}, b\right)$. Hence

$$|f(b) - f(a)| = \left(\frac{b-a}{2}\right)^2 \frac{1}{2}|f''(x_2) - f''(x_1)| \leq \left(\frac{b-a}{2}\right)^2 |f''(c)|,$$

where $|f''(c)| = \max\{|f''(x_1)|, |f''(x_2)|\}$.

2.3.25. It follows from Taylor's formula that

$$1 = f(1) = \frac{1}{2}f''(0) + \frac{f'''(x_1)}{3!} \quad \text{and} \quad 0 = f(-1) = \frac{1}{2}f''(0) - \frac{f'''(x_2)}{3!}$$

with some $x_1 \in (0,1)$ and $x_2 \in (-1,0)$. Thus

$$f'''(x_1) + f'''(x_2) = 6,$$

which implies $f'''(x_1) \geq 3$ or $f'''(x_2) \geq 3$.

It is worth noting here that the equality is attained, for example, for $f(x) = \frac{1}{2}(x^3 + x^2)$.

2.3. Taylor's Formula and L'Hospital's Rule 255

2.3.26. Write

(1) $$f(t) = f(x) + (t-x)Q(t).$$

Differentiating both sides of this equality with respect to t gives

(2) $$f'(t) = Q(t) + (t-x)Q'(t).$$

Substituting x_0 for t, we get

(3) $$f(x) = f(x_0) + (x-x_0)f'(x_0) + (x-x_0)^2 Q'(x_0).$$

Differentiating (2) with respect to t, next taking $t = x_0$, and aided by (3), we obtain

$$f(x) = f(x_0) + (x-x_0)f'(x_0) + \frac{1}{2}f''(x_0)(x-x_0)^2 + \frac{1}{2}(x-x_0)^3 Q''(x_0).$$

To get the desired equality it is enough to repeat the above procedure n times.

2.3.27. By Taylor's formula given in 2.3.1,

$$f(y_n) = f(0) + f'(0)y_n + o(y_n),$$
$$f(x_n) = f(0) + f'(0)x_n + o(x_n).$$

Hence

(1) $$f'(0) = \frac{f(y_n) - f(x_n)}{y_n - x_n} - \frac{o(y_n) - o(x_n)}{y_n - x_n}.$$

(a) Since $x_n < 0 < y_n$, we see that

$$\left| \frac{o(y_n) - o(x_n)}{y_n - x_n} \right| \le \frac{|o(y_n)|}{y_n - x_n} + \frac{|o(x_n)|}{y_n - x_n} \le \frac{|o(y_n)|}{y_n} + \frac{|o(x_n)|}{-x_n}.$$

Consequently,

$$\lim_{n \to \infty} \frac{o(y_n) - o(x_n)}{y_n - x_n} = 0,$$

which along with (1) shows that $\lim_{n \to \infty} D_n = f'(0)$.

(b) By (1) it is enough to show that $\lim_{n \to \infty} \frac{o(y_n) - o(x_n)}{y_n - x_n} = 0$. We have

$$\lim_{n \to \infty} \frac{o(y_n) - o(x_n)}{y_n - x_n}$$
$$= \lim_{n \to \infty} \left(\frac{o(y_n)}{y_n} \cdot \frac{y_n}{y_n - x_n} - \frac{o(x_n)}{x_n} \cdot \frac{x_n}{y_n - x_n} \right) = 0,$$

where the last equality follows from the boundedness of $\{\frac{y_n}{y_n-x_n}\}$ and $\{\frac{x_n}{y_n-x_n}\}$.

(c) By the mean value theorem, $D_n = f'(\theta_n)$, where $x_n < \theta_n < y_n$. The desired result follows from the continuity of f' at zero.

2.3.28. Observe first that P is a polynomial of degree at most m. Differentiating the equality

$$(1-y)^{m+1} = \sum_{k=0}^{m+1} \binom{m+1}{k}(-1)^k y^k,$$

we get

(1) $\quad -(m+1)(1-y)^m = \sum_{k=1}^{m+1} \binom{m+1}{k}(-1)^k k y^{k-1}.$

Taking $y = 1$ gives

(2) $\quad 0 = \sum_{k=1}^{m+1} \binom{m+1}{k}(-1)^k k.$

It follows from this equality that $P^{(m-1)}(0) = 0$. Next, differentiating (1) and then putting $y = 1$, we see by (2) that $P^{(m-2)}(0) = 0$. Continuing this process, one can show that $P^{(j)}(0) = 0$ for $j = 0, 1, 2, .., m-1$. Moreover, $P^{(m)}(0) = 0$, because

$$0 = (1-1)^{m+1} = \sum_{k=0}^{m+1} \binom{m+1}{k}(-1)^k.$$

Now it follows from Taylor's formula that $P(x) \equiv 0$.

2.3.29 [E. I. Poffald, Amer. Math. Monthly 97 (1990), 205-213]. We will need the following mean value theorem for integrals.

Theorem. *Let f and g be continuous on $[a,b]$, and let g have constant sign on $[a,b]$. Then there is $\xi \in (a,b)$ such that*

$$\int_a^b f(x)g(x)dx = f(\xi)\int_a^b g(x)dx.$$

2.3. Taylor's Formula and L'Hospital's Rule

Proof. Set

$$m = \min\{f(x) : x \in [a,b]\} \quad \text{and} \quad M = \max\{f(x) : x \in [a,b]\}.$$

Assume, e.g., that $g(x) > 0$. Then $mg(x) \leq f(x)g(x) \leq Mg(x)$. Integrating both sides of this inequality yields

$$m \int_a^b g(x)dx \leq \int_a^b f(x)g(x)dx \leq M \int_a^b g(x)dx.$$

Hence

$$m \leq \frac{\int_a^b f(x)g(x)dx}{\int_a^b g(x)dx} \leq M.$$

Since f enjoys the intermediate value property on $[a,b]$, the assertion follows.

Now following E. I. Poffald, we start the proof of the given formula. By Taylor's formula with integral remainder (see, e.g., 2.3.4),

$$f^{(n)}\left(\frac{x}{n+1}\right) = f^{(n)}(0) + f^{(n+1)}(0)\frac{x}{n+1}$$
$$+ \int_0^{\frac{x}{n+1}} f^{(n+2)}(t)\left(\frac{x}{n+1} - t\right) dt.$$

Hence

$$f(0) + \frac{f'(0)}{1!}x + \cdots + \frac{f^{(n-1)}(0)}{(n-1)!}x^{n-1} + \frac{f^{(n)}\left(\frac{x}{n+1}\right)}{n!}x^n$$

$$= f(0) + \frac{f'(0)}{1!}x + \cdots + \frac{f^{(n-1)}(0)}{(n-1)!}x^{n-1}$$
$$+ \left(f^{(n)}(0) + f^{(n+1)}(0)\frac{x}{n+1} + \int_0^{\frac{x}{n+1}} f^{(n+2)}(t)\left(\frac{x}{n+1} - t\right) dt\right)\frac{x^n}{n!}$$

$$= f(0) + \frac{f'(0)}{1!}x + \cdots + \frac{f^{(n+1)}(0)}{(n+1)!}x^{n+1}$$
$$+ \int_0^{\frac{x}{n+1}} f^{(n+2)}(t)\left(\frac{x}{n+1} - t\right) dt \frac{x^n}{n!}.$$

On the other hand, by Taylor's formula with integral remainder,

$$f(x) = f(0) + \frac{f'(0)}{1!}x + \cdots + \frac{f^{(n+1)}(0)}{(n+1)!}x^{n+1}$$
$$+ \frac{1}{(n+1)!}\int_0^x f^{(n+2)}(t)(x-t)^{n+1}dt.$$

Consequently,

$$f(x) - \left(f(0) + \frac{f'(0)}{1!}x + \cdots + \frac{f^{(n-1)}(0)}{(n-1)!}x^{n-1} + \frac{f^{(n)}\left(\frac{x}{n+1}\right)}{n!}x^n\right)$$
$$= \frac{1}{(n+1)!}\int_0^x f^{(n+2)}(t)(x-t)^{n+1}dt$$
$$- \int_0^{\frac{x}{n+1}} f^{(n+2)}(t)\left(\frac{x}{n+1} - t\right)dt\frac{x^n}{n!}$$
$$= \frac{1}{n!}\int_0^{\frac{x}{n+1}} f^{(n+2)}(t)\left(\frac{(x-t)^{n+1}}{n+1} - x^n\left(\frac{x}{n+1} - t\right)\right)dt$$
$$+ \frac{1}{(n+1)!}\int_{\frac{x}{n+1}}^x f^{(n+2)}(t)(x-t)^{n+1}dt.$$

Consider g defined by

$$g(t) = \frac{(x-t)^{n+1}}{n+1} - x^n\left(\frac{x}{n+1} - t\right) \quad \text{for} \quad t \in [0,x].$$

It easy to check that $g'(t) > 0$ for $t \in (0, x)$ and $g(0) = 0$. Thus g is positive on the open interval $(0, x)$. By the mean value theorem for integrals (proved at the beginning of the solution), we get

$$\int_0^{\frac{x}{n+1}} f^{(n+2)}(t)\left(\frac{(x-t)^{n+1}}{n+1} - x^n\left(\frac{x}{n+1} - t\right)\right)dt$$
$$= f^{(n+2)}(\xi_1)\int_0^{\frac{x}{n+1}} g(t)dt$$

and

$$\int_{\frac{x}{n+1}}^x f^{(n+2)}(t)(x-t)^{n+1}dt = f^{(n+2)}(\xi_2)\int_{\frac{x}{n+1}}^x (x-t)^{n+1}dt.$$

2.3. Taylor's Formula and L'Hospital's Rule

It follows from the above that

$$f(x) - \left(f(0) + \frac{f'(0)}{1!}x + \cdots + \frac{f^{(n-1)}(0)}{(n-1)!}x^{n-1} + \frac{f^{(n)}\left(\frac{x}{n+1}\right)}{n!}x^n\right)$$
$$= \frac{1}{n!}f^{(n+2)}(\xi_1)\int_0^{\frac{x}{n+1}}g(t)dt + \frac{1}{(n+1)!}f^{(n+2)}(\xi_2)\int_{\frac{x}{n+1}}^x (x-t)^{n+1}dt.$$

Setting

$$\lambda_1 = \int_0^{\frac{x}{n+1}} g(t)dt \quad \text{and} \quad \lambda_2 = \int_{\frac{x}{n+1}}^x \frac{(x-t)^{n+1}}{n+1}dt$$

we see that

$$\lambda_1 + \lambda_2 = \frac{x^{n+2}n}{2(n+1)^2(n+2)}.$$

By the intermediate value property,

$$\frac{f^{(n+2)}(\xi_1)\int_0^{\frac{x}{n+1}} g(t)dt + f^{(n+2)}(\xi_2)\int_{\frac{x}{n+1}}^x \frac{(x-t)^{n+1}}{n+1}dt}{\frac{x^{n+2}n}{2(n+1)^2(n+2)}} = f^{(n+2)}(\xi),$$

where ξ is between ξ_1 and ξ_2. Consequently,

$$f(x) - \left(f(0) + \frac{f'(0)}{1!}x + \cdots + \frac{f^{(n-1)}(0)}{(n-1)!}x^{n-1} + \frac{f^{(n)}\left(\frac{x}{n+1}\right)}{n!}x^n\right)$$
$$= \frac{1}{n!}f^{(n+2)}(\xi)\frac{x^{n+2}n}{2(n+1)^2(n+2)} = \frac{n}{2(n+1)}f^{(n+2)}(\xi)\frac{x^{n+2}}{(n+2)!}.$$

To end the proof it is enough to set $\theta = \frac{\xi}{x}$.

2.3.30. It follows from the assumption and Taylor's formula applied to $f^{(n)}$ that

$$f^{(n)}(x_0 + \theta(x)(x-x_0))$$
$$= f^{(n)}(x_0) + \frac{f^{(n+p)}(x_0 + \theta_1\theta(x)(x-x_0))}{p!}(\theta(x)(x-x_0))^p.$$

Hence

$$f(x) = f(x_0) + \frac{f'(x_0)}{1!}(x-x_0) + \cdots + \frac{f^{(n)}(x_0)}{n!}(x-x_0)^n$$
$$+ \frac{f^{(n+p)}(x_0+\theta_1\theta(x)(x-x_0))}{n!p!}(x-x_0)^{n+p}(\theta(x))^p.$$

On the other hand,

$$f(x) = f(x_0) + \frac{f'(x_0)}{1!}(x-x_0) + \cdots + \frac{f^{(n)}(x_0)}{n!}(x-x_0)^n$$
$$+ \frac{f^{(n+p)}(x_0+\theta_2(x-x_0))}{(n+p)!}(x-x_0)^{n+p}.$$

Upon putting the last two equalities together, we get

$$\frac{f^{(n+p)}(x_0+\theta_1\theta(x)(x-x_0))}{n!p!}(\theta(x))^p = \frac{f^{(n+p)}(x_0+\theta_2(x-x_0))}{(n+p)!}.$$

Since $f^{(n+p)}$ is continuous at x_0 and $f^{(n+p)}(x_0) \neq 0$, upon passage to the limit as $x \to x_0$, we obtain $\frac{1}{(n+p)!} = \frac{1}{n!p!} \lim\limits_{x \to x_0} (\theta(x))^p$.

Note that if $p = 1$, then we get the result in 2.3.15.

2.3.31. By Taylor's formula,

(1)
$$\sum_{k=1}^{\left[\frac{1}{\sqrt{x}}\right]} f(kx) = \sum_{k=1}^{\left[\frac{1}{\sqrt{x}}\right]} \left(f'(0)kx + \frac{1}{2}f''(\theta kx)k^2x^2 \right)$$
$$= f'(0)x \frac{\left[\frac{1}{\sqrt{x}}\right]\left(\left[\frac{1}{\sqrt{x}}\right]+1\right)}{2} + \eta(x),$$

where

(2)
$$\eta(x) = \frac{1}{2} \sum_{k=1}^{\left[\frac{1}{\sqrt{x}}\right]} f''(\theta kx)k^2x^2.$$

Since f'' is bounded in a neighborhood of zero and

$$\sum_{k=1}^{\left[\frac{1}{\sqrt{x}}\right]} k^2 = \frac{\left[\frac{1}{\sqrt{x}}\right]\left(\left[\frac{1}{\sqrt{x}}\right]+1\right)\left(2\left[\frac{1}{\sqrt{x}}\right]+1\right)}{6},$$

2.3. Taylor's Formula and L'Hospital's Rule

(2) implies that $\lim\limits_{x\to 0+} \eta(x) = 0$. Now (1) shows that

$$\lim_{x\to 0+} \sum_{k=1}^{\left[\frac{1}{\sqrt{x}}\right]} f(kx) = \frac{f'(0)}{2}.$$

2.3.32. It follows from the Bolzano-Weierstrass theorem (see, e.g., I, 2.4.30) that the set of zeros of f has at least one limit point, say p, in $[c,d]$. Clearly, $f(p) = 0$. Let $\{x_n\}$ be a sequence of zeros of f converging to p. By Rolle's theorem, between two zeros of f there is at least one zero of f'. Therefore p is also a limit point of the set of zeros of f'. Since f' is continuous, $f'(p) = 0$, and inductively, $f^{(k)}(p) = 0$ for $k \in \mathbb{N}$. Consequently, by Taylor's formula,

$$f(x) = \frac{f^{(n)}(p + \theta(x-p))}{n!}(x-p)^n$$

with some $\theta \in (0,1)$. Since $\sup\{|f^{(n)}(x)| : x \in (a,b)\} = O(n!)$, there is $M > 0$ such that $|f(x)| \leq M|x-p|^n$ for sufficiently large n. So, if $x \in (a,b)$ and $|x-p| < 1$, then $f(x) = 0$.

2.3.33. As in the solution of the foregoing problem, one can show that $f^{(k)}(0) = 0$ for $k \in \mathbb{N}$. By Taylor's formula,

$$f(x) = \frac{f^{(n)}(\theta x)}{n!}x^n, \quad n \in \mathbb{N}.$$

Since given x, $\lim\limits_{n\to\infty} \frac{x^n}{n!} = 0$, we see that $f(x) = 0$ for $x \in \mathbb{R}$.

2.3.34.

(a) 1.

(b) $-e/2$.

(c) $1/e$.

(d) 1.

(e) $e^{-\frac{1}{6}}$.

2.3.35. To show that

$$\lim_{x\to+\infty} \left(f\left(\frac{a}{\sqrt{x}}\right)\right)^x = \lim_{x\to+\infty} e^{x\ln f\left(\frac{a}{\sqrt{x}}\right)} = e^{-\frac{a^2}{2}},$$

one can use Taylor's formula with the Peano form for the remainder (see, e.g., 2.3.1), which gives

$$f(x) = 1 - \frac{x^2}{2} + o(x^2),$$

and next apply 1.1.17(a). One can also use l'Hospital's rule in the following way:

$$\lim_{x \to +\infty} x \ln f\left(\frac{a}{\sqrt{x}}\right) = \lim_{t \to 0^+} \frac{\ln f(a\sqrt{t})}{t} = \lim_{t \to 0^+} \frac{af'(a\sqrt{t})}{2\sqrt{t}f(a\sqrt{t})}$$
$$= \lim_{t \to 0^+} \frac{a^2 f''(a\sqrt{t})}{2f(a\sqrt{t}) + 2a\sqrt{t}f'(a\sqrt{t})} = -\frac{a^2}{2}.$$

2.3.36. Assume first that $a > 1$. Then by l'Hospital's rule,

$$\lim_{x \to +\infty} \left(\frac{a^x - 1}{x(a-1)}\right)^{\frac{1}{x}} = a.$$

If $0 < a < 1$, then

$$\lim_{x \to +\infty} \left(\frac{a^x - 1}{x(a-1)}\right)^{\frac{1}{x}} = 1.$$

2.3.37.

(a) Since $\lim\limits_{x \to \infty} \frac{1 - \cos x}{2 + \cos x}$ does not exist, one cannot apply l'Hospital's rule. Clearly, the limit is $1/2$.

(b) We are not allowed to apply l'Hospital's, because the derivative of the function in the denominator vanishes at $\frac{\pi}{2} + 2n\pi$, $n \in \mathbb{N}$. On the other hand, it is easy to show that the limit does not exist.

(c) To find the limit of $f(x)^{g(x)}$ when $x \to 0^+$ it is enough to determine $\lim\limits_{x \to 0^+} \frac{\ln f(x)}{\frac{1}{g(x)}}$. However, this limit cannot be calculated using l'Hospital's rule because the assumption that the limit of the quotient of derivatives exists is not satisfied. It is shown in 1.1.23(a) that the limit is 1.

(d) The limit is 1 (see, e.g., 1.1.23(b)). However, to find it one cannot apply l'Hospital's rule.

2.3. Taylor's Formula and L'Hospital's Rule

2.3.38. We have

$$\lim_{x\to 0}\frac{\frac{1}{x\ln 2}-\frac{1}{2^x-1}-\frac{1}{2}}{x} = \lim_{t\to 0}\frac{\frac{1}{\ln(1+t)}-\frac{1}{t}-\frac{1}{2}}{\frac{\ln(1+t)}{\ln 2}}$$

$$= \lim_{t\to 0}\frac{\ln 2(2t - 2\ln(1+t) - t\ln(1+t))}{2t\ln^2(1+t)}$$

$$= -\frac{\ln 2}{12},$$

where the last equality can be obtained applying l'Hospital's rule several times in succession. Hence $f'(0) = -\frac{\ln 2}{12}$.

2.3.39. As in the solution of 2.3.28, one can show that

$$\sum_{k=0}^{n}(-1)^k\binom{n}{k}k^r = \begin{cases} 0 & \text{if } r = 0, 1, \ldots, n-1, \\ n! & \text{if } r = n. \end{cases}$$

Now to get the desired equality it is enough to apply l'Hospital's rule n times successively.

2.3.40. Assume first that $\lim_{x\to a^+} g(x) = +\infty$ and $L \in \mathbb{R}$. By (iii), given $\varepsilon > 0$, there is a_1, such that for $x \in (a, a_1)$,

(1) $$L - \varepsilon < \frac{f'(x)}{g'(x)} < L + \varepsilon.$$

Since g' enjoys the intermediate value property, (i) implies that g' has the same sign on (a, b). Consequently, g is strictly monotonic on (a, b). For $x, y \in (a, a_1)$, $x < y$, by the generalized mean value theorem,

$$\frac{f(x) - f(y)}{g(x) - g(y)} = \frac{f'(x_0)}{g'(x_0)}$$

with some $x_0 \in (x, y) \subset (a, a_1)$. Fix y for a moment. Then by (1),

$$L - \varepsilon < \frac{\frac{f(x)}{g(x)} - \frac{f(y)}{g(x)}}{1 - \frac{g(y)}{g(x)}} < L + \varepsilon.$$

Thus if, for example, g is strictly decreasing in (a, b), then

$$(L-\varepsilon)\left(1 - \frac{g(y)}{g(x)}\right) + \frac{f(y)}{g(x)} < \frac{f(x)}{g(x)} < (L+\varepsilon)\left(1 - \frac{g(y)}{g(x)}\right) + \frac{f(y)}{g(x)}.$$

Letting $x \to a^+$ gives
$$L - \varepsilon \leq \lim_{x \to a^+} \frac{f(x)}{g(x)} \leq L + \varepsilon,$$
which ends the proof in this case. The other cases can be proved analogously.

Note that a similar result holds for the left-hand limit at b.

2.3.41.

(a) By l'Hospital's rule (see 2.3.40), we have
$$\lim_{x \to +\infty} f(x) = \lim_{x \to +\infty} \frac{e^{ax} f(x)}{e^{ax}} = \lim_{x \to +\infty} \frac{e^{ax}(af(x) + f'(x))}{ae^{ax}} = \frac{L}{a}.$$

(b) Likewise,
$$\lim_{x \to +\infty} f(x) = \lim_{x \to +\infty} \frac{e^{a\sqrt{x}} f(x)}{e^{a\sqrt{x}}} = \lim_{x \to +\infty} \frac{e^{a\sqrt{x}} \left(f'(x) + \frac{a}{2\sqrt{x}} f(x)\right)}{\frac{a}{2\sqrt{x}} e^{a\sqrt{x}}}$$
$$= \lim_{x \to +\infty} \frac{1}{a}(af(x) + 2\sqrt{x} f'(x)) = \frac{L}{a}.$$

To see that statements (a) and (b) are not true for negative a, consider the functions $f(x) = e^{-ax}$ and $f(x) = e^{-a\sqrt{x}}$, respectively.

2.3.42. Using l'Hospital's rule proved in 2.3.40, we get
$$\lim_{x \to \infty} \left(1 - \frac{f'(x)}{xf''(x)}\right) = \lim_{x \to \infty} \frac{\left(x - \frac{f'(x)}{f''(x)}\right)'}{x'} = \lim_{x \to \infty} \frac{f'(x) f'''(x)}{(f''(x))^2} = c.$$
Hence
$$\lim_{x \to \infty} \frac{f'(x)}{xf''(x)} = 1 - c.$$
By assumption, we therefore see that $c \leq 1$. Clearly, if $c \neq 1$, then

(1) $$\lim_{x \to \infty} \frac{xf''(x)}{f'(x)} = \frac{1}{1-c}.$$

Now we prove that $\lim_{x \to \infty} f(x) = +\infty$. By Taylor's formula we have
$$f(x + h) = f(x) + f'(x)h + f''(\zeta)\frac{h^2}{2}, \quad h > 0.$$

2.3. Taylor's Formula and L'Hospital's Rule

Hence $f(x + h) > f(x) + f'(x)h$. Letting $h \to +\infty$ shows that $\lim_{x \to \infty} f(x) = +\infty$. Now, by l'Hospital's rule again,

$$\lim_{x \to \infty} \frac{xf'(x)}{f(x)} = \lim_{x \to \infty} \frac{f'(x) + xf''(x)}{f'(x)} = 1 + \frac{1}{1-c},$$

which combined with (1) yields

$$\lim_{x \to \infty} \frac{f(x)f''(x)}{(f'(x))^2} = \lim_{x \to \infty} \frac{xf''(x)}{f'(x)} \cdot \frac{f(x)}{xf'(x)} = \frac{1}{2-c}.$$

2.3.43. For $x \neq 0$, by the Leibniz formula, we get

(1)
$$g^{(n)}(x) = \sum_{k=0}^{n} \binom{n}{k} f^{(k)}(x) \left(\frac{1}{x}\right)^{(n-k)}$$
$$= \sum_{k=0}^{n} (-1)^{n-k} \frac{n!}{k!} f^{(k)}(x) \frac{1}{x^{n+1-k}}.$$

Set $g(0) = f'(0)$. Then an application of l'Hospital's rule gives

$$g'(0) = \lim_{x \to 0} \frac{g(x) - f'(0)}{x} = \lim_{x \to 0} \frac{f(x) - xf'(0)}{x^2}$$
$$= \lim_{x \to 0} \frac{f'(x) - f'(0)}{2x} = \frac{f''(0)}{2}.$$

Thus $g'(0)$ exists. Now we show that g' is continuous at zero. By (1) and l'Hospital's rule,

$$\lim_{x \to 0} g'(x) = \lim_{x \to 0} \frac{f'(x) - g(x)}{x} = \lim_{x \to 0} \frac{xf'(x) - f(x)}{x^2}$$
$$= \lim_{x \to 0} \frac{xf''(x)}{2x} = g'(0).$$

So, g is in $C^1(-1, 1)$. Now we proceed by induction. Suppose that $g^{(n)}(0) = \frac{f^{(n+1)}(0)}{n+1}$ and $g \in C^n(-1, 1)$. Then by (1) and l'Hospital's rule we obtain

$$g^{(n+1)}(0) = \lim_{x \to 0} \frac{g^{(n)}(x) - g^{(n)}(0)}{x}$$

$$= \lim_{x \to 0} \frac{\sum_{k=0}^{n}(-1)^{n-k}\frac{n!}{k!}f^{(k)}(x)x^k - x^{n+1}g^{(n)}(0)}{x^{n+2}}$$

$$= \lim_{x \to 0} \frac{\sum_{k=0}^{n}(-1)^{n-k}\frac{n!}{k!}f^{(k)}(x)x^k - x^{n+1}\frac{f^{(n+1)}(0)}{n+1}}{x^{n+2}}$$

$$= \lim_{x \to 0} \left\{ \frac{n!(-1)^n f'(x) + \sum_{k=1}^{n}(-1)^{n-k}\frac{n!}{k!}f^{(k+1)}(x)x^k}{(n+2)x^{n+1}} \right.$$

$$+ \left. \frac{\sum_{k=1}^{n}(-1)^{n-k}\frac{n!}{(k-1)!}f^{(k)}(x)x^{k-1} - x^n f^{(n+1)}(0)}{(n+2)x^{n+1}} \right\}$$

$$= \lim_{x \to 0} \left\{ \frac{n!(-1)^n f'(x) + \sum_{k=1}^{n}(-1)^{n-k}\frac{n!}{k!}f^{(k+1)}(x)x^k}{(n+2)x^{n+1}} \right.$$

$$+ \left. \frac{\sum_{k=0}^{n-1}(-1)^{n-1-k}\frac{n!}{k!}f^{(k+1)}(x)x^k - x^n f^{(n+1)}(0)}{(n+2)x^{n+1}} \right\}$$

$$= \lim_{x \to 0} \frac{x^n(f^{(n+1)}(x) - f^{(n+1)}(0))}{(n+2)x^{n+1}} = \frac{f^{(n+2)}(0)}{n+2}.$$

Our task is now to show that $g^{(n+1)}$ is continuous at zero. By (1) and l'Hospital's rule again,

$$\lim_{x \to 0} g^{(n+1)}(x) = \lim_{x \to 0} \frac{\sum_{k=0}^{n+1}(-1)^{n+1-k}\frac{(n+1)!}{k!}f^{(k)}(x)x^k}{x^{n+2}} =$$

2.4. Convex Functions

$$\lim_{x \to 0} \frac{\sum_{k=0}^{n+1} \frac{(-1)^{n+1-k}(n+1)!}{k!} f^{(k+1)}(x) x^k + \sum_{k=1}^{n+1} \frac{(-1)^{n+1-k}(n+1)!}{(k-1)!} f^{(k)}(x) x^{k-1}}{(n+2) x^{n+1}}$$

$$= \lim_{x \to 0} \frac{f^{(n+2)}(x)}{n+2} = g^{(n+1)}(0).$$

Summing up the above, we see that the extended g defined above is in C^∞ on $(-1, 1)$, and $g^{(n)}(0) = \frac{f^{(n+1)}(0)}{n+1}$, $n = 0, 1, 2, \ldots$.

2.4. Convex Functions

2.4.1. Assume that f is convex on **I**. Then for $x_1 < x < x_2$ we have

(1) $$\frac{f(x) - f(x_1)}{x - x_1} \leq \frac{f(x_2) - f(x_1)}{x_2 - x_1}$$

(see (1) in the solution of 1.2.33). On the other hand, since

$$x = \frac{x_2 - x}{x_2 - x_1} x_1 + \frac{x - x_1}{x_2 - x_1} x_2,$$

we see that

$$f(x) \leq \frac{x_2 - x}{x_2 - x_1} f(x_1) + \frac{x - x_1}{x_2 - x_1} f(x_2),$$

and consequently,

$$\frac{f(x_2) - f(x_1)}{x_2 - x_1} \leq \frac{f(x_2) - f(x)}{x_2 - x}.$$

This combined with (1) gives

(2) $$\frac{f(x) - f(x_1)}{x - x_1} \leq \frac{f(x_2) - f(x)}{x_2 - x}.$$

Upon passage to the limit as $x \to x_1^+$ we see that

$$f'(x_1) \leq \frac{f(x_2) - f(x_1)}{x_2 - x_1}.$$

Similarly, letting $x \to x_2^-$ in (2) yields

$$f'(x_2) \geq \frac{f(x_2) - f(x_1)}{x_2 - x_1},$$

Consequently, $f'(x_1) \leq f'(x_2)$, which shows that f' is increasing.

Assume now that f' is increasing on **I**. Let $x_1 < x < x_2$. By the mean value theorem,

$$\frac{f(x)-f(x_1)}{x-x_1} = f'(\xi_1), \quad \frac{f(x_2)-f(x)}{x_2-x} = f'(\xi_2),$$

where $x_1 < \xi_1 < x < \xi_2 < x_2$. Therefore by the monotonicity of f' we get inequality (2). Now we show that (2) implies the convexity of f. To this end set $x = \lambda x_1 + (1-\lambda)x_2$, where $x_1 < x_2$ and $\lambda \in (0,1)$. Then $x \in (x_1, x_2)$,

$$x - x_1 = (1-\lambda)(x_2 - x_1) \quad \text{and} \quad x_2 - x = \lambda(x_2 - x_1).$$

So (2) gives $f(x) \leq \lambda f(x_1) + (1-\lambda)f(x_2)$. It is worth noting here that (2) is in fact equivalent to the convexity of f.

Note also that if f' is strictly increasing on **I**, then f is strictly convex on **I**.

2.4.2. It suffices to observe that the condition $f''(x) \geq 0$ for $x \in$ **I** is equivalent to the fact that f' increases, and to apply the result in the foregoing problem.

2.4.3. We proceed by induction. The condition with $n = 2$ is the definition of convexity of f on **I**. We therefore assume the inequality to be proved is true for some $n \geq 2$ and show that it is also true for $n+1$. Let $\lambda_1, \lambda_2, \ldots, \lambda_n, \lambda_{n+1}$ be nonnegative and such that $\lambda_1 + \lambda_2 + \cdots + \lambda_n + \lambda_{n+1} = 1$. Since $\lambda_n x_n + \lambda_{n+1} x_{n+1}$ can be written in the form $(\lambda_n + \lambda_{n+1})\left(\frac{\lambda_n}{\lambda_n + \lambda_{n+1}} x_n + \frac{\lambda_{n+1}}{\lambda_n + \lambda_{n+1}} x_{n+1}\right)$, by the induction hypothesis we get

$$f(\lambda_1 x_1 + \lambda_2 x_2 + \cdots + \lambda_{n+1} x_{n+1})$$
$$\leq \lambda_1 f(x_1) + \lambda_2 f(x_2)$$
$$+ \cdots + (\lambda_n + \lambda_{n+1}) f\left(\frac{\lambda_n}{\lambda_n + \lambda_{n+1}} x_n + \frac{\lambda_{n+1}}{\lambda_n + \lambda_{n+1}} x_{n+1}\right).$$

Now we only have to apply the definition of the convexity of f to the last summand.

2.4. Convex Functions

2.4.4. Since $\ln''(x) = -\frac{1}{x^2}$, the function $x \mapsto \ln x$ is concave on $(0, \infty)$. Thus

$$\ln\left(\frac{x^p}{p} + \frac{y^q}{q}\right) \geq \frac{1}{p}\ln x^p + \frac{1}{q}\ln y^q = \ln(xy).$$

2.4.5. Since $x \mapsto \ln x$ is concave on $(0, \infty)$, we get

$$\ln\left(\frac{x_1}{n} + \frac{x_2}{n} + \cdots + \frac{x_n}{n}\right) \geq \frac{1}{n}(\ln x_1 + \ln x_2 + \cdots + \ln x_n)$$
$$= \frac{1}{n}\ln(x_1 \cdot x_2 \cdots x_n).$$

2.4.6. The function $x \mapsto e^x$ is strictly convex on $\mathbb{R}$.

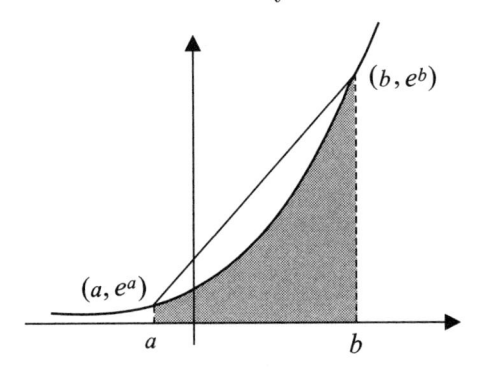

If, for example, $a < b$, then the area under the graph of $y = e^x$ from $x = a$ to $x = b$ is less than the area of the trapezoid with vertices $(a, 0), (b, 0), (a, e^a)$ and (b, e^b). Therefore

$$e^b - e^a = \int_a^b e^t dt < (b-a)\frac{e^a + e^b}{2}.$$

2.4.7. Consider the function given by $f(x) = x\ln x$, $x > 0$. Then $f''(x) = \frac{1}{x} > 0$. Thus f is convex. Consequently,

$$\frac{x+y}{2}\ln\frac{x+y}{2} \leq \frac{x}{2}\ln x + \frac{y}{2}\ln y.$$

2.4.8. Use the fact that $x \mapsto x^\alpha$, $\alpha > 1$, is convex on $(0, \infty)$.

2.4.9.

(a) The function $f(x) = \ln\left(\frac{1}{x} + 1\right)$ is convex on $(0, \infty)$, because $f''(x) > 0$ on that interval. So the result follows from the Jensen inequality (see, e.g., 2.4.3).

Observe that if $p_k = \frac{1}{n}$ for $k = 1, 2, \ldots, n$, and if $\sum_{k=1}^{n} x_k = 1$, then we get the inequality given in I, 1.2.43(a).

(b) It is enough to apply the Jensen inequality to the function

$$f(x) = \ln\left(\frac{1+x}{1-x}\right), \quad 0 < x < 1.$$

Observe that if $p_k = \frac{1}{n}$ for $k = 1, 2, \ldots, n$, and if $\sum_{k=1}^{n} x_k = 1$, then we get the inequality given I,1.2.45.

2.4.10.

(a) Define $f(x) = \ln(\sin x)$ for $x \in (0, \pi)$. Since $f''(x) = -\frac{1}{\sin^2 x} < 0$, we see that f is concave on $(0, \pi)$. Now it suffices to apply the Jensen inequality (see, e.g., 2.4.3) to $-f$.

(b) Consider the function defined by

$$f(x) = \ln(\sin x) - \ln x, \quad x \in (0, \pi),$$

and observe that $f''(x) = -\frac{1}{\sin^2 x} + \frac{1}{x^2} < 0$, and apply the Jensen inequality (see, e.g., 2.4.3) to $-f$.

2.4.11. Note that the function $f(x) = \left(x + \frac{1}{x}\right)^a$ is convex on $(0, \infty)$, because

$$f''(x) = a\left(x + \frac{1}{x}\right)^{a-2}\left((a-1)\left(1 - \frac{1}{x^2}\right)^2 + \frac{2}{x^3}\left(x + \frac{1}{x}\right)\right) > 0.$$

By Jensen's inequality (see, e.g., 2.4.3),

$$\left(\frac{n^2+1}{n}\right)^a = \left(\frac{1}{n}\sum_{k=1}^{n} x_k + \frac{1}{\frac{1}{n}\sum_{k=1}^{n} x_k}\right)^a \leq \frac{1}{n}\sum_{k=1}^{n}\left(x_k + \frac{1}{x_k}\right)^a.$$

2.4. Convex Functions

Hence
$$\sum_{k=1}^{n}\left(x_k + \frac{1}{x_k}\right)^a \geq \frac{(n^2+1)^a}{n^{a-1}}.$$

2.4.12. Applying Jensen's inequality to $x \mapsto -\ln x$, $x > 0$, we get

$$\frac{1}{n}\left(\ln 1 + \ln\frac{2^2-1}{2} + \ln\frac{2^3-1}{2^2} + \cdots + \ln\frac{2^n-1}{2^{n-1}}\right)$$
$$\leq \ln\left(\frac{1}{n}\left(1 + \frac{2^2-1}{2} + \cdots + \frac{2^n-1}{2^{n-1}}\right)\right) = \ln\left(2 - \frac{2}{n} + \frac{1}{n2^{n-1}}\right).$$

2.4.13.

(a) Applying Jensen's inequality to $f(x) = \frac{1}{x}$, $x > 0$, we obtain

$$\frac{1}{\frac{1}{n}x_1 + \frac{1}{n}x_2 + \cdots + \frac{1}{n}x_n} \leq \frac{1}{n}\cdot\frac{1}{x_1} + \frac{1}{n}\cdot\frac{1}{x_2} + \cdots + \frac{1}{n}\cdot\frac{1}{x_n}.$$

Hence
$$\frac{n^2}{x_1 + x_2 + \cdots + x_n} \leq \frac{1}{x_1} + \frac{1}{x_2} + \cdots + \frac{1}{x_n}.$$

(b) By Jensen's inequality applied to $f(x) = -\ln x$, $x > 0$, we have

$$\ln(x_1^{\alpha_1} x_2^{\alpha_2} \cdots x_n^{\alpha_n}) = \alpha_1 \ln x_1 + \alpha_2 \ln x_2 + \cdots + \alpha_n \ln x_n$$
$$\leq \ln(\alpha_1 x_1 + \alpha_2 x_2 + \cdots + \alpha_n x_n).$$

Consequently,

(1) $\qquad x_1^{\alpha_1} x_2^{\alpha_2} \cdots x_n^{\alpha_n} \leq \alpha_1 x_1 + \alpha_2 x_2 + \cdots + \alpha_n x_n.$

If in (1) we replace x_k by $\frac{1}{x_k}$, we get the left inequality.

(c) If one of x_k or y_k is zero, then the inequality is obvious. So we can assume that $x_k, y_k > 0$ for $k = 1, 2, \ldots, n$. Then the inequality in question can be rewritten as

$$\frac{x_1^{\alpha_1} x_2^{\alpha_2} \cdots x_n^{\alpha_n} + y_1^{\alpha_1} y_2^{\alpha_2} \cdots y_n^{\alpha_n}}{(x_1+y_1)^{\alpha_1}(x_2+y_2)^{\alpha_2}\cdots(x_n+y_n)^{\alpha_n}} \leq 1.$$

Now by (b) we get

$$\frac{x_1^{\alpha_1} x_2^{\alpha_2} \cdots x_n^{\alpha_n} + y_1^{\alpha_1} y_2^{\alpha_2} \cdots y_n^{\alpha_n}}{(x_1+y_1)^{\alpha_1}(x_2+y_2)^{\alpha_2}\cdots(x_n+y_n)^{\alpha_n}}$$
$$\leq \alpha_1 \frac{x_1}{x_1+y_1} + \cdots + \alpha_n \frac{x_n}{x_n+y_n} + \alpha_1 \frac{y_1}{x_1+y_1}$$
$$+ \cdots + \alpha_n \frac{y_n}{x_n+y_n} = 1.$$

(d) The inequality follows from (c) by induction on m.

2.4.14. Suppose, contrary to our claim, that f is not constant on $\mathbb{R}$. Then there exist $x_1 < x_2$ such that $f(x_1) < f(x_2)$ or $f(x_1) > f(x_2)$. Let x be such that $x_1 < x_2 < x$. Then we have

$$f(x_2) = f\left(\frac{x-x_2}{x-x_1} x_1 + \frac{x_2-x_1}{x-x_1} x\right) \leq \frac{x-x_2}{x-x_1} f(x_1) + \frac{x_2-x_1}{x-x_1} f(x).$$

Thus

(1) $$f(x) \geq \frac{x-x_1}{x_2-x_1} f(x_2) - \frac{x-x_2}{x_2-x_1} f(x_1).$$

If $f(x_2) = f(x_1) + A$ with some positive A, then (1) implies that $f(x) > A\frac{x-x_1}{x_2-x_1} + f(x_1)$, contrary to the assumption that f is bounded above. Similarly, if $f(x_1) > f(x_2)$, then $f(x_1) = A + f(x_2)$ with some $A > 0$. Now taking $x < x_1 < x_2$ we get

$$f(x) \geq A\frac{x_2-x}{x_2-x_1} + f(x_2),$$

again contrary to the assumption that f is bounded above.

2.4.15. No. It is enough to consider $f(x) = e^{-x}$, $x \in (a, \infty)$; and $f(x) = e^x$, $x \in (-\infty, a)$.

2.4.16. Suppose that f is not monotonic. Then there are $a < x_1 < x_2 < x_3 < b$ such that

$$f(x_1) > f(x_2) \quad \text{and} \quad f(x_2) < f(x_3),$$

or

$$f(x_1) < f(x_2) \quad \text{and} \quad f(x_2) > f(x_3).$$

Since f is convex, $f(x_2) \leq \max\{f(x_1), f(x_3)\}$. So the latter case cannot hold. The continuity of f (see, e.g., 1.2.33) implies that there

2.4. Convex Functions

is $c \in [x_1, x_3]$ such that $f(c) = \min\{f(x) : x \in [x_1, x_3]\}$. By the convexity of f we see that $f(x_1) \leq \max\{f(x), f(c)\}$ for $x \in (a, x_1)$. Consequently, since $f(c) \leq f(x_1)$, we have $f(x_1) \leq f(x)$. It then follows that if $x, y \in (a, c]$, then

- $x < y < x_1$ implies $f(y) \leq \max\{f(x), f(x_1)\} = f(x)$,
- $x < x_1 \leq y$ implies $f(y) \leq \max\{f(x_1), f(c)\} = f(x_1) \leq f(x)$,
- $x_1 \leq x < y$ implies $f(y) \leq \max\{f(x), f(c)\} = f(x)$.

So we have shown that f is decreasing on $(a, c]$. In an entirely similar manner one can show that f increases on $[c, b)$.

2.4.17. This is an immediate consequence of the result in the preceding problem.

2.4.18. Since f is bounded, by the foregoing result, one-sided limits of f at a and b exist and are finite. Consequently, the assertion follows from 1.2.33 and 1.5.7.

2.4.19. Let $x_1 < x_2$ be two points in (a, b). Then for $a < y < x_1 < x < x_2$ we have (see (1) and (2) in the solution of 2.4.1)

$$(*) \qquad \frac{f(y) - f(x_1)}{y - x_1} \leq \frac{f(x) - f(x_1)}{x - x_1} \leq \frac{f(x_2) - f(x_1)}{x_2 - x_1}.$$

This means that the function $x \mapsto \frac{f(x) - f(x_1)}{x - x_1}$ is increasing and bounded below in (x_1, b). Consequently, the right-hand derivative $f'_+(x_1)$ exists and

$$(**) \qquad f'_+(x_1) \leq \frac{f(x_2) - f(x_1)}{x_2 - x_1}.$$

Now note that for $x_1 < x_2 < t < b$,

$$\frac{f(x_2) - f(x_1)}{x_2 - x_1} \leq \frac{f(t) - f(x_2)}{t - x_2},$$

which gives

$$\frac{f(x_2) - f(x_1)}{x_2 - x_1} \leq f'_+(x_2).$$

This combined with $(**)$ shows that $f'_+(x_1) \leq f'_+(x_2)$. To show that the left-hand derivative exists and is increasing on (a, b), similar reasoning can be applied. Moreover, it follows from $(*)$ that for

$x_1 \in (a,b)$ we have $f'_-(x_1) \leq f'_+(x_1)$. Recall that, by (2) in the solution of 2.4.1, if $x_1 < x < x_2$, then

$$\frac{f(x) - f(x_1)}{x - x_1} \leq \frac{f(x_2) - f(x)}{x_2 - x}.$$

This implies that

$$f'_+(x_1) \leq f'_-(x_2).$$

Summing up, we get

$$f'_-(x_1) \leq f'_+(x_1) \leq f'_-(x_2) \leq f'_+(x_2) \quad \text{for} \quad x_1 < x_2.$$

This shows that if one of the one-sided derivatives is continuous at a point in (a,b), then both derivatives are equal at this point. Since a monotone function can have only a countable number of discontinuities (see, e.g., 1.2.29), the one sided derivatives are equal except on a countable set.

An analogous statement is also true for concave functions.

2.4.20. Since f' is strictly increasing, the inverse function $(f')^{-1}$ exists and

$$\xi(x) = (f')^{-1}\left(\frac{f(b+x) - f(a-x)}{b - a + 2x}\right).$$

It then follows that ξ is differentiable on $(0, \infty)$. Upon differentiating the equality given in the problem, we obtain

(1) $$\frac{f'(b+x) + f'(a-x) - 2f'(\xi)}{b - a + 2x} = f''(\xi)\xi'(x).$$

Now note that $f'' \geq 0$, and since f'' is strictly increasing, f' is strictly convex (see, e.g., 2.4.1). Hence (see the figure below)

$$(b - a + 2x)\frac{f'(b+x) + f'(a-x)}{2} > \int_{a-x}^{b+x} f'(t)dt = f(b+x) - f(a-x).$$

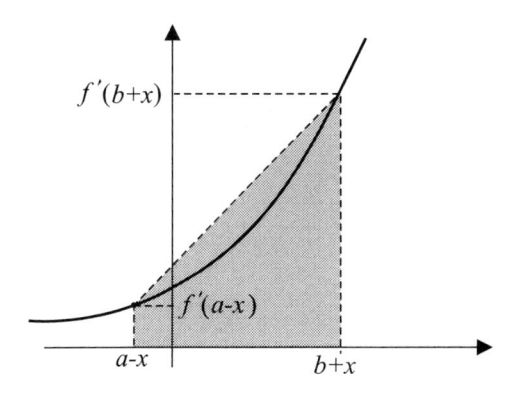

2.4. Convex Functions

Thus
$$\frac{f'(b+x)+f'(a-x)}{2} > f'(\xi).$$
Consequently, by (1), we see that $\xi'(x) > 0$ for $x > 0$.

2.4.21. Without loss of generality we may assume that $\sum\limits_{i=1}^{n}|x_i| > 0$ and $\sum\limits_{i=1}^{n}|y_i| > 0$. By 2.4.4 we get

$$\frac{|x_i|}{\left(\sum\limits_{i=1}^{n}|x_i|^p\right)^{\frac{1}{p}}} \cdot \frac{|y_i|}{\left(\sum\limits_{i=1}^{n}|y_i|^q\right)^{\frac{1}{q}}} \leq \frac{1}{p}\left(\frac{|x_i|}{\left(\sum\limits_{i=1}^{n}|x_i|^p\right)^{\frac{1}{p}}}\right)^p + \frac{1}{q}\left(\frac{|y_i|}{\left(\sum\limits_{i=1}^{n}|y_i|^q\right)^{\frac{1}{q}}}\right)^q.$$

Upon summing both sides of this inequality from $i=1$ to $i=n$, we obtain

$$\frac{\sum\limits_{i=1}^{n}|x_iy_i|}{\left(\sum\limits_{i=1}^{n}|x_i|^p\right)^{\frac{1}{p}}\left(\sum\limits_{i=1}^{n}|y_i|^q\right)^{\frac{1}{q}}}$$
$$\leq \frac{1}{p}\sum_{i=1}^{n}\left(\frac{|x_i|}{\left(\sum\limits_{i=1}^{n}|x_i|^p\right)^{\frac{1}{p}}}\right)^p + \frac{1}{q}\sum_{i=1}^{n}\left(\frac{|y_i|}{\left(\sum\limits_{i=1}^{n}|y_i|^q\right)^{\frac{1}{q}}}\right)^q$$

$$= \frac{1}{p}\frac{\sum\limits_{i=1}^{n}|x_i|^p}{\sum\limits_{i=1}^{n}|x_i|^p} + \frac{1}{q}\frac{\sum\limits_{i=1}^{n}|y_i|^q}{\sum\limits_{i=1}^{n}|y_i|^q} = \frac{1}{p} + \frac{1}{q} = 1.$$

2.4.22. For $p = 1$ the inequality is obvious. For $p > 1$, let q be such that $\frac{1}{p} + \frac{1}{q} = 1$. So, $q = \frac{p}{p-1}$. Hence

$$\sum_{i=1}^{n}|x_i + y_i|^p = \sum_{i=1}^{n}|x_i + y_i||x_i + y_i|^{p-1}$$
$$\leq \sum_{i=1}^{n}|x_i||x_i + y_i|^{p-1} + \sum_{i=1}^{n}|y_i||x_i + y_i|^{p-1}$$
$$\leq \left(\sum_{i=1}^{n}|x_i|^p\right)^{\frac{1}{p}}\left(\sum_{i=1}^{n}|x_i + y_i|^{(p-1)q}\right)^{\frac{1}{q}}$$
$$+ \left(\sum_{i=1}^{n}|y_i|^p\right)^{\frac{1}{p}}\left(\sum_{i=1}^{n}|x_i + y_i|^{(p-1)q}\right)^{\frac{1}{q}}$$
$$= \left(\left(\sum_{i=1}^{n}|x_i|^p\right)^{\frac{1}{p}} + \left(\sum_{i=1}^{n}|y_i|^p\right)^{\frac{1}{p}}\right)\left(\sum_{i=1}^{n}|x_i + y_i|^p\right)^{\frac{1}{q}}.$$

Consequently,

$$\left(\sum_{i=1}^{n}|x_i + y_i|^p\right)^{\frac{1}{p}} \leq \left(\sum_{i=1}^{n}|x_i|^p\right)^{\frac{1}{p}} + \left(\sum_{i=1}^{n}|y_i|^p\right)^{\frac{1}{p}}.$$

2.4.23. By Hölder's inequality we have

$$\sum_{n=1}^{N}\frac{|a_n|}{n^{\frac{4}{5}}} \leq \left(\sum_{n=1}^{N}a_n^4\right)^{\frac{1}{4}}\left(\sum_{n=1}^{N}\frac{1}{n^{\frac{16}{15}}}\right)^{\frac{3}{4}}.$$

2.4.24. Set $s_1 = x_1 + x_2 + \cdots + x_n$, $s_2 = y_1 + y_2 + \cdots + y_n$ and $S = (s_1^p + s_2^p)^{\frac{1}{p}}$. Then

$$S^p = s_1^p + s_2^p = (x_1 s_1^{p-1} + y_1 s_2^{p-1}) + (x_2 s_1^{p-1} + y_2 s_2^{p-1})$$
$$+ \cdots + (x_n s_1^{p-1} + y_n s_2^{p-1}).$$

2.4. Convex Functions

It then follows by Hölder's inequality that

$$S^p \leq (x_1^p + y_1^p)^{\frac{1}{p}}(s_1^p + s_2^p)^{\frac{1}{q}} + (x_2^p + y_2^p)^{\frac{1}{p}}(s_1^p + s_2^p)^{\frac{1}{q}}$$
$$+ \cdots + (x_n^p + y_n^p)^{\frac{1}{p}}(s_1^p + s_2^p)^{\frac{1}{q}}$$
$$= S^{\frac{p}{q}}\left((x_1^p + y_1^p)^{\frac{1}{p}} + (x_2^p + y_2^p)^{\frac{1}{p}} + \cdots + (x_n^p + y_n^p)^{\frac{1}{p}}\right),$$

which implies the desired inequality.

2.4.25. Set

$$s_i = \sum_{j=1}^{m} x_{i,j} \quad \text{and} \quad S = \left(\sum_{i=1}^{n} s_i^p\right)^{\frac{1}{p}}.$$

Then by Hölder's inequality,

$$S^p = \sum_{i=1}^{n} s_i s_i^{p-1} = \sum_{i=1}^{n}\sum_{j=1}^{m} x_{i,j} s_i^{p-1} = \sum_{j=1}^{m}\sum_{i=1}^{n} x_{i,j} s_i^{p-1}$$
$$\leq \sum_{j=1}^{m}\left(\sum_{i=1}^{n} x_{i,j}^p\right)^{\frac{1}{p}}\left(\sum_{i=1}^{n} s_i^p\right)^{\frac{p-1}{p}} = S^{p-1}\sum_{j=1}^{m}\left(\sum_{i=1}^{n} x_{i,j}^p\right)^{\frac{1}{p}},$$

which implies the desired inequality.

2.4.26. Let $x, y \in \mathbf{I}$ and assume that $x < y$. For $n = 0, 1, 2, \ldots$, set $\mathbf{T}_n = \left\{\frac{i}{2^n} : i = 0, 1, \ldots, 2^n\right\}$. We show by induction that for $n = 0, 1, \ldots$ and for $s \in \mathbf{T}_n$,

$$f((1-s)x + sy) \leq (1-s)f(x) + sf(y).$$

Clearly, if $n = 0$, then $s = 0$ or $s = 1$, and therefore the above inequality is obvious. Assuming the inequality to hold for an arbitrarily chosen $n \in \{0, 1, \ldots\}$ and for $s \in \mathbf{T}_n$, we will prove it for $n + 1$. Suppose that $s \in \mathbf{T}_{n+1}$. Clearly, it suffices to consider the case $s \notin \mathbf{T}_n$. Since there exist $\xi, \eta \in \mathbf{T}_n$ such that $s = \frac{\xi + \eta}{2}$, we see that

$$(1-s)x + sy = \left(1 - \frac{\xi + \eta}{2}\right)x + \frac{\xi + \eta}{2}y$$
$$= \frac{(1-\xi) + (1-\eta)}{2}x + \frac{\xi + \eta}{2}y$$
$$= \frac{((1-\xi)x + \xi y) + ((1-\eta)x + \eta y)}{2}.$$

By the midpoint-convexity of f,
$$f((1-s)x+sy) \leq \frac{f((1-\xi)x+\xi y)+f((1-\eta)x+\eta y)}{2}.$$

By the induction hypothesis,
$$\begin{aligned}
f((1-s)x+sy) &\leq \frac{(1-\xi)f(x)+\xi f(y)+(1-\eta)f(x)+\eta f(y)}{2} \\
&= \left(1-\frac{\xi+\eta}{2}\right)f(x)+\frac{\xi+\eta}{2}f(y) \\
&= (1-s)f(x)+sf(y).
\end{aligned}$$

Let t be an arbitrarily chosen point in $[0,1]$. Since the set
$$\mathbf{T} = \bigcup_{n=0}^{\infty} \mathbf{T}_n$$
is dense in $[0,1]$, there is a sequence $\{s_n\}$ of points in $\mathbf{T}$ such that $t = \lim_{n\to\infty} s_n$. Hence, by the continuity of f,
$$\begin{aligned}
f((1-t)x+ty) &= \lim_{n\to\infty} f((1-s_n)x+s_n y) \\
&\leq \lim_{n\to\infty}((1-s_n)f(x)+s_n f(y)) \\
&= (1-t)f(x)+tf(y).
\end{aligned}$$

2.4.27. There are functions $f:\mathbb{R}\to\mathbb{R}$ which are additive and not continuous (see, e.g., 1.6.31). If f is such a function, then for any $x\in\mathbb{R}$,
$$f(x)=f\left(\frac{x}{2}+\frac{x}{2}\right)=f\left(\frac{x}{2}\right)+f\left(\frac{x}{2}\right)=2f\left(\frac{x}{2}\right).$$

Thus $f\left(\frac{x}{2}\right)=\frac{1}{2}f(x)$. Consequently, for $x,y\in\mathbb{R}$,
$$f\left(\frac{x+y}{2}\right)=f\left(\frac{x}{2}+\frac{y}{2}\right)=\frac{1}{2}f(x)+\frac{1}{2}f(y)=\frac{f(x)+f(y)}{2}.$$

If f were convex on $\mathbb{R}$, then it would be continuous (see, e.g., 1.2.33), a contradiction.

2.4. Convex Functions

2.4.28. Suppose, for example, that $x < y$. For $t \in (0,1)$ set $z = (1-t)x + ty$. Then $x < z < y$ and there are $a \in (x,z)$ and $b \in (z,y)$ such that $z = \frac{a+b}{2}$. Analogously, there are $t_a \in (0,t)$ and $t_b \in (t,1)$ such that

$$a = (1-t_a)x + t_a y \quad \text{and} \quad b = (1-t_b)x + t_b y.$$

Since $z = \frac{a+b}{2}$, we have $t = \frac{t_a+t_b}{2}$. We know, by the result in 2.4.26, that f is convex on **I**. Now we show that f is strictly convex. We have

$$\begin{aligned}
f((1-t)x + ty) = f(z) &< \frac{f(a) + f(b)}{2} \\
&= \frac{f((1-t_a)x + t_a y) + f((1-t_b)x + t_b y)}{2} \\
&\leq \frac{(1-t_a)f(x) + t_a f(y) + (1-t_b)f(x) + t_b f(y)}{2} \\
&= \left(1 - \frac{t_a + t_b}{2}\right) f(x) + \frac{t_a + t_b}{2} f(y) \\
&= (1-t)f(x) + tf(y).
\end{aligned}$$

2.4.29. Since f is continuous on **I** (see, e.g., 1.2.33), it is locally bounded. Let x_0 be in **I** and let $\varepsilon > 0$ be so small that the interval $[x_0 - 2\varepsilon, x_0 + 2\varepsilon]$ is contained in **I**. By the local boundedness of f there is $M > 0$ such that

(1) $\qquad |f(x)| \leq M \quad \text{for} \quad x \in [x_0 - 2\varepsilon, x_0 + 2\varepsilon].$

Take two distinct points x_1 and x_2 in $[x_0 - \varepsilon, x_0 + \varepsilon]$. Then $x_3 = x_2 + \frac{\varepsilon}{|x_2 - x_1|}(x_2 - x_1)$ is in $[x_0 - 2\varepsilon, x_0 + 2\varepsilon]$ and

$$x_2 = \frac{\varepsilon}{|x_2 - x_1| + \varepsilon} x_1 + \frac{|x_2 - x_1|}{|x_2 - x_1| + \varepsilon} x_3.$$

Since f is convex, we see that

$$f(x_2) \leq \frac{\varepsilon}{|x_2 - x_1| + \varepsilon} f(x_1) + \frac{|x_2 - x_1|}{|x_2 - x_1| + \varepsilon} f(x_3).$$

Consequently,
$$f(x_2) - f(x_1) \le \frac{|x_2 - x_1|}{|x_2 - x_1| + \varepsilon}(f(x_3) - f(x_1))$$
$$\le \frac{|x_2 - x_1|}{\varepsilon}|f(x_3) - f(x_1)|,$$

which combined with (1) shows that $f(x_2) - f(x_1) \le \frac{2M}{\varepsilon}|x_2 - x_1|$. Finally, since the roles of x_1 and x_2 can be interchanged, we get $|f(x_2) - f(x_1)| \le \frac{2M}{\varepsilon}|x_2 - x_1|$.

2.4.30. Let $x_1 < x_2$ be chosen arbitrarily in $(0, \infty)$. If $0 < x < x_1$, then
$$x_1 = \frac{x_2 - x_1}{x_2 - x}x + \frac{x_1 - x}{x_2 - x}x_2.$$
It follows from the convexity of f that
$$f(x_1) \le \frac{x_2 - x_1}{x_2 - x}f(x) + \frac{x_1 - x}{x_2 - x}f(x_2).$$
Upon passage to the limit as $x \to 0^+$, we see that
$$f(x_1) \le \frac{x_1}{x_2}f(x_2).$$

2.4.31.

(a) It follows from the monotonicity of $x \mapsto \frac{f(x)}{x}$ that for $x_1, x_2 \ge 0$,
$$f(x_1 + x_2) = x_1\frac{f(x_1 + x_2)}{x_1 + x_2} + x_2\frac{f(x_1 + x_2)}{x_1 + x_2} \le f(x_1) + f(x_2).$$

(b) Suppose that $0 < a < b$, and set $p = \frac{a}{b}$ and $q = 1 - p$. By the convexity and subadditivity of f we see that
$$f(b) = f(pa + q(a + b)) \le pf(a) + qf(a + b)$$
$$\le pf(a) + q(f(a) + f(b)) = f(a) + \left(1 - \frac{a}{b}\right)f(b).$$
Thus
$$\frac{f(b)}{b} \le \frac{f(a)}{a}.$$

2.4. Convex Functions

2.4.32. Assume, contrary to our claim, that f is neither strictly convex nor concave. Then there are points $\alpha < \beta$ in (a,b) such that the line through $(\alpha, f(\alpha))$ and $(\beta, f(\beta))$ meets the graph of f at a point $(\gamma, f(\gamma))$, where $\alpha < \gamma < \beta$. By assumption, there are unique $\zeta_1 \in (\alpha, \gamma)$ and $\zeta_2 \in (\gamma, \beta)$ such that

$$\frac{f(\gamma) - f(\alpha)}{\gamma - \alpha} = f'(\zeta_1) \quad \text{and} \quad \frac{f(\beta) - f(\gamma)}{\beta - \gamma} = f'(\zeta_2).$$

Since $(\alpha, f(\alpha))$, $(\beta, f(\beta))$ and $(\gamma, f(\gamma))$ are collinear, we see that $f'(\zeta_1) = f'(\zeta_2)$, which contradicts the hypothesis.

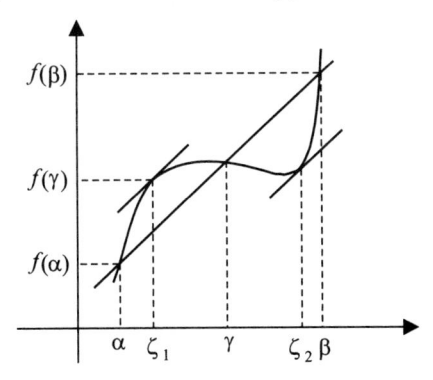

2.4.33. Note first that the so-called *Dini derivatives*

$$D^+ f(x) = \overline{\lim_{t \to 0^+}} \frac{f(x+t) - f(x)}{t}, \quad D_+ f(x) = \underline{\lim_{t \to 0^+}} \frac{f(x+t) - f(x)}{t},$$

$$D^- f(x) = \overline{\lim_{t \to 0^-}} \frac{f(x+t) - f(x)}{t}, \quad D_- f(x) = \underline{\lim_{t \to 0^-}} \frac{f(x+t) - f(x)}{t}$$

always exist (finite or infinite). Moreover, since g_d is differentiable, we get (see, e.g., 1.4.10)

$$D^+ f(x+d) = D^+ f(x) + g'_d(x), \quad D_+ f(x+d) = D_+ f(x) + g'_d(x),$$
$$D^- f(x+d) = D^- f(x) + g'_d(x), \quad D_- f(x+d) = D_- f(x) + g'_d(x)$$

for $x \in \mathbb{R}$. So each of the Dini derivatives of f at $x + d$ is a translate of the corresponding derivative at x. Next, if $a < b$ are arbitrarily chosen, we set $m = (f(b) - f(a))/(b-a)$ and define $F(x) = f(x) - m(x-a)$. Then $F(a) = F(b) = f(a)$, and therefore F attains its maximum or minimum value on $[a,b]$ at some point $c \in (a,b)$. We may assume without loss of generality that $F(c)$ is the maximum of F. So if $c+t \in (a,b)$, then $F(c+t) \leq F(c)$, or in other words, $f(c+t) - f(c) \leq mt$. This implies that $D^+f(c) \leq m \leq D_-f(c)$. Since each of the Dini derivatives of f at x is a translate of the corresponding derivative at c, we see that $D^+f(x) \leq D_-f(x)$ for all x. If f is concave on $[a,b]$, then f is differentiable except on an at most countable set (see, e.g., 2.4.19). It then follows that f is differentiable on (a,b). If f is not concave on $[a,b]$, then f attains also its minimum value on $[a,b]$ at some point in the open interval (a,b). Then, as above, one can show that $D^-f(x) \leq D_+f(x)$ for all x. Consequently, $D^+f(x) \leq D_-f(x) \leq D^-f(x) \leq D_+f(x)$ for all x. So the differentiability of f on $\mathbb{R}$ is proved.

Our task is now to show that f' is continuous. Assume, contrary to our claim, that there is x_0 at which f' is not continuous. Then there is a sequence $\{z_n\}$ converging to x_0 such that $\{f'(z_n)\}$ converges to $f'(x_0) + r$ with some $r \neq 0$, or $\{f'(z_n)\}$ is unbounded. In the latter case, we can find a sequence $\{y_n\}$ such that $\{f'(y_n)\}$ diverges, say, to $+\infty$. Then
$$\frac{f(x_0) - f(y_n)}{x_0 - y_n} = f'(y_n) + o(1),$$
and upon passage to the limit as $n \to \infty$, we get $f'(x_0) = +\infty$, a contradiction. In the former case, since f' enjoys the intermediate value property (see, e.g., 2.2.31), there is a sequence $\{y_n\}$ such that $f'(y_n) = f'(x_0) + r/2$. Clearly, we can assume that the sequence approaches x_0 from one side, say, from above. By the hypothesis, for every x we can find such a sequence with the same r. Indeed, since $x = x_0 + (x - x_0) = x_0 + d$ and
$$f'(z_n + d) - g'_d(z_n) = f'(z_n) = f'(x_0) + r = f'(x_0 + d) - g'_d(x_0) + r,$$
upon passage to the limit,
$$\lim_{n \to \infty} f'(z_n + d) = f'(x) + r.$$

By the intermediate value property of f' again, there is $\{\tilde{z}_n\}$ such that $f'(\tilde{z}_n) = f'(x) + r/2$. Now we construct a sequence $\{x_n\}$ as follows. Let x be arbitrarily chosen and let x_1 be such that $x <$

2.4. Convex Functions

$x_1 < x + 2^{-1}$ and $f'(x_1) = f'(x) + r/2$. Next let x_2 be such that $x_1 < x_2 < x_1 + 2^{-2}$ and $f'(x_2) = f'(x_1) + r/2$. Continuing this procedure, we get a sequence $\{x_n\}$ with

$$x_n < x_{n+1} < x_n + 2^{-n} \quad \text{and} \quad f'(x_{n+1}) = f'(x_n) + \frac{r}{2}.$$

Hence the sequence converges, say, to a. Moreover, it follows from the last equality that $f'(x_n) = f'(x_1) + (n-1)r/2$. Consequently, $\{f'(x_n)\}$ diverges to $+\infty$ or $-\infty$, contradicting differentiability of f at a. This ends the proof of the continuity of f'. Thus f' satisfies the same hypothesis as f, so it too is continuously differentiable, and by induction so are all derivatives.

2.4.34. For $n = 2$ we get the obvious equality. So, assume that $n > 2$ and that $\{a_n\}$ is a decreasing sequence. Set

$$S_n = (f(a_n)a_1 - f(a_1)a_n) + \sum_{k=1}^{n-1}(f(a_k)a_{k+1} - f(a_{k+1})a_k).$$

Our task is to show that $S_n \geq 0$. Since f is convex, we see that

$$f(a_n) = f\left(\frac{a_n - a_{n+1}}{a_1 - a_{n+1}}a_1 + \frac{a_1 - a_n}{a_1 - a_{n+1}}a_{n+1}\right)$$

$$\leq \frac{a_n - a_{n+1}}{a_1 - a_{n+1}}f(a_1) + \frac{a_1 - a_n}{a_1 - a_{n+1}}f(a_{n+1}).$$

Hence

$$(a_{n+1} - a_1)f(a_n) + (a_n - a_{n+1})f(a_1) + (a_1 - a_n)f(a_{n+1}) \geq 0,$$

which means that $S_{n+1} - S_n \geq 0$. Consequently, $S_n \geq S_2 = 0$.

2.4.35 [M. Kuczma, A. Smajdor, Amer. Math. Monthly 74 (1967), 401-402]. Since f is strictly increasing and $a < f(x) < x$, we get

$$a < f^{n+1}(x) < f^n(x) < x \quad \text{for} \quad n \in \mathbb{N} \quad \text{and} \quad x \in (a,b).$$

Consequently, the sequence $\{f^n(x)\}$ converges to an l (the case $l = -\infty$ is possible). Now we show that $l = a$. To this end, recall that, by the result in 1.2.33, each f^n is continuous. So, if $l \in (a,b)$, then we would get

$$f(l) = f\left(\lim_{n \to \infty} f^n(x)\right) = \lim_{n \to \infty} f^{n+1}(x) = l,$$

which would contradict the assumption $f(x) < x$ for $x \in (a,b)$. Therefore $l = a$ for each $x \in (a,b)$. By 2.4.19 the right-hand derivative of f exists and is decreasing on (a,b). So, if $a < t_1 < t_0 < b$, then (see the solution of 2.4.19)

(1) $$f'_+(t_0) \leq \frac{f(t_1) - f(t_0)}{t_1 - t_0} \leq f'_+(t_1).$$

Taking $t_0 = f^n(x)$ and $t_1 = f^{n+1}(x)$, we get

$$f'_+(f^n(x)) \leq \frac{f^{n+2}(x) - f^{n+1}(x)}{f^{n+1}(x) - f^n(x)} \leq f'_+(f^{n+1}(x)).$$

It then follows from $\lim_{x \to a^+} f'_+(x) = 1$ that

$$\lim_{n \to \infty} \frac{f^{n+2}(x) - f^{n+1}(x)}{f^{n+1}(x) - f^n(x)} = 1,$$

and consequently, for $k \in \mathbb{N}$,

(2) $$\lim_{n \to \infty} \frac{f^{n+k+1}(x) - f^{n+k}(x)}{f^{n+1}(x) - f^n(x)}$$
$$= \lim_{n \to \infty} \prod_{i=0}^{k-1} \frac{f^{n+i+2}(x) - f^{n+i+1}(x)}{f^{n+i+1}(x) - f^{n+i}(x)} = 1.$$

Since f'_+ decreases, the equality $\lim_{x \to a^+} f'_+(x) = 1$ implies $f'_+(x) \leq 1$. Thus by (1) the function $x \mapsto f(x) - x$ decreases on (a,b). Since $f(v) - v < 0$,

$$\frac{f(u) - u}{f(v) - v} \geq 1 \quad \text{for} \quad v < u, \; u,v \in (a,b).$$

Assume that $a < y < x < b$, and put $u = f^n(x)$ and $v = f^n(y)$. Then

$$\frac{f^{n+1}(x) - f^n(x)}{f^{n+1}(y) - f^n(y)} \geq 1.$$

On the other hand, there is $k \in \mathbb{N}$ such that $f^k(x) < y < x$. This implies that $f^{n+k}(x) < f^n(y)$. Since the function $x \mapsto f(x) - x$ decreases, we see that

$$f^{n+1}(y) - f^n(y) \leq f^{n+k+1}(x) - f^{n+k}(x).$$

2.5. Applications of Derivatives

Consequently,

$$1 \leq \frac{f^{n+1}(x) - f^n(x)}{f^{n+1}(y) - f^n(y)} \leq \frac{f^{n+1}(x) - f^n(x)}{f^{n+k+1}(x) - f^{n+k}(x)},$$

which combined with (2) gives the desired equality.

2.5. Applications of Derivatives

2.5.1.

(a) Applying the generalized mean value theorem to the functions $f(x) = 1 - \cos x$ and $g(x) = \frac{x^2}{2}$, we get

$$\frac{1 - \cos x}{\frac{x^2}{2}} = \frac{\sin \theta}{\theta} < 1$$

for $x \neq 0$.

(b) For $x \geq 0$, consider $f(x) = x - \sin x$ and $g(x) = \frac{x^3}{3!}$. The generalized mean value theorem combined with (a) shows that

$$\frac{x - \sin x}{\frac{x^3}{3!}} = \frac{1 - \cos \theta}{\frac{\theta^2}{2!}} < 1,$$

which implies that the inequality holds true. Note that for negative x we have $\sin x < x - \frac{x^3}{3!}$.

(c) Apply the generalized mean value theorem to the functions

$$f(x) = \cos x - 1 + \frac{x^2}{2}, \quad g(x) = \frac{x^4}{4!}$$

on the interval with the endpoints 0 and x, and use (b).

(d) Apply the generalized mean value theorem to the functions

$$f(x) = \sin x - x + \frac{x^3}{3!}, \quad g(x) = \frac{x^5}{5!}, \quad x \geq 0,$$

and use (c).

2.5.2. Use induction and reasoning similar to that in the solution of the foregoing problem.

2.5.3. Applying the generalized mean value theorem to the functions f and $g(x) = x$, $g(x) = x^2$ and $g(x) = x^3$ in succession, we see that
$$\frac{f(b)-f(a)}{b-a} = \frac{f'(x_1)}{1}, \quad \frac{f(b)-f(a)}{b^2-a^2} = \frac{f'(x_2)}{2x_2},$$
$$\frac{f(b)-f(a)}{b^3-a^3} = \frac{f'(x_3)}{3x_3^2}.$$

2.5.4. Set $f(x) = f_1(x) + if_2(x)$ and $\alpha = a + ib$, $a > 0$. It follows from $\lim\limits_{x\to+\infty}(\alpha f(x) + f'(x)) = 0$ that

(1) $$\lim_{x\to+\infty}(af_1(x) + f_1'(x) - bf_2(x)) = 0$$

and

(2) $$\lim_{x\to+\infty}(af_2(x) + f_2'(x) + bf_1(x)) = 0.$$

Observe now that
$$\lim_{x\to+\infty} e^{ibx} f(x) = \lim_{x\to+\infty} \frac{e^{ax+ibx}f(x)}{e^{ax}}$$
$$= \lim_{x\to+\infty} \frac{e^{ax}(f_1(x)\cos bx - f_2(x)\sin bx)}{e^{ax}}$$
$$+ i \lim_{x\to+\infty} \frac{e^{ax}(f_2(x)\cos bx + f_1(x)\sin bx)}{e^{ax}}.$$

Using l'Hospital's rule (given in 2.3.40), by (1) and (2) we get
$$\lim_{x\to+\infty} \frac{e^{ax}(f_1(x)\cos bx - f_2(x)\sin bx)}{e^{ax}}$$
$$= \lim_{x\to+\infty} \frac{\cos bx(af_1(x)+f_1'(x)-bf_2(x)) - \sin bx(af_2(x)+f_2'(x)+bf_1(x))}{a}$$
$$= 0.$$

Similarly, one can show that
$$\lim_{x\to+\infty} \frac{e^{ax}(f_2(x)\cos bx + f_1(x)\sin bx)}{e^{ax}} = 0.$$

So, we see that $\lim\limits_{x\to+\infty} e^{ibx} f(x) = 0$, which immediately implies that $\lim\limits_{x\to+\infty} f(x) = 0$.

2.5. Applications of Derivatives

Finally, let us remark that the proved result can be generalized as follows. If $\lim\limits_{x\to+\infty}(\alpha f(x)+f'(x))=L$, then $\lim\limits_{x\to+\infty}f(x)=L/\alpha$. Indeed, in this case we have $\lim\limits_{x\to+\infty}(\alpha(f(x)-L/\alpha)+(f(x)-L/\alpha)')=0$ and, by what we have proved, $\lim\limits_{x\to+\infty}(f(x)-L/\alpha)=0$.

2.5.5. Take $\alpha_1=\frac{1}{2}-\frac{\sqrt{3}}{2}i$ and $\alpha_2=\frac{1}{2}+\frac{\sqrt{3}}{2}i$. Then

$$f(x)+f'(x)+f''(x)=\alpha_2\alpha_1 f(x)+(\alpha_2+\alpha_1)f'(x)+f''(x)$$
$$=\alpha_2(\alpha_1 f(x)+f'(x))+(\alpha_1 f(x)+f'(x))'.$$

So, by the result in the foregoing problem (see the final remark in the solution), we get $\lim\limits_{x\to+\infty}(\alpha_1 f(x)+f'(x))=L/\alpha_2$ and $\lim\limits_{x\to+\infty}f(x)=L/(\alpha_2\alpha_1)=L$.

2.5.6. No. Consider, for example, the function $f(x)=\cos x$, $x>0$.

2.5.7.

(a) Set $g(x)=f(x)-e^{-x}$, $x\geq 0$. Then $g(0)=0$, $g(x)\leq 0$ and $\lim\limits_{x\to\infty}g(x)=0$. If $g(x)\equiv 0$, then $f'(x)=-e^{-x}$ for $x\in(0,\infty)$. So, suppose that there is $a>0$ such that $g(a)<0$. Then for sufficiently large x, say $x>M$, we have $g(x)>\frac{1}{2}g(a)$. Consequently, g attains its minimum value at some x_0 in $(0,M)$. Thus $g'(x_0)=0$.

(b) Apply reasoning analogous to that in (a).

2.5.8. We have

$$\left(\frac{f(x)}{g(x)}\right)'=\frac{g'(x)}{g(x)}\left(\frac{f'(x)}{g'(x)}-\frac{f(x)-f(0)}{g(x)-g(0)}\right).$$

By the generalized mean value theorem,

$$\left(\frac{f(x)}{g(x)}\right)'=\frac{g'(x)}{g(x)}\left(\frac{f'(x)}{g'(x)}-\frac{f'(\theta)}{g'(\theta)}\right),$$

where $0<\theta<x\leq a$. Since f'/g' monotonically increases, we see that

$$\left(\frac{f(x)}{g(x)}\right)'>0 \quad\text{for}\quad x>0.$$

2.5.9. Setting $f(x) = \sin(\cos x) - x$, we see that $f(0) = \sin 1$ and $f(\pi/2) = -\pi/2$. The intermediate value property implies that there is an $x_1 \in (0, \pi/2)$ such that $f(x_1) = 0$. Since $f'(x) < 0$ in $(0, \pi/2)$, there are no other zeros of f in this interval. In an entirely similar manner one can show that there is a unique root in $(0, \pi/2)$, say x_2, of the equation $\cos(\sin x) = x$.

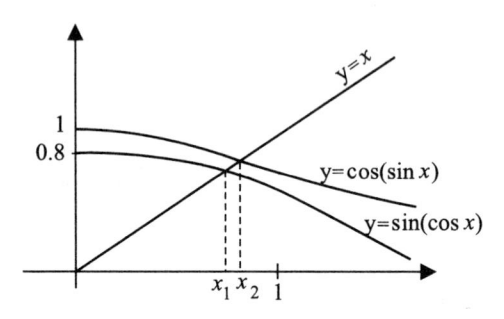

Moreover, we have

$$x_1 = \sin(\cos x_1) < \cos x_1, \quad x_2 = \cos(\sin x_2) > \cos x_2.$$

Therefore $x_1 < x_2$.

2.5.10. Suppose, contrary to our claim, that there is $x_1 \in (a, b]$ such that $f(x_1) \neq 0$. Then the continuity of f implies that $f(x) \neq 0$ for $x \in (\alpha, \beta)$. We may assume, for example, that $f(x) > 0$ for $x \in (\alpha, \beta)$, $f(\alpha) = 0$, $\alpha \geq a$, and $f(\beta) > 0$. Then by the mean value theorem, for $0 < \varepsilon < \beta - \alpha$,

$$|\ln f(\beta) - \ln f(\alpha + \varepsilon)| = \left|\frac{f'(\theta)}{f(\theta)}\right|(\beta - \alpha - \varepsilon) \leq C(\beta - \alpha - \varepsilon).$$

Letting $\varepsilon \to 0^+$, we obtain $+\infty \leq C(\beta - \alpha)$, a contradiction.

2.5.11. Let $0 < p < q$ and let x be positive. It follows from the mean value theorem that

$$\frac{\ln\left(1 + \frac{x}{q}\right)}{\frac{x}{q}} = \frac{1}{1 + \zeta_0} > \frac{1}{1 + \zeta_1} = \frac{\ln\left(1 + \frac{x}{p}\right) - \ln\left(1 + \frac{x}{q}\right)}{\frac{x}{p} - \frac{x}{q}},$$

2.5. Applications of Derivatives

where $\zeta_0 \in \left(0, \frac{x}{q}\right), \zeta_1 \in \left(\frac{x}{q}, \frac{x}{p}\right)$. Hence

$$\left(\frac{x}{p} - \frac{x}{q}\right)\ln\left(1 + \frac{x}{q}\right) > \frac{x}{q}\left(\ln\left(1 + \frac{x}{p}\right) - \ln\left(1 + \frac{x}{q}\right)\right).$$

Consequently,

$$\frac{x}{p}\ln\left(1 + \frac{x}{q}\right) > \frac{x}{q}\ln\left(1 + \frac{x}{p}\right),$$

or in other words,

$$q\ln\left(1 + \frac{x}{q}\right) > p\ln\left(1 + \frac{x}{p}\right).$$

2.5.12. The inequality $e^x \geq 1 + x$, $x \in \mathbb{R}$, follows, e.g., from the mean value theorem. Indeed, we have

$$\frac{e^x - 1}{x} = e^\zeta > 1 \quad \text{for} \quad x > 0,$$

and

$$\frac{e^x - 1}{x} = e^\zeta < 1 \quad \text{for} \quad x < 0.$$

If $x = 0$ we get the equality.

Let

$$A_n = \frac{1}{n}\sum_{k=1}^{n} a_k \quad \text{and} \quad G_n = \sqrt[n]{\prod_{k=1}^{n} a_k}$$

denote the arithmetic and geometric mean of nonnegative numbers $a_1, a_2, \ldots, a_n$. If $A_n \neq 0$, then

$$e^{\frac{a_k}{A_n} - 1} \geq \frac{a_k}{A_n} \geq 0 \quad \text{for} \quad k = 1, 2, \ldots, n.$$

Thus

$$1 = e^0 = e^{\sum_{k=1}^{n}\left(\frac{a_k}{A_n} - 1\right)} = \prod_{k=1}^{n} e^{\frac{a_k}{A_n} - 1} \geq \prod_{k=1}^{n} \frac{a_k}{A_n} = \frac{G_n^n}{A_n^n},$$

which gives $A_n \geq G_n$. If $A_n = 0$, then $A_n = G_n = 0$. Since in $e^x \geq 1 + x$ the equality is attained only for $x = 0$, we see that $A_n = G_n$ if and only if $a_1 = a_2 = \cdots = a_n$.

2.5.13. If in the inequality $e^t \geq 1+t$ we replace t by $x-z$, we obtain

$$xe^z \leq e^x + e^z(z-1) \quad \text{for} \quad x, z \in \mathbb{R}.$$

So the desired result follows if we replace z by $\ln y$.

2.5.14. By the mean value theorem, there is $a \in (-2, 0)$ such that

$$|f'(a)| = \frac{|f(0) - f(-2)|}{2} \leq \frac{|f(0)| + |f(-2)|}{2} \leq 1.$$

Similarly, there is $b \in (0, 2)$ such that $|f'(b)| \leq 1$. Set $F(x) = (f(x))^2 + (f'(x))^2$. The function F attains its maximum on $[a,b]$, say, at an x_0. Since $F(0) = 4$, $F(a) \leq 2$, and $F(b) \leq 2$, x_0 is in the open interval (a,b). Then $F'(x_0) = 2f'(x_0)(f(x_0) + f''(x_0)) = 0$. Note that $f'(x_0) \neq 0$, because $f'(x_0) = 0$ would give $F(x_0) = f(x_0)^2 \leq 1$, a contradiction. Therefore $f(x_0) + f''(x_0) = 0$.

2.5.15.

(a) The inequality to be proved is equivalent to

$$f(x) = (x^2 + 1)\arctan x - x > 0, \quad x > 0.$$

Since $f'(x) = 2x \arctan x + 1 - 1 > 0$, we see that $f(x) > f(0) = 0$ for $x > 0$.

(b) By Taylor's formula with the Lagrange form for the remainder,

$$(1) \quad 2\tan x = 2x + \frac{2}{3}x^3 + 2\left(\frac{\sin^3 \xi_1}{\cos^5 \xi_1} + \frac{2}{3}\frac{\sin \xi_1}{\cos^3 \xi_1}\right)x^4 > 2x + \frac{2}{3}x^3$$

and

$$(2) \quad \sinh x = x + \frac{x^3}{6} + \frac{1}{4!}\frac{e^{\xi_2} - e^{-\xi_2}}{2}x^4 < x + \frac{x^3}{6} + \frac{1}{4!}\frac{e^{\frac{\pi}{2}} - e^{-\frac{\pi}{2}}}{2}x^4.$$

We now show that $e^{\pi/2} < 8$. To this end note that (see, e.g., I, 2.5.3) $\ln(1+x) > \frac{2x}{x+2}$ for $x > 0$. This implies that $\ln 8 = 3\ln 2 = 3\ln(1+1) > 2$. Consequently, $8 > e^2 > e^{\pi/2}$. It then follows that

$$\frac{e^{\frac{\pi}{2}} - e^{-\frac{\pi}{2}}}{2} < \frac{e^{\pi/2}}{2} < 4,$$

2.5. Applications of Derivatives

which combined with (1) and (2) gives

$$\sinh x < x + \frac{x^3}{6} + \frac{x^4}{6} < 2x + \frac{2}{3}x^3,$$

because $x + \frac{1}{2}x^3 - \frac{1}{6}x^4 > 0$ for $0 < x < 2$.

(c) Set $f(x) = \ln x - \frac{x}{e}$ for $x > 0$, $x \neq e$. We have $f'(x) = \frac{e-x}{xe}$. Thus $f'(x) > 0$ if $0 < x < e$, and $f'(x) < 0$ if $x > e$. Consequently, $f(x) < f(e) = 0$ if $x \neq e$.

(d) For $x > 1$ the inequality to be proved is equivalent to

$$f(x) = 2x \ln x - x^2 + 1 < 0.$$

Since $f'(x) = 2 \ln x + 2 - 2x$ and $f''(x) = \frac{2}{x} - 2 < 0$, we get $f'(x) < f'(1) = 0$, and consequently, $f(x) < f(1) = 0$.
For $0 < x < 1$ the inequality to be proved is equivalent to $f(x) = 2x \ln x - x^2 + 1 > 0$. Since $f''(x) = \frac{2}{x} - 2 > 0$, we get $f'(x) < f'(1) = 0$, and therefore $f(x) > f(1) = 0$.

2.5.16.

(a) By (c) in the foregoing problem we get $\ln \pi < \frac{\pi}{e}$, which means that $e^\pi > \pi^e$.

(b) By (d) in the foregoing problem we see that $\sqrt{2} \ln \sqrt{2} < \frac{1}{2}$. This gives $2^{\sqrt{2}} < e$.

(c) The inequality $\ln 8 > 2$ is proved in the solution of (b) in the foregoing problem.

2.5.17.

(a) The inequality to be proved is equivalent to

$$\left(1 + \frac{b}{x}\right)^{\ln\left(1 + \frac{x}{a}\right)} < e^{\frac{b}{a}}.$$

Since $\ln(1 + t) < t$ for $t > 0$,

$$\left(1 + \frac{b}{x}\right)^{\ln\left(1 + \frac{x}{a}\right)} < \left(1 + \frac{b}{x}\right)^{\frac{x}{a}} = \left(\left(1 + \frac{b}{x}\right)^{\frac{x}{b}}\right)^{\frac{b}{a}} < e^{\frac{b}{a}}.$$

(b) For positive integers m and n, define the function f by setting
$$f(x) = \left(1 + \frac{x}{m}\right)^m \left(1 - \frac{x}{n}\right)^n, \quad |x| < \min\{m,n\}.$$
Then $f'(x) < 0$ if $x > 0$ and $f'(x) > 0$ if $x < 0$. Therefore $f(x) < f(0) = 1$ for $x \neq 0$, $|x| < \min\{m,n\}$, which implies the desired result.

(c) Set $f(x) = \ln(\sqrt{1+x^2} + 1) - \frac{1}{x} - \ln x$, $x > 0$. Then
$$f'(x) = \frac{(1-x)(\sqrt{1+x^2}+1) + x^2}{x^2(\sqrt{1+x^2}+1+x^2)}.$$
Clearly, $f'(x) > 0$ if $0 < x \leq 1$. If $x > 1$, then
$$(1-x)(\sqrt{1+x^2}+1) + x^2 > 0$$
if and only if
$$x^2 > (x-1)(\sqrt{1+x^2}+1).$$
The last inequality is equivalent to
$$\frac{x^2}{x-1} - 1 > \sqrt{1+x^2},$$
which can be proved by squaring both sides. So, $f'(x) > 0$ for all positive x. Moreover, since
$$\lim_{x \to \infty} \ln\left(\frac{1+\sqrt{1+x^2}}{x}\right) = 0,$$
we see that $\lim_{x \to \infty} f(x) = 0$. Consequently, $f(x) < 0$ for $x > 0$.

2.5.18.

(a) Set
$$f(x) = \ln(1+x) - \frac{x}{\sqrt{1+x}}, \quad x > 0.$$
Then
$$f'(x) = \frac{2\sqrt{1+x} - 2 - x}{2(1+x)\sqrt{1+x}} < 0,$$
because $\sqrt{1+x} < 1 + \frac{x}{2}$, $x > 0$. Thus $f(x) < f(0) = 0$.

2.5. Applications of Derivatives 293

(b) For $x > 1$ the inequality follows from (a). Indeed, it suffices to replace x by $x - 1$. If $x \in (0,1)$, then we apply the proved inequality to $\frac{1}{x} > 1$.

2.5.19.

(a) It suffices to apply Taylor's formula to $f(x) = (1+x)\ln(1+x)$.

(b) By Taylor's formula,

$$\ln(1+\cos x) = \ln 2 - \frac{x^2}{4} - \frac{\sin \zeta}{(1+\cos \zeta)^2} \cdot \frac{x^3}{3!} < \ln 2 - \frac{x^2}{4}.$$

2.5.20.

(a) Set $f(x) = e^x - 1 - xe^x$. We have $f'(x) = -xe^x < 0$. Hence $f(x) < f(0) = 0$.

(b) Setting $f(x) = e^x - 1 - x - x^2 e^x$, we get
$$f'(x) = e^x - 1 - 2xe^x - x^2 e^x$$
$$< 1 + xe^x - 1 - 2xe^x - x^2 e^x = -xe^x(1+x) < 0,$$
where the first inequality follows from (a).

(c) If $f(x) = xe^{\frac{x}{2}} - e^x + 1$, then
$$f'(x) = e^{\frac{x}{2}}\left(1 + \frac{x}{2} - e^{\frac{x}{2}}\right) < 0,$$
because $e^x > 1 + x$ for $x > 0$.

(d) The inequality to be proved is equivalent to the easily verifiable inequality
$$x < (1+x)\ln(1+x).$$

(e) We will prove the equivalent inequality $(x+1)(\ln(1+x) - \ln 2) \leq x \ln x$. To this end consider $f(x) = (x+1)(\ln(1+x) - \ln 2) - x \ln x$. This function attains its global maximum at $x = 1$. Therefore $f(x) \leq f(1) = 0$.

2.5.21. Taking logarithms of both sides, we rewrite the inequality to be proved in the form $(e-x)\ln(e+x) > (e+x)\ln(e-x)$. Now consider the function f defined by $f(x) = (e-x)\ln(e+x) - (e+x)\ln(e-x)$. We have $f''(x) > 0$ for $x \in (0,e)$. Consequently, $f'(x) > f'(0) = 0$, which implies $f(x) > f(0) = 0$.

2.5.22. Setting $f(x) = e^{x-1} + \ln x - 2x + 1$, we get

$$f'(x) = e^{x-1} + \frac{1}{x} - 2.$$

So for $x > 1$ we have $f''(x) = e^{x-1} - \frac{1}{x^2} > 0$, because $e^{x-1} > 1$ and $\frac{1}{x^2} < 1$.

2.5.23.
(a) Set $f(x) = \frac{1}{3}\tan x + \frac{2}{3}\sin x - x$. Then

$$f'(x) = \frac{2(1-\cos x)^2 \left(\cos x + \frac{1}{2}\right)}{3\cos^2 x} > 0 \quad \text{for} \quad x \in \left(0, \frac{\pi}{2}\right).$$

This means that f is strictly increasing. So $f(x) > f(0)$ for $0 < x < \frac{\pi}{2}$.

(b) We define $f(x) = x - \frac{3\sin x}{2+\cos x}$. Then

$$f'(x) = \frac{(\cos x - 1)^2}{(2+\cos x)^2} \geq 0.$$

(c) Putting $f(x) = \frac{\sin x}{\sqrt{\cos x}} - x$ for $0 < x < \frac{\pi}{2}$, we see that

$$f'(x) = \frac{1 + \cos^2 x - 2\sqrt{\cos x}\cos x}{2\sqrt{\cos x}\cos x} > \frac{(1-\cos x)^2}{2\sqrt{\cos x}\cos x} > 0.$$

2.5.24. Let $f(x) = x^\alpha + (1-x)^\alpha$. Then f' vanishes only at $x = \frac{1}{2}$. Moreover, the function attains its global minimum value $\frac{1}{2^{\alpha-1}}$ at this point and its global maximum value 1 at the endpoints of $[0, 1]$.

2.5.25. Dividing both sides by x^α, we see that it suffices to prove

$$(1+t)^\alpha < 1 + t^\alpha \quad \text{for} \quad t > 0.$$

If $f(t) = (1+t)^\alpha - 1 - t^\alpha$, then $f'(t) < 0$. Thus $f(t) < f(0)$ for $t > 0$.

2.5.26. Consider the function

$$f(x) = (1+x)^\alpha - 1 - \alpha x - \frac{\alpha(\alpha-1)}{8}x^2, \quad x \in [-1, 1].$$

2.5. Applications of Derivatives

We have $f(0) = 0$, $f'(0) = 0$ and, for $x \in (-1,1)$,

$$f''(x) = -\alpha(1-\alpha)(1+x)^{\alpha-2} + \frac{\alpha(1-\alpha)}{4}$$
$$< -2^{\alpha-2}\alpha(1-\alpha) + \frac{\alpha(1-\alpha)}{4}$$
$$= \frac{1}{4}\alpha(1-\alpha)(1-2^\alpha) < 0.$$

Consequently, f' decreases on the interval $(-1,1)$. Thus $f'(x) > 0$ for $x \in (-1,0)$, and $f'(x) < 0$ for $x \in (0,1)$. It then follows that f attains its maximum at zero. Since $f(0) = 0$, we see that $f(x) \leq 0$ for $x \in [-1,1]$.

2.5.27 [D.S. Mitrinović, J.E. Pečarić, Rendiconti del Circolo Mat. di Palermo 42 (1993), 317-337]. Consider the function

$$f(x) = (1+x)^\alpha - 1 - \alpha x - \frac{\alpha(\alpha-1)}{2}(1+B)^{\alpha-2}x^2, \quad x \in [-1,B].$$

We have

$$f'(x) = \alpha(1+x)^{\alpha-1} - \alpha - \alpha(\alpha-1)(1+B)^{\alpha-2}x$$

and

$$f''(x) = \alpha(\alpha-1)\left((1+x)^{\alpha-2} - (1+B)^{\alpha-2}\right).$$

(a) If $0 < \alpha < 1$ and $x \in (-1,B)$, then $f''(x) < 0$, which means that f' decreases. Thus $0 = f'(0) < f'(x)$ if $x \in (-1,0)$, and $0 = f'(0) > f'(x)$ if $x \in (0,B]$. Therefore f attains its maximum at zero. Hence $f(x) \leq f(0) = 0$ for $x \in [-1,B]$. Finally, since $(1+B)^\alpha \geq 1$, we see that

$$(1+x)^\alpha - 1 - \alpha x - \frac{\alpha(\alpha-1)}{2(1+B)^2}x^2 \leq f(x) \leq f(0) = 0.$$

(b) As in (a), if $1 < \alpha < 2$ and $x \in [-1,B]$, then $f(x) \geq f(0) = 0$, and consequently,

$$(1+x)^\alpha - 1 - \alpha x - \frac{\alpha(\alpha-1)}{2(1+B)^2}x^2 \geq f(x) \geq f(0) = 0.$$

2.5.28.

(a) Define
$$f(x) = \begin{cases} \frac{\sin x}{x} & \text{if } x \in (0, \frac{\pi}{2}], \\ 1 & \text{if } x = 0. \end{cases}$$
We will show that f decreases on $[0, \frac{\pi}{2}]$. By the mean value theorem
$$f'(x) = \frac{x \cos x - \sin x}{x^2} = \frac{\cos x - \frac{\sin x}{x}}{x} = \frac{\cos x - \cos \theta}{x},$$
where $0 < \theta < x$. This implies that $f'(x) < 0$ on $(0, \frac{\pi}{2}]$. Since $f\left(\frac{\pi}{2}\right) = \frac{2}{\pi}$, the desired inequality follows.

(b) The inequality to be proved can be written in the form
$$\sin x \geq \frac{3}{\pi} x - 4\frac{x^3}{\pi^3}.$$
Set
$$f(x) = \sin x - \frac{3}{\pi} x + 4\frac{x^3}{\pi^3}, \quad \mathbf{I} = \left[0, \frac{\pi}{4}\right] \quad \text{and} \quad \mathbf{J} = \left[\frac{\pi}{4}, \frac{\pi}{2}\right].$$
We have $f(0) = f\left(\frac{\pi}{2}\right) = 0$ and $f\left(\frac{\pi}{4}\right) > 0$. Moreover, $f''(0) = 0$, $f''\left(\frac{\pi}{4}\right) < 0$ and $f^{(4)}(x) \geq 0$ for $x \in \mathbf{I}$. This implies that $f'' \leq 0$ on $\mathbf{I}$, which means that f is concave on $\mathbf{I}$. Since $f(0) = 0$ and $f\left(\frac{\pi}{4}\right) > 0$, we see that $f(x) \geq 0$ for $x \in \mathbf{I}$. Now we show that f' is convex on $\mathbf{J}$. Indeed, since $f^{(3)}\left(\frac{\pi}{4}\right) > 0$ and $f^{(4)}(x) > 0$ for $x \in \left(\frac{\pi}{4}, \frac{\pi}{2}\right)$, we see that the third derivative is positive on $\mathbf{J}$. Moreover, we have $f'\left(\frac{\pi}{4}\right) < 0$ and $f'\left(\frac{\pi}{2}\right) = 0$. It then follows that $f'(x) \leq 0$ for $x \in \mathbf{J}$, which together with $f\left(\frac{\pi}{2}\right) = 0$ gives $f \geq 0$ on $\mathbf{J}$.

2.5.29. Assume first that $x \in (0, \frac{1}{2})$. Then $\frac{\pi^3 x^3}{3!} < \pi x^2$. Hence the inequality $\sin \pi x > \pi x - \frac{\pi^3 x^3}{3!}$ (see, e.g., 2.5.1(b)) shows that $\sin \pi x > \pi x - \frac{\pi^3 x^3}{3!} > \pi x(1 - x)$. To prove the other inequality we consider the function defined by $f(x) = 4x - 4x^2 - \sin \pi x$, $x \in [0, \frac{1}{2}]$. We have $f'(x) = 4 - 8x - \pi \cos \pi x$ and $f''(x) = -8 + \pi^2 \sin \pi x$. Thus
$$f''(x_0) = 0 \quad \text{if and only if} \quad x_0 = \frac{1}{\pi} \arcsin \frac{8}{\pi^2},$$

2.5. Applications of Derivatives

and $f''(0) = -8$, and $f''(\frac{1}{2}) = \pi^2 - 8 > 0$. Hence $f''(x) < 0$ for $x \in (0, x_0)$ and $f''(x) > 0$ for $x \in (x_0, \frac{1}{2})$, or in other words, f' strictly decreases on $(0, x_0)$ and strictly increases on $(x_0, \frac{1}{2})$. Moreover, since $f'(0) = 4 - \pi > 0$ and $f'(\frac{1}{2}) = 0$, we see that $f'(x) < 0$ for $x \in (x_0, \frac{1}{2})$, so also $f'(x_0) < 0$. This implies that there is $x_1 \in (0, x_0)$ such that $f'(x_1) = 0$. It follows from the monotonicity of f' that f increases on $(0, x_1)$ and decreases on $(x_1, \frac{1}{2})$. Since $f(0) = f(\frac{1}{2}) = 0$, we get $f(x) \geq 0$ if $x \in (0, \frac{1}{2})$. So we have proved that the inequalities hold for $x \in (0, \frac{1}{2})$. It is easy to check that they also hold for $x = \frac{1}{2}$. Finally, note that the inequalities are not changed if we replace x by $1 - x$. Therefore they are satisfied for all $x \in (0, 1)$.

2.5.30. Define $f(x) = e^x - \sum_{k=0}^{n} \frac{x^k}{k!} - \frac{x}{n}(e^x - 1)$, $x > 0$. Then

$$f'(x) = e^x - \sum_{k=1}^{n} \frac{x^{k-1}}{(k-1)!} - \frac{x}{n}e^x - \frac{1}{n}e^x + \frac{1}{n},$$

and

$$f^{(l)}(x) = e^x - \sum_{k=l}^{n} \frac{x^{k-l}}{(k-l)!} - \frac{x}{n}e^x - \frac{l}{n}e^x, \quad l = 2, 3, \ldots, n.$$

Moreover, $f^{(l)}(0) = -\frac{l}{n} < 0$, $l = 2, 3, \ldots, n$, and $f'(0) = 0$, and $f(0) = 0$. Since $f^{(n)}(x) < 0$ for $x > 0$, the derivative $f^{(n-1)}$ strictly decreases, which implies that $f^{(n-1)}(x) < f^{(n-1)}(0) < 0$. This in turn implies the monotonicity of $f^{(n-2)}$ and $f^{(n-2)}(x) < 0$ for $x > 0$. By repeating the reasoning we obtain $f(x) < f(0) = 0$, $x > 0$

2.5.31. Since $f'(x) = -\frac{x^n}{n!}e^{-x}$, we see that the derivative vanishes only at zero. Moreover, if n is even, then $f'(x) < 0$ for $x \neq 0$. So in this case f does not have any local extrema. On the other hand, if n is odd, then $f'(x) > 0$ for $x < 0$ and $f'(x) < 0$ for $x > 0$. So in the case where n is odd, $f(0) = 1$ is a (global) maximum value of f.

2.5.32. The derivative $f'(x) = (m+n)x^{m-1}(1-x)^{n-1}\left(\frac{m}{m+n} - x\right)$ vanishes only at $x_0 = 0$ (if $m > 1$), $x_1 = 1$ (if $n > 1$), and at $x_2 = \frac{m}{m+n}$. It is easy to verify that $f(x_2) = \frac{m^m n^n}{(m+n)^{m+n}}$ is the local maximum value of f. Moreover, if m is even, then $f(x_0) = 0$ is the local minimum value of f. On the other hand, if m is odd, any local extremum of f does not occur at zero. Analogous analysis shows that if n is even, then $f(x_1) = 0$ is a local minimum value, and if n is odd no local extremum of f occurs at x_1.

2.5.33. It follows from the result in the foregoing problem that the maximum value $\frac{m^m n^n}{(m+n)^{m+n}}$ of f is attained at x satisfying the equation

$$\sin^2 x = \frac{m}{m+n}.$$

2.5.34. For $x \neq 0, 1$,

$$f'(x) = \frac{1}{9} \cdot \frac{\frac{1}{3} - x}{\sqrt[3]{x^2(1-x)}}.$$

So f' vanishes at $x = \frac{1}{3}$. Moreover, $f'(x) > 0$ if $x \in (0, \frac{1}{3})$ and $f'(x) < 0$ if $x \in (\frac{1}{3}, 1)$. Thus $f(\frac{1}{3}) = \frac{\sqrt[3]{4}}{3}$ is a local maximum value of f. The function is not differentiable at 0 and at 1. Since $f(x) > 0$ for $x \in (0, 1)$ and $f(x) < 0$ for $x < 0$, no local extremum of f occurs at zero. But $f(1) = 0$ is a local minimum value of f, because $f(x) > 0 = f(1)$ for $x > 1$ and for $x \in (0, 1)$.

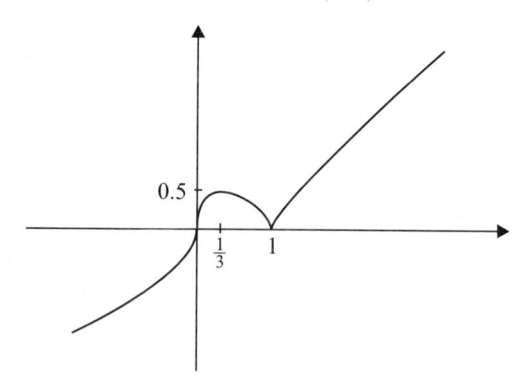

2.5.35. We have $f'(x) = \arcsin x$. Thus zero is the only critical point of f. Since $f(0) = 1$, and $f(-1) = f(1) = \frac{\pi}{2}$, $\frac{\pi}{2}$ is the global maximum value and 1 is the global minimum value of f in $[-1, 1]$.

2.5. Applications of Derivatives

2.5.36. For $x > 1$ the derivative f' is negative, and consequently, $f(x) < f(1) = \frac{3}{2}$. For $x \in (0,1)$ we have $f'(\frac{1}{2}) = 0$, $f'(x) < 0$ if $x \in (0, \frac{1}{2})$, and $f'(x) > 0$ if $x \in (\frac{1}{2}, 1)$. Thus $f(\frac{1}{2}) = \frac{4}{3}$ is the local minimum value of the function. For $x < 0$ the derivative f' is positive and therefore $\frac{3}{2} = f(0) > f(x)$.

So the global maximum value of f is $f(0) = f(1) = \frac{3}{2}$. On the other hand, since $\lim\limits_{x \to \infty} f(x) = \lim\limits_{x \to -\infty} f(x) = 0$ and $f(x) > 0$ for all $x \in \mathbb{R}$, the greatest lower bound of $f(\mathbb{R})$ is zero, but the function is not minimized over $\mathbb{R}$.

2.5.37.

(a) The global maximum value of the function $x \mapsto xe^{-x}$, $x \geq 0$, is $f(1) = \frac{1}{e}$. Therefore

$$\frac{1}{n} \sum_{k=1}^{n} a_k e^{-a_k} \leq \frac{1}{n} \cdot n \cdot \frac{1}{e} = \frac{1}{e}.$$

(b) As in (a), it is enough to find the global maximum value of

$$x \mapsto x^2 e^{-x}, \quad x \geq 0.$$

(c) If one of the numbers a_k is zero, the inequality is obvious. So suppose that all a_k are positive. Then, taking logarithms of both sides, we obtain the equivalent form of the inequality:

$$\frac{1}{n} \sum_{k=1}^{n} \left(\ln a_k - \frac{a_k}{3} \right) \leq \ln 3 - 1.$$

Now it suffices to find the global maximum value of

$$x \mapsto \ln x - \frac{x}{3}, \quad x > 0.$$

2.5.38. We have

$$f'(x) = \begin{cases} \frac{1}{x^2} e^{-\frac{1}{|x|}} \left(\left(\sqrt{2} + \sin \frac{1}{x} \right) \operatorname{sgn} x - \cos \frac{1}{x} \right) & \text{if } x \neq 0, \\ 0 & \text{if } x = 0. \end{cases}$$

Since

$$\left| \sin \frac{1}{x} \pm \cos \frac{1}{x} \right| \leq \sqrt{2},$$

we see that $f'(x) \geq 0$ if $x > 0$, and $f'(x) \leq 0$ if $x < 0$. Therefore no local extremum of f occurs at $x \neq 0$. Moreover, $0 = f(0)$ is a global minimum value of f, because $f(x) > f(0) = 0$ for $x \neq 0$.

2.5.39. Note first that $f(x) > f(0) = 0$ for $x \neq 0$. Moreover,

$$f'(x) = \begin{cases} x^2 \left(8x + 4x \sin \frac{1}{x} - \cos \frac{1}{x}\right) & \text{if } x \neq 0, \\ 0 & \text{if } x = 0. \end{cases}$$

Consequently, if $n \in \mathbb{Z} \setminus \{0, 1\}$, then

$$f'\left(\frac{1}{2n\pi}\right) = \frac{1}{4n^2\pi^2}\left(\frac{4}{n\pi} - 1\right) < 0,$$

and if $n \in \mathbb{Z} \setminus \{-1\}$, then

$$f'\left(\frac{1}{(2n+1)\pi}\right) = \frac{1}{(2n+1)^2\pi^2}\left(\frac{8}{(2n+1)\pi} + 1\right) > 0.$$

2.5.40. Observe that $\sinh x > 0$ and $\tanh x > 0$ for $x > 0$. Thus the inequality

$$\frac{\sinh x}{\sqrt{\sinh^2 x + \cosh^2 x}} < \tanh x$$

can be rewritten in the following equivalent form:

$$\frac{1}{\sqrt{\sinh^2 x + \cosh^2 x}} < \frac{1}{\cosh x}.$$

This inequality is obvious. The other inequalities can be proved by standard arguments.

2.5.41. For $0 < a < b$, set $x = \ln\sqrt{\frac{b}{a}}$. Then

$$\frac{b-a}{2\sqrt{\frac{a^2+b^2}{2}}} < \frac{b-a}{b+a} < \ln\sqrt{\frac{b}{a}} < \frac{b-a}{2\sqrt{ab}} < \frac{1}{2} \cdot \frac{b^2-a^2}{2ab}.$$

Dividing by $\frac{b-a}{2}$ gives the desired inequalities.

2.5. Applications of Derivatives

2.5.42.

(a) We have
$$\lim_{p \to 0} \left(\frac{x^p + y^p}{2} \right)^{\frac{1}{p}} = \lim_{p \to 0} e^{\frac{1}{p} \ln \frac{x^p+y^p}{2}} = e^{\frac{1}{2} \ln xy} = \sqrt{xy},$$

because by l'Hospital's rule,
$$\lim_{p \to 0} \frac{1}{p} \ln \frac{x^p + y^p}{2} = \lim_{p \to 0} \frac{\left(\ln \frac{x^p+y^p}{2} \right)'}{p'} = \frac{1}{2} \ln xy.$$

(b) For $p \neq 0$, set $f(p) = \left(\frac{x^p+y^p}{2} \right)^{1/p}$. It suffices to show that
$$F(p) = \ln f(p) = \frac{1}{p} \ln \frac{x^p + y^p}{2}$$

is a strictly increasing function. We have
$$F'(p) = \frac{1}{p^2} \left(\frac{p}{x^p + y^p} \left(x^p \ln x + y^p \ln y \right) - \ln \frac{x^p + y^p}{2} \right).$$

Now let
$$G(p) = \frac{p}{x^p + y^p} (x^p \ln x + y^p \ln y) - \ln \frac{x^p + y^p}{2}.$$

Then
$$G'(p) = \frac{p \left[\left(x^p \ln^2 x + y^p \ln^2 y \right)(x^p + y^p) - \left(x^p \ln x + y^p \ln y \right)^2 \right]}{(x^p + y^p)^2}.$$

Our aim is to show that
$$\left(x^p \ln^2 x + y^p \ln^2 y \right)(x^p + y^p) - \left(x^p \ln x + y^p \ln y \right)^2 \geq 0.$$

Applying the Cauchy inequality
$$(x_1 y_1 + x_2 y_2)^2 \leq (x_1^2 + x_2^2)(y_1^2 + y_2^2)$$

with
$$x_1 = x^{\frac{p}{2}}, \quad x_2 = y^{\frac{p}{2}}, \quad y_1 = x^{\frac{p}{2}} \ln x, \quad y_2 = y^{\frac{p}{2}} \ln y,$$

we obtain
$$\left(x^{\frac{p}{2}} \cdot x^{\frac{p}{2}} \ln x + y^{\frac{p}{2}} \cdot y^{\frac{p}{2}} \ln y\right)^2 \leq (x^p + y^p)\left(x^p \ln^2 x + y^p \ln^2 y\right).$$
Hence
$$(x^p \ln x + y^p \ln y)^2 \leq (x^p + y^p)\left(x^p \ln^2 x + y^p \ln^2 y\right).$$
This means that $G'(p) \geq 0$ for $p > 0$. Consequently, in this case we get $G(p) = p^2 F'(p) > G(0) = 0$. If $p < 0$, then $G'(p) < 0$, and so $G(p) = p^2 F'(p) > G(0)$. Summing up, we see that the function $p \mapsto f(p)$ is strictly increasing on each of the intervals $(-\infty, 0)$ and $(0, \infty)$. It then follows by the definition of $M_0(x, y)$ (see 2.5.42) that f is strictly increasing on $\mathbb{R}$.

2.5.43. For $\lambda \geq 1$, define
$$f(\lambda) = \frac{x^n + y^n + \lambda((x+y)^n - x^n - y^n)}{2 + \lambda(2^n - 2)}.$$
Using the inequality
$$(x+y)^n \leq 2^{n-1}(x^n + y^n),$$
one can show that $f'(\lambda) \leq 0$. So f is decreasing on $[1, \infty)$. Since $f(1) = (x+y)^n/2^n$, the right inequality follows. To prove the left inequality it is enough to show that $\lim\limits_{\lambda \to \infty} f(\lambda) \geq (\sqrt{xy})^n$. By the arithmetic-geometric mean inequality,
$$\lim_{\lambda \to \infty} f(\lambda) = \frac{(x+y)^n - x^n - y^n}{2^n - 2}$$
$$= \frac{\binom{n}{1} x y^{n-1} + \binom{n}{2} x^2 y^{n-2} + \cdots + \binom{n}{n-1} x^{n-1} y}{2^n - 2}$$
$$\geq \sqrt[2^n - 2]{(xy^{n-1})^{\binom{n}{1}} (x^2 y^{n-2})^{\binom{n}{2}} \cdots (x^{n-1} y)^{\binom{n}{n-1}}} = (\sqrt{xy})^n,$$
where the last equality follows from the identity
$$\binom{n}{1} + 2\binom{n}{2} + \cdots + (n-1)\binom{n}{n-1} = n(2^{n-1} - 1),$$
which in turn can be proved using the fact that
$$k \binom{n}{k} = n \binom{n-1}{k-1}, \quad k \geq 1.$$

2.5. Applications of Derivatives 303

2.5.44.

(a) Set $f(x) = \sin(\tan x) - x$ for $x \in \left[0, \frac{\pi}{4}\right]$. Then

$$f(0) = 0 \quad \text{and} \quad f'(x) = \cos(\tan x)\frac{1}{\cos^2 x} - 1,$$

and therefore,

$$f'(x) \geq 0 \quad \text{if and only if} \quad \cos(\tan x) \geq \cos^2 x.$$

Note now that $\cos(\tan x) \geq 1 - \frac{\tan^2 x}{2}$ (see, e.g., 2.5.1(a)). So it suffices to show that $1 - \frac{\tan^2 x}{2} \geq \cos^2 x$ for $x \in \left[0, \frac{\pi}{4}\right]$. The last inequality can be rewritten in the form

$$2\cos^4 x - 3\cos^2 x + 1 \leq 0,$$

which is clearly satisfied for all $x \in \left[0, \frac{\pi}{4}\right]$.

(b) For $x \in \left[0, \frac{\pi}{3}\right]$, define $f(x) = \tan(\sin x) - x$. Then

$$f(0) = 0 \quad \text{and} \quad f'(x) = \frac{\cos x}{\cos^2(\sin x)} - 1.$$

Consequently, $f'(x) \geq 0$ if and only if

$$\cos x \geq \cos^2(\sin x) = \frac{1 + \cos(2\sin x)}{2}.$$

So it suffices to prove the last inequality. To this end note that, by 2.5.1(c),

$$1 + \cos(2\sin x) \leq 2 - 2\sin^2 x + \frac{2}{3}\sin^4 x \leq 2\cos x.$$

To see that the last inequality holds for $x \in \left[0, \frac{\pi}{3}\right]$, standard arguments can be used.

2.5.45. Define $f(x) = \frac{1}{\sin^2 x} - \frac{1}{x^2}$. Then for $x \in (0, \pi/2)$ we have $f'(x) > 0$ if and only if

$$\frac{1}{x^3} > \frac{\cos x}{\sin^3 x},$$

or in other words, if and only if

$$\frac{\sin x}{\sqrt[3]{\cos x}} - x > 0.$$

Now, if we put
$$g(x) = \frac{\sin x}{\sqrt[3]{\cos x}} - x,$$
then
$$g'(x) = (\cos x)^{\frac{2}{3}} + \frac{1}{3}(\cos x)^{-\frac{4}{3}} \sin^2 x - 1$$
and
$$g''(x) = \frac{4}{9}(\cos x)^{-\frac{7}{3}} \sin^3 x.$$
Since $g''(x) > 0$ for $x \in (0, \pi/2)$, we see that $g'(x) > g'(0) = 0$. Consequently, $g(x) > g(0) = 0$ for $x \in (0, \pi/2)$. This in turn implies that f increases on that interval, and so $f(x) \leq f\left(\frac{\pi}{2}\right) = 1 - \frac{4}{\pi^2}$.

2.5.46. It is enough to observe that
$$\left(\arctan x - \frac{3x}{1 + 2\sqrt{1+x^2}}\right)' = \frac{\left(\sqrt{1+x^2} - 1\right)^2}{(1+x^2)\left(1 + 2\sqrt{1+x^2}\right)^2} > 0.$$

2.5.47. If $a_k = b_k$ for all k, then the assertion is clear. So assume that $a_k \neq b_k$ for at least one k, and put
$$f(x) = \prod_{k=1}^{n}(xa_k + (1-x)b_k) \quad \text{and} \quad g(x) = \ln f(x).$$
Then
$$g'(x) = \sum_{k=1}^{n} \frac{a_k - b_k}{xa_k + (1-x)b_k} \quad \text{and} \quad g''(x) = -\sum_{k=1}^{n}\left(\frac{a_k - b_k}{xa_k + (1-x)b_k}\right)^2.$$
Since $g''(x) < 0$, the function g (so also f) attains its maximum over $[0, 1]$ at one of the endpoints if and only if $g'(0)$ and $g'(1)$ have the same sign, that is, if $g'(0)g'(1) \geq 0$. The last inequality means that
$$\left(\sum_{k=1}^{n} \frac{a_k - b_k}{a_k}\right)\left(\sum_{k=1}^{n} \frac{a_k - b_k}{b_k}\right) \geq 0.$$

2.5. Applications of Derivatives

2.5.48. By 2.5.1(a) and (c),
$$1 - \frac{x^2}{2} \leq \cos x \leq 1 - \frac{x^2}{2} + \frac{x^4}{24}, \quad x \in \mathbb{R}.$$
Consequently, to prove our inequality it is enough to show that
$$1 - \frac{x^2}{2} + \frac{x^4}{24} + 1 - \frac{y^2}{2} + \frac{y^4}{24} \leq 1 + 1 - \frac{x^2 y^2}{2},$$
or equivalently,
$$x^4 + y^4 + 12x^2 y^2 - 12(x^2 + y^2) \leq 0 \quad \text{for} \quad x^2 + y^2 \leq \pi.$$
The last inequality can be written in polar coordinates r, θ as follows:
(1) $\quad r^2(2 + 5\sin^2 2\theta) \leq 24 \quad \text{for} \quad r^2 \leq \pi \quad \text{and} \quad \theta \in [0, 2\pi].$

Since
$$r^2(2 + 5\sin^2 2\theta) \leq 7\pi < 24,$$
we see that (1) is true.

2.5.49. The inequality is obvious if $x \geq 1$ or $y \geq 1$. So assume that $x, y \in (0,1)$ and write $y = tx$. In view of the symmetry, it is enough to consider the case where $0 < t \leq 1$. We have
$$x^y + y^x = x^{tx} + (tx)^x = (x^x)^t + t^x x^x.$$
Since the function $x \mapsto x^x$ attains its minimum value $e^{-1/e} = a$ at $\frac{1}{e}$ and since $t^x \geq t$, we see that $x^y + y^x \geq a^t + ta$. Moreover, $F(t) = a^t + ta$, $t \in \mathbb{R}$, has only one local minimum $t_0 = 1 - e < 0$, and F is strictly increasing on (k_0, ∞), and $F(0) = 1$. It then follows that $x^y + y^x > 1$.

2.5.50. For $0 < x < 1$, the inequality to be proved can be rewritten in the form
$$1 - 2x^n + x^{n+1} < (1-x^n)\sqrt[n]{1-x^n},$$
or
$$\frac{1-x^n}{1-x} < \frac{1-(1-x^n)}{1-\sqrt[n]{1-x^n}}.$$
Since the function $t \mapsto \frac{1-t^n}{1-t}$ is strictly increasing on $(0,1)$, it is enough to show that $x < \sqrt[n]{1-x^n}$, or equivalently, $0 < x < \frac{1}{\sqrt[n]{2}}$. Finally, note that $\left(1 + \frac{1}{n}\right)^n > 2$ for $n \geq 2$, which means that $\frac{1}{\sqrt[n]{2}} > \frac{n}{n+1}$.

2.5.51. For $0 < x < 1$, consider the function
$$g(x) = \frac{f(x)}{x} = 1 - \frac{x^2}{6} + \frac{x^3}{24}\sin\frac{1}{x}.$$
Since $g'(x) < 0$ for $0 < x < 1$, we see that g is strictly increasing on $(0,1)$. Therefore $g(y+z) < g(y)$ and $g(y+z) < g(z)$, and consequently,
$$yg(y+z) + zg(y+z) < yg(y) + zg(z),$$
which means that $f(y+z) < f(y) + f(z)$.

2.5.52. We start with the well-known binomial formula
$$(1) \qquad (x+y)^n = \sum_{k=0}^{n} \binom{n}{k} x^k y^{n-k}.$$
Differentiating (1) with respect to x and multiplying the resulting equality by x, we get
$$(2) \qquad nx(x+y)^{n-1} = \sum_{k=0}^{n} k\binom{n}{k} x^k y^{n-k}.$$
Now differentiating (1) twice and multiplying the result by x^2, we get
$$(3) \qquad n(n-1)x^2(x+y)^{n-2} = \sum_{k=0}^{n} k(k-1)\binom{n}{k} x^k y^{n-k}.$$
If in (1), (2) and (3) we replace y by $1-x$, we obtain
$$1 = \sum_{k=0}^{n} \binom{n}{k} x^k (1-x)^{n-k},$$
$$nx = \sum_{k=0}^{n} k\binom{n}{k} x^k (1-x)^{n-k},$$
and
$$n(n-1)x^2 = \sum_{k=0}^{n} k(k-1)\binom{n}{k} x^k (1-x)^{n-k}.$$
It then follows that
$$\sum_{k=0}^{n} (k-nx)^2 \binom{n}{k} x^k (1-x)^{n-k} = nx(1-x) \leq \frac{n}{4}.$$

2.5. Applications of Derivatives

2.5.53. By assumption the equation $f(x) = 0$ has a unique zero, say ξ, in $[a,b]$.

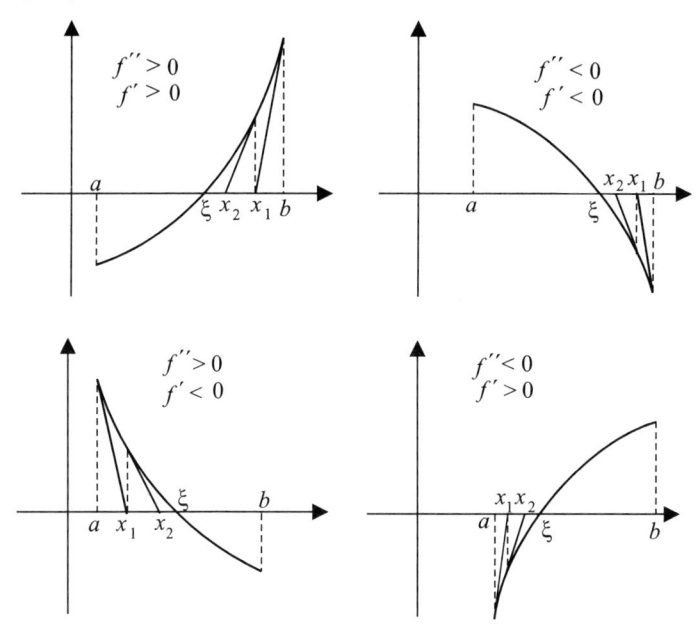

Suppose, for example, that $f'(x) > 0$ and $f''(x) < 0$ for $x \in [a,b]$. So we set $x_0 = a$. By Taylor's formula with the Lagrange form for the remainder,

$$0 = f(\xi) = f(x_n) + f'(x_n)(\xi - x_n) + \frac{1}{2}f''(c_n)(\xi - x_n)^2,$$

where c_n is an element of the open interval with the endpoints x_n, ξ. By the definition of $\{x_n\}$,

$$\xi - x_{n+1} = -\frac{f''(c_n)}{2f'(x_n)}(\xi - x_n)^2 > 0.$$

So $\{x_n\}$ is bounded above by ξ. Consequently, $f(x_n) < 0$. Hence

$$\xi - x_{n+1} = \xi - x_n + \frac{f(x_n)}{f'(x_n)} < \xi - x_n,$$

which means that $\{x_n\}$ is a strictly increasing sequence. Thus it converges, and $\lim_{n \to \infty} x_n = \xi$. The other cases can be proved analogously.

2.5.54. Clearly, m and M are positive. It follows from the solution of the foregoing problem that

$$0 = f(\xi) = f(x_n) + f'(x_n)(\xi - x_n) + \frac{1}{2}f''(c_n)(\xi - x_n)^2,$$

where ξ is a unique root of the equation $f(x) = 0$ in $[a, b]$, and c_n is an element of the open interval with the endpoints x_n, ξ. Hence

$$|x_{n+1} - \xi| = \left| x_n - \xi - \frac{f(x_n)}{f'(x_n)} \right| = \frac{|f''(c_n)|}{2|f'(x_n)|}(\xi - x_n)^2 \leq \frac{M}{2m}(\xi - x_n)^2.$$

2.5.55. We will show that $\sup\{2^{-x} + 2^{-\frac{1}{x}} : x > 0\} = 1$. Define $f(x) = 2^{-x} + 2^{-\frac{1}{x}}$, $x > 0$. Clearly, $f(1) = 1$ and $f(x) = f\left(\frac{1}{x}\right)$. Therefore it suffices to show that if $x > 1$, then $f(x) < 1$, or in other words,

(1) $$\frac{1}{2^{\frac{1}{x}}} < 1 - \frac{1}{2^x} \quad \text{for} \quad x > 1.$$

By 2.3.7(a) we get

$$\left(1 - \frac{1}{2^x}\right)^x > 1 - \frac{x}{2^x}.$$

Now we show that

(2) $$1 - \frac{x}{2^x} \geq \frac{1}{2} \quad \text{for} \quad x \geq 2.$$

To this end we write (2) in the form $g(x) = 2^{x-1} - x \geq 0$ and note that g is strictly increasing on $[2, \infty)$, and $g(2) = 0$. So, inequality (1) is proved for $x \geq 2$. Thus $f(x) < 1$ for $x \geq 2$. Our task is now to prove that $f(x) < 1$ for $x \in (1, 2)$. To this end we define the function h by

$$h(x) = \ln f(x) = \ln\left(2^x + 2^{\frac{1}{x}}\right) - \left(x + \frac{1}{x}\right) \ln 2.$$

Since

$$h'(x) = \ln 2 \frac{-2^{\frac{1}{x}} + \frac{1}{x^2} 2^x}{2^x + 2^{\frac{1}{x}}},$$

it follows that $h'(x) < 0$ if and only if $(x^2 - 1) \ln 2 < 2x \ln x$. To prove the last inequality, consider

$$k(x) = (x^2 - 1) \ln 2 - 2x \ln x, \quad x \in (1, 2).$$

2.5. Applications of Derivatives

Then $k'(x) = 2x\ln 2 - 2\ln x - 2$ and $k''(x) = 2(\ln 2 - 1/x)$. Thus $k''(x) < 0$ if $x \in (1, 1/\ln 2)$, and $k''(x) > 0$ if $x \in (1/\ln 2, 2)$. Since $k'(1) = k'(2) < 0$, we get $k'(x) < 0$ for all $x \in (1,2)$. This means that k decreases on this interval; that is, $k(x) < k(1) = 0$. So $h'(x) < 0$ if $x \in (1,2)$, and therefore $h(x) < h(1) = 0$, or $f(x) < 1$ for $x \in (1,2)$. Thus we conclude that the inequality (1) holds for all x in $(1, \infty)$.

2.5.56 [5]. The proof is based on Baire's category theorem. For $n \in \mathbb{N}$, define $\mathbf{A}_n = \{x \in [0,1] : f^{(n)}(x) = 0\}$. By assumption, $[0,1]$ is a union of $\mathbf{A}_n$. So, by Baire's theorem not every $\mathbf{A}_n$ is nowhere dense. Therefore there are a closed interval $\mathbf{I}$ and an n such that $\mathbf{I} \subset \overline{\mathbf{A}_n}$. Since $f^{(n)}$ is continuous, we have $f^{(n)}(x) = 0$ for $x \in \mathbf{I}$, and consequently, f coincides on $\mathbf{I}$ with a polynomial. If $\mathbf{I} = [0,1]$, the proof is complete. If not, we can repeat the reasoning on the remaining intervals of $[0,1]$. Continuing this procedure, we get a collection of intervals whose union is dense in $[0,1]$. Moreover, f coincides on each of these intervals with a polynomial. Our task is now to show that f coincides with the same polynomial in all the intervals. To this end, consider the set $\mathbf{B}$ which is left when we remove the interiors of the intervals in the collection. Clearly, $\mathbf{B}$ is closed. Moreover, if $\mathbf{B}$ is not empty, then each element of $\mathbf{B}$ is also a limit point of $\mathbf{B}$. Indeed, if an $x_0 \in \mathbf{B}$ were not a limit point of $\mathbf{B}$, then x_0 would be a common endpoint of two intervals, say $\mathbf{I}_1$ and $\mathbf{I}_2$, such that $f^{(n_1)}(x) = 0$ for $x \in \mathbf{I}_1$ and $f^{(n_2)}(x) = 0$ for $x \in \mathbf{I}_2$. Thus $f^{(n)}(x) = 0$ for $x \in \mathbf{I}_1 \cup \mathbf{I}_2$ and $n \geq \max\{n_1, n_2\}$. Since $f^{(n)}$ is continuous, f would coincide with one polynomial on the union of $\mathbf{I}_1$ and $\mathbf{I}_2$, and consequently, x_0 would not belong to $\mathbf{B}$. A contradiction. Since $\mathbf{B}$ is closed, if it is not empty we can apply Baire's theorem again. So there is $\mathbf{A}_n$ such that $\mathbf{A}_n \cap \mathbf{B}$ is dense in $\mathbf{B} \cap \mathbf{J}$, where $\mathbf{J}$ is an interval. This implies that $f^{(n)}$ vanishes on $\mathbf{B} \cap \mathbf{J}$. On the other hand, there is a subinterval $\mathbf{K}$ of $\mathbf{J}$ which is complementary to $\mathbf{B}$. Therefore there is an integer m such that $f^{(m)}(x) = 0$ for $x \in \mathbf{K}$. If $m \leq n$, then $f^{(n)}(x) = 0$ for $x \in \mathbf{K}$. If $m > n$, then $f^{(n+1)}(x) = f^{(n+2)}(x) = \cdots = f^{(m)}(x) = \cdots = 0$ for $x \in \mathbf{B} \cap \mathbf{J}$, because each point of $\mathbf{B}$ is also its limit point. In particular, $f^{(n+1)}(x) = f^{(n+2)}(x) = \cdots = f^{(m)}(x) = \cdots = 0$ at the endpoints of

K, say a and b. So, for each $x \in \mathbf{K}$ we have

$$0 = \int_a^x f^{(m)}(t)dt = f^{(m-1)}(x) - f^{(m-1)}(a) = f^{(m-1)}(x).$$

Repeating the process, we get $f^{(n)}(x) = 0$ for $x \in \mathbf{K}$ also in the case where $m > n$. Of course, this reasoning applies to every subinterval **K** of **J** which is complementary to **B**. It then follows that $f^{(n)}(x) = 0$ for $x \in \mathbf{J}$, and consequently, there are no points of **B** in **J**, a contradiction. Thus **B** is empty, which means that $\mathbf{I} = [0,1]$ was the only interval to begin with.

2.5.57. Let

$$f(x) = \begin{cases} 0 & \text{if } x \in [0, 1/2], \\ \left(x - \tfrac{1}{2}\right)^2 & \text{if } x \in (1/2, 1]. \end{cases}$$

Then $f'(x) = 0$ for $x \in [0, 1/2]$ and $f^{(3)}(x) = 0$ for $x \in (1/2, 1]$.

The following example shows that the conclusion of 2.5.56 is not true if $\lim_{n \to \infty} f^{(n)}(x) = 0$ for each $x \in [0,1]$:

$$f(x) = \sin \frac{x}{2}, \quad x \in [0,1].$$

2.6. Strong Differentiability and Schwarz Differentiability

2.6.1. It suffices to set $x_2 = a$ in Definition 1. The converse does not hold (see, e.g., 2.1.13).

2.6.2 [M. Esser, O. Shisha, Amer. Math. Monthly 71 (1964), 904-906]. Let $\varepsilon > 0$ be arbitrarily chosen and let $\delta > 0$ be such that

$$\mathbf{B} = \{x : |x - a| < \delta\} \subset \mathbf{A}$$

and if $x_1, x_2 \in \mathbf{B}$, $x_1 \neq x_2$, then

$$\left| \frac{f(x_2) - f(x_1)}{x_2 - x_1} - f^*(a) \right| < \varepsilon.$$

2.6. Strong and Schwarz Differentiability

Now if $x \in \mathbf{A}^1$ (that is, if $f'(x)$ exists) and if $|x - a| < \frac{\delta}{2}$, then for all x_2 such that $|x_2 - x| < \frac{\delta}{2}$,

$$\left| \frac{f(x_2) - f(x)}{x_2 - x} - f^*(a) \right| < \varepsilon.$$

Letting $x_2 \to x$ gives $|f'(x) - f^*(a)| \le \varepsilon$. So, $\lim_{\substack{x \to a \\ x \in \mathbf{A}^1}} f'(x) = f^*(a) = f'(a)$. Since $\mathbf{A}^* \subset \mathbf{A}^1$, it then follows that

$$\lim_{\substack{x \to a \\ x \in \mathbf{A}^*}} f^*(x) = f^*(a) = f'(a).$$

2.6.3 [M. Esser, O. Shisha, Amer. Math. Monthly 71 (1964), 904-906]. Since f' is continuous at a, the mean value theorem yields

$$\lim_{\substack{(x_1,x_2)\to(a,a) \\ x_1 \ne x_2}} \frac{f(x_1) - f(x_2)}{x_1 - x_2} = \lim_{\substack{(x_1,x_2)\to(a,a) \\ x_1 \ne x_2}} f'(x_1 + \theta(x_2 - x_1)) = f'(a).$$

2.6.4 [M. Esser, O. Shisha, Amer. Math. Monthly 71 (1964), 904-906]. No. Consider the function f defined on the interval $(-1, 1)$ by

$$f(x) = \int_0^x g(t)dt,$$

where

$$g(t) = \begin{cases} 0 & \text{if } t \in (-1, 0] \cup \bigcup_{k=1}^\infty \left[\frac{1}{2k+1}, \frac{1}{2k}\right), \\ t & \text{if } t \in \bigcup_{k=1}^\infty \left[\frac{1}{2k}, \frac{1}{2k-1}\right). \end{cases}$$

Then f is continuous on $(-1, 1)$, and

$$\lim_{\substack{(x_1,x_2)\to(0,0) \\ x_1 \ne x_2}} \frac{f(x_1) - f(x_2)}{x_1 - x_2} = \lim_{\substack{(x_1,x_2)\to(0,0) \\ x_1 \ne x_2}} \frac{1}{x_1 - x_2} \int_{x_2}^{x_1} g(t)dt = 0.$$

The last equality follows from the fact that

$$0 \le \int_{x_2}^{x_1} g(t)dt \le \frac{x_1^2 - x_2^2}{2} \quad \text{for} \quad x_2 < x_1.$$

So, f is strongly differentiable at zero. On the other hand, f' does not exist at $\frac{1}{n}$, $n = 3, 4, 5, \ldots$.

2.6.5. The result follows immediately from 2.6.2 and 2.6.3.

2.6.6 [C.L. Belna, M.J. Evans, P.D. Humke, Amer. Math. Monthly 86 (1979), 121-123]. Note first that f' is in the first class of Baire, because
$$f'(x) = \lim_{n\to\infty} \frac{f\left(x+\frac{1}{n}\right) - f(x)}{\frac{1}{n}}.$$
Therefore the set of points of discontinuity of f' is of the first category (see, e.g., 1.7.20). So, the assertion follows from the result in 2.6.3.

2.6.7. Let α be a real number such that $f(a) < \alpha < f(b)$, and denote $c = \inf\{x \in (a,b) : f(x) > \alpha\}$. Clearly, $c \neq a$ and $c \neq b$. By the definition of the greatest lower bound, $f(x) \leq \alpha$ for $x \in [a,c]$, and there exists a positive sequence $\{h_n\}$ converging to zero such that $f(c+h_n) > \alpha$. Since f is Schwarz differentiable at c,
$$f^s(c) = \lim_{n\to\infty} \frac{f(c+h_n) - f(c-h_n)}{2h_n} \geq 0.$$
It is worth noting here that in much the same way one can show that if $f(a) > f(b)$, then there is $c \in (a,b)$ such that $f^s(c) \leq 0$.

2.6.8 [C.E. Aull, Amer. Math. Monthly 74 (1967), 708-711]. If f is identically zero on $[a,b]$, then the statement is obvious. So assume that there is $c \in (a,b)$ such that, for example, $f(c) > 0$. Then it follows from the foregoing result that there are x_1 and x_2 such that $a < x_1 < c < x_2 < b$, $f^s(x_1) \geq 0$, and $f^s(x_2) \leq 0$.

2.6.9 [C.E. Aull, Amer. Math. Monthly 74 (1967), 708-711]. It is easy to see that the auxiliary function
$$F(x) = f(x) - f(a) - \frac{f(b) - f(a)}{b-a}(x-a)$$
satisfies the assumptions of the foregoing problem.

2.6.10 [C.E. Aull, Amer. Math. Monthly 74 (1967), 708-711]. Since f^s is bounded on (a,b), there is $M \geq 0$ such that $|f^s(x)| \leq M$ for all $x \in (a,b)$. It follows from the result in the foregoing problem that
$$-M \leq \frac{f(x) - f(t)}{x - t} \leq M \quad \text{if} \quad x,t \in (a,b),\ x \neq t.$$

2.6. Strong and Schwarz Differentiability

Consequently, $|f(x) - f(t)| \leq M|x - t|$.

2.6.11 [C.E. Aull, Amer. Math. Monthly 74 (1967), 708-711]. By 2.6.9, there are x_1 and x_2 between x and $x+h$ $(x, x+h \in (a,b))$ such that
$$f^s(x_2) \leq \frac{f(x+h) - f(x)}{h} \leq f^s(x_1).$$
On the other hand, by the continuity of f^s there is x_3 between x and $x + h$ such that $f^s(x_3) = \frac{f(x+h)-f(x)}{h}$. Upon letting $h \to 0$, we get $f^s(x) = f'(x)$.

2.6.12. If x, z are in **I** and $x < z$, then by 2.6.9 there is $x_2 \in (x, z)$ such that
$$\frac{f(z) - f(x)}{z - x} \geq f^s(x_2) \geq 0.$$

2.6.13. As above, it suffices to apply the result given in 2.6.9.

2.6.14. No. Consider, for example, the function f given by $f(x) = x - 2|x|$, $x \in (-1, 1)$. It is easy to check that $f^s(0) = 1$ and $f(0)$ is the maximum value of f on $(-1, 1)$.

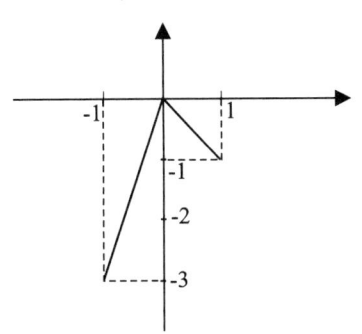

2.6.15 [C. Belna, M. Evans, P. Humke, Amer. Math. Monthly 86 (1979), 121-123]. It suffices to show that there is a residual set on which the first equality holds, because the second equality can be obtained by replacing f by $-f$ in the first one. By definition, $D_s f(x) \geq D_* f(x)$. Our task is to show that the set
$$\mathbf{A}(f) = \{x : D_s f(x) > D_* f(x)\}$$

is of the first category. Observe that $\mathbf{A}(f)$ is a countable union of the sets
$$\mathbf{A}(f,\alpha) = \{x : D_s f(x) > \alpha > D_* f(x)\}, \quad \alpha \in \mathbb{Q}.$$
So, it is sufficient to show that each of these sets is of the first category. Since $\mathbf{A}(f,\alpha) = \mathbf{A}(g,0)$, where $g(x) = f(x) - \alpha x$, it is enough to show that $\mathbf{A}(f,0)$ is of the first category. To this end, note that

$$\mathbf{A}(f,0) = \bigcup_{n=1}^{\infty} \mathbf{A}_n(f,0)$$
$$= \bigcup_{n=1}^{\infty} \left(\left\{ x : f(x-h) \leq f(x+h) \text{ for } 0 < h < \frac{1}{n} \right\} \cap \mathbf{A}(f,0) \right).$$

Thus it suffices to show that all the sets $\mathbf{A}_n(f,0)$ are of the first category. Suppose, contrary to our claim, that there is $n \in \mathbb{N}$ such that $\mathbf{A}_n(f,0)$ is of the second category. Then there is an open interval $\mathbf{I}$ such that $\mathbf{A}_n(f,0)$ is also of the second category on every open subinterval of $\mathbf{I}$. Assume additionally that the length of $\mathbf{I}$ is less than $\frac{1}{n}$ and that $a, b \in \mathbf{I}$ with $a < b$. Let $\mathbf{S} \subset \mathbb{R}$ be a residual set such that $f_{|\mathbf{S}}$ is continuous, and choose $c \in \mathbf{S} \cap (a,b)$. Let $\varepsilon > 0$ be chosen arbitrarily. Then there is an open subinterval $\mathbf{J}$ of the interval (a,b) such that $c \in \mathbf{J}$ and

(1) $$f(x) > f(c) - \varepsilon \quad \text{for} \quad x \in \mathbf{S} \cap \mathbf{J}.$$

Now let $\mathbf{K}$ be an open subinterval of (a,b) such that $\tilde{\mathbf{K}} = 2\mathbf{K} - b = \{y : y = 2x - b, x \in \mathbf{K}\} \subset \mathbf{J}$. Since the set

$$\mathbf{S}_n = \left\{ x : f(x-h) \leq f(x+h) \text{ for } 0 < h < \frac{1}{n} \right\}$$

is of the second category on $\mathbf{K}$ and $\mathbf{S}$ is residual on $\tilde{\mathbf{K}}$, we see that the set $(2\mathbf{S}_n - b) \cap (\mathbf{S} \cap \tilde{\mathbf{K}})$ is of the second category, hence nonempty. We can pick an $\tilde{x} \in \mathbf{S} \cap \tilde{\mathbf{K}}$ such that $\frac{\tilde{x}+b}{2} \in \mathbf{S}_n$. Consequently, taking $h = \frac{b-\tilde{x}}{2}$ (clearly, $0 < h < 1/n$), we see that $f(\tilde{x}) \leq f(b)$. Moreover, (1) implies that $f(c) - \varepsilon < f(\tilde{x})$. In view of the arbitrariness of $\varepsilon > 0$, we get $f(c) \leq f(b)$. In an entirely similar manner one can show that $f(a) \leq f(c)$. So we have proved that f increases on $\mathbf{I}$. Consequently, $D_* f(x) \geq 0$ for $x \in \mathbf{I}$. Thus $A(f,0) \cap \mathbf{I} = \emptyset$, a contradiction.

2.6. Strong and Schwarz Differentiability 315

2.6.16. The result follows immediately from the foregoing problem. Note that this is a generalization of 2.6.6.

2.6.17 [J. Swetits, Amer. Math. Monthly 75 (1968), 1093-1095]. We may assume that f is locally bounded on $[x_1, x_0)$ and $x_0 - x_1 < \delta < 1$. Let x_2 be the midpoint of $[x_1, x_0)$. Then there is $M > 0$ such that $|f(x)| \leq M$ for $x \in [x_1, x_2]$. Choose h, $0 < h < \frac{\delta}{2}$, such that

$$|f(x_2 + h)| > 1 + M + |f^s(x_2)|.$$

Then
$$\left|\frac{f(x_2 + h) - f(x_2 - h)}{2h} - f^s(x_2)\right|$$
$$\geq \left|\frac{f(x_2 + h) - f(x_2 - h)}{2h}\right| - |f^s(x_2)|$$
$$\geq |f(x_2 + h)| - |f(x_2 - h)| - |f^s(x_2)|$$
$$\geq |f(x_2 + h)| - M - |f^s(x_2)| > 1.$$

So, f is not uniformly Schwarz differentiable on $[a, b]$.

2.6.18. One can use the result in 2.6.9 and proceed as in the solution of 2.2.26.

2.6.19. Consider the function defined as follows:
$$f(x) = \begin{cases} 0 & \text{if } x \in \mathbb{R} \setminus \{0\}, \\ 1 & \text{if } x = 0. \end{cases}$$

Then f^s is identically zero on $\mathbb{R}$, so continuous, but f is not uniformly Schwarz differentiable on any interval containing zero.

2.6.20 [J. Swetits, Amer. Math. Monthly 75 (1968), 1093-1095]. Assume first that f is uniformly Schwarz differentiable on every $[a, b] \subset \mathbf{I}$. Let $x_0 \in (a, b)$ and let $\delta_1 > 0$ be such that $[x_0 - \delta_1, x_0 + \delta_1] \subset (a, b)$. Put $\mathbf{I}_1 = (x_0 - \delta_1, x_0 + \delta_1)$. Since f is locally bounded on $\mathbf{I}$, there is $M > 0$ such that
$$|f(x)| \leq M \quad \text{for} \quad x \in \mathbf{I}_1.$$
Let $\delta > 0$ be such that
$$\left|\frac{f(x + h) - f(x - h)}{2h} - f^s(x)\right| < 1$$

for $|h| < \delta$ and $x \in [a,b]$. Then, for $x \in \mathbf{I}_2 = \left(x_0 - \frac{\delta_1}{2}, x_0 + \frac{\delta_1}{2}\right)$ and $|h_1| < \min\{\delta, \delta_1/2\}$,

$$|f^s(x)| < 1 + \left|\frac{f(x+h_1) - f(x-h_1)}{2h_1}\right| < 1 + \frac{2M}{2|h_1|}.$$

Thus f^s is locally bounded on $\mathbf{I}$. We now show that f is continuous on $\mathbf{I}$. Suppose, contrary to our claim, that f is discontinuous at an $x_0 \in [a,b] \subset \mathbf{I}$. Then there is $\varepsilon > 0$ such that for every $\delta > 0$ there is $x' \in [a,b] \cap (x_0 - \delta, x_0 + \delta)$ for which $|f(x') - f(x_0)| > \varepsilon$. Since f^s is locally bounded, there is $M_1 > 0$ such that $|f^s(x)| \leq M_1$ for x in the interval with endpoints x' and x_0. Consequently,

$$\left|\frac{f(x') - f(x_0)}{x' - x_0} - f^s\left(\frac{x' + x_0}{2}\right)\right| \geq \frac{\varepsilon}{|x' - x_0|} - M_1,$$

which contradicts uniform Schwarz differentiability of f on $[a,b]$. So we have shown that f is continuous on $\mathbf{I}$, and, by the result in 2.6.18, so is f^s. This combined with 2.6.11 shows that f' exists and is continuous on $\mathbf{I}$. Sufficiency follows immediately from the result in 2.6.18.

Chapter 3

Sequences and Series of Functions

3.1. Sequences of Functions, Uniform Convergence

3.1.1. Suppose first that $f_n \underset{\mathbf{B}}{\rightrightarrows} f$. Then, given $\varepsilon > 0$, there is n_0 such that
$$|f_n(x) - f(x)| < \varepsilon$$
for all $n \geq n_0$ and all $x \in \mathbf{B}$. Hence, for $n \geq n_0$,
$$d_n = \sup\{|f_n(x) - f(x)| : x \in \mathbf{B}\} \leq \varepsilon,$$
and consequently, $\lim\limits_{n \to \infty} d_n = 0$.

Suppose now that $\lim\limits_{n \to \infty} d_n = 0$. Then
$$|f_n(x) - f(x)| \leq \sup\{|f_n(x) - f(x)| : x \in \mathbf{B}\} < \varepsilon$$
for sufficiently large n and for all $x \in \mathbf{B}$, which means that $\{f_n\}$ is uniformly convergent on $\mathbf{B}$ to f.

3.1.2. Given $\varepsilon > 0$, we get
$$|f_n(x) - f(x)| < \frac{\varepsilon}{2} \quad \text{and} \quad |g_n(x) - g(x)| < \frac{\varepsilon}{2}$$

317

for n large enough and for all $x \in \mathbf{A}$. Thus
$$|f_n(x) + g_n(x) - (f(x) + g(x))| \leq |f_n(x) - f(x)| + |g_n(x) - g(x)| < \varepsilon$$
for n large enough and for all $x \in \mathbf{A}$.

To see that the analogous assertion does not hold for the product of two uniformly convergent sequences, consider the following functions:
$$f_n(x) = x\left(1 - \frac{1}{n}\right) \quad \text{and} \quad g_n(x) = \frac{1}{x^2}, \quad x \in (0,1).$$

We have $f_n \underset{(0,1)}{\rightrightarrows} f$ and $g_n \underset{(0,1)}{\rightrightarrows} g$, where $f(x) = x$ and $g(x) = \frac{1}{x^2}$. On the other hand,
$$f_n(x) g_n(x) = \frac{1}{x}\left(1 - \frac{1}{n}\right).$$
So $\{f_n g_n\}$ is pointwise convergent on $(0,1)$ to the function $x \mapsto \frac{1}{x}$. Since
$$d_n = \sup\left\{\left|f_n(x)g_n(x) - \frac{1}{x}\right| : x \in (0,1)\right\} = +\infty, \quad n \in \mathbb{N},$$
the convergence is not uniform.

3.1.3. Note first that if $\{g_n\}$ converges uniformly on $\mathbf{A}$ to a bounded function g, then there is $C > 0$ such that for sufficiently large n,
$$|g_n(x)| \leq C \quad \text{for all } x \in \mathbf{A}.$$
Given $\varepsilon > 0$, by the uniform convergence of $\{f_n\}$ and $\{g_n\}$, we get
$$|f_n(x) - f(x)| < \frac{\varepsilon}{2C} \quad \text{and} \quad |g_n(x) - g(x)| < \frac{\varepsilon}{2M}$$
for sufficiently large n and for all $x \in \mathbf{A}$. Hence, for sufficiently large n and for all $x \in \mathbf{A}$,
$$|f_n(x) \cdot g_n(x) - f(x) \cdot g(x)|$$
$$\leq |f_n(x) - f(x)||g_n(x)| + |g_n(x) - g(x)||f(x)| < \varepsilon.$$

3.1. Sequences of Functions, Uniform Convergence

3.1.4. It follows from the Cauchy criterion for convergence of sequences of real numbers that $\{f_n\}$ is pointwise convergent on $\mathbf{A}$, say, to f. Our task is now to show that the convergence is uniform. Let $\varepsilon > 0$ be arbitrarily chosen. By assumption, there is n_0 such that if $n, m > n_0$, then
$$|f_n(x) - f_m(x)| < \frac{1}{2}\varepsilon$$
for every $x \in \mathbf{A}$. By the continuity of the absolute value function we get
$$\lim_{m \to \infty} |f_n(x) - f_m(x)| = |f_n(x) - f(x)| \leq \frac{1}{2}\varepsilon < \varepsilon$$
for every $x \in \mathbf{A}$ and for all $n > n_0$.

3.1.5. Let $\{f_n\}$ be a sequence of bounded functions uniformly convergent on $\mathbf{A}$ to f. Then, given $\varepsilon > 0$, there is $n_0 \in \mathbb{N}$ such that
$$|f(x)| \leq |f_{n_0}(x) - f(x)| + |f_{n_0}(x)| < \varepsilon + |f_{n_0}(x)|$$
for all $x \in \mathbf{A}$. Since f_{n_0} is bounded on $\mathbf{A}$, so is f.

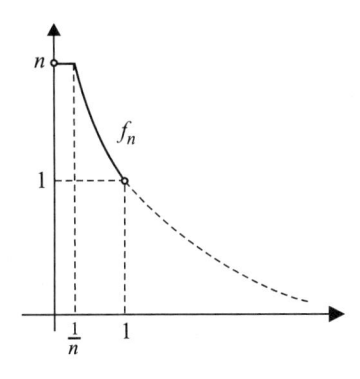

The limit function of a pointwise convergent sequence of bounded functions need not be bounded. To see this take, for example,
$$f_n(x) = \min\left\{\frac{1}{x}, n\right\}, \quad x \in (0,1), \ n \in \mathbb{N}.$$
The sequence $\{f_n\}$ converges to the unbounded function $x \mapsto 1/x$, $x \in (0,1)$.

3.1.6. For $x \in \mathbb{R}$, $\lim\limits_{n\to\infty} f_n(x) = 0$. The convergence is not uniform on $\mathbb{R}$ because $d_{2n} = +\infty$. Clearly, the subsequence $\{f_{2n-1}\}$ is uniformly convergent on $\mathbb{R}$.

3.1.7. The proof runs as in 3.1.4.

3.1.8.

(a) We have
$$\frac{1}{1+(nx-1)^2} \xrightarrow[n\to\infty]{} f(x),$$
where
$$f(x) = \begin{cases} 0 & \text{for } x \in (0,1], \\ \frac{1}{2} & \text{for } x = 0. \end{cases}$$
Since the limit function is not continuous, the convergence is not uniform (see, e.g., 1.2.34).

(b) We have
$$\frac{x^2}{x^2+(nx-1)^2} \xrightarrow[n\to\infty]{} 0$$
and $d_n = \sup\{|f_n(x) - 0| : x \in [0,1]\} = f_n\left(\frac{1}{n}\right) = 1$. By 3.1.1 the convergence is not uniform.

(c) Since
$$x^n(1-x) \xrightarrow[n\to\infty]{} 0$$
and $d_n = \sup\{|f_n(x) - 0| : x \in [0,1]\} = f_n\left(\frac{n}{n+1}\right) = \frac{n^n}{(n+1)^{n+1}}$, we see that $\{f_n\}$ converges uniformly on $[0,1]$.

(d) The convergence is not uniform because
$$d_n = \frac{n^{n+1}}{(n+1)^{n+1}} \xrightarrow[n\to\infty]{} \frac{1}{e}.$$

(e) Since $d_n = f\left(\frac{n}{n+4}\right) \xrightarrow[n\to\infty]{} 0$, the convergence is uniform.

(f) The sequence is uniformly convergent because
$$d_n = \sup\{|f_n(x) - x| : x \in [0,1]\} = 1 - f_n(1) = \frac{1}{n+1} \xrightarrow[n\to\infty]{} 0.$$

3.1. Sequences of Functions, Uniform Convergence

(g) The sequence is pointwise convergent to
$$f(x) = \begin{cases} 1 & \text{for } x \in [0,1), \\ \frac{1}{2} & \text{for } x = 1. \end{cases}$$
So the limit function is not continuous and therefore the convergence is not uniform (see, e.g., 1.2.34).

3.1.9.

(a) One can easily see that $f_n(x) \xrightarrow[n \to \infty]{} 0$ and $d_n = \frac{1}{4}$. Thus the convergence is not uniform on **A**. On the other hand,
$$\sup\{|f_n(x)| : x \in \mathbf{B}\} = \left(\frac{1}{\sqrt{2}}\right)^n \left(1 - \left(\frac{1}{\sqrt{2}}\right)^n\right), \quad n \geq 2,$$
and therefore the sequence converges uniformly on **B**.

(b) The sequence converges uniformly on $\mathbb{R}$ to the zero function, and so it also converges uniformly on each subset of $\mathbb{R}$.

3.1.10.

(a) Since $d_n = \arctan \frac{1}{\sqrt{n^3}}$, $\{f_n\}$ converges uniformly on $\mathbb{R}$ to zero.

(b) $f_n(x) \xrightarrow[n \to \infty]{} x^2$, and since $f_n(\sqrt{n}) - n = n(\ln 2 - 1)$, the convergence cannot be uniform on $\mathbb{R}$.

(c) We have $f_n(x) \xrightarrow[n \to \infty]{} \frac{1}{x}$. The sequence cannot converge uniformly on $(0, \infty)$, because $f_n\left(\frac{1}{n}\right) - n = n(\ln 2 - 1)$.

(d) $f_n(x) \xrightarrow[n \to \infty]{} f(x)$, where
$$f(x) = \begin{cases} 1 & \text{if } |x| \leq 1, \\ |x| & \text{if } |x| > 1. \end{cases}$$
Set $u_n = \sqrt[2n]{1 + x^{2n}}$. Then, for $x > 1$,
$$\sqrt[2n]{1 + x^{2n}} - x = u_n - x = \frac{u_n^{2n} - x^{2n}}{u_n^{2n-1} + u_n^{2n-2}x + \cdots + x^{2n-1}}$$
$$= \frac{1}{u_n^{2n-1} + u_n^{2n-2}x + \cdots + x^{2n-1}} \leq \frac{1}{2n}.$$
It then follows that
$$d_n \leq \sup_{x \in [0,1]} |f_n(x) - f(x)| + \sup_{x \in [1, \infty)} |f_n(x) - f(x)| \leq \sqrt[2n]{2} - 1 + \frac{1}{2n},$$

which shows that $\{f_n\}$ converges uniformly on $\mathbb{R}$.

(e) As in (d), one can show that the sequence is uniformly convergent on $\mathbb{R}$ to
$$f(x) = \begin{cases} 2 & \text{if } |x| \leq 2, \\ |x| & \text{if } |x| > 2. \end{cases}$$

(f) We have
$$d_n = \sup_{x \in \mathbb{R}} |\sqrt{n+1} \sin^n x \cos x| = \left(\sqrt{\frac{n}{n+1}}\right)^n \xrightarrow[n \to \infty]{} \frac{1}{\sqrt{e}}.$$
Thus the convergence cannot be uniform on $\mathbb{R}$.

(g) The sequence is pointwise convergent to $\ln x$ (see, e.g., I, 2.5.4). By Taylor's formula,
$$d_n = \sup_{x \in [1,a]} |n(\sqrt[n]{x} - 1) - \ln x| = \sup_{x \in [1,a]} \left| n\left(e^{\frac{1}{n}\ln x} - 1\right) - \ln x \right|$$
$$= \sup_{x \in [1,a]} \left| n\left(1 + \frac{1}{n}\ln x - \frac{\ln^2 x}{2n^2} e^{\zeta_n} - 1\right) - \ln x \right| \leq \frac{\ln^2 a}{2n} a^{\frac{1}{n}},$$
because $0 < \zeta_n < \frac{\ln a}{n}$. Consequently, $\lim_{n \to \infty} d_n = 0$, which shows that $f_n \underset{[1,a]}{\rightrightarrows} f$, where $f(x) = \ln x$.

3.1.11. We have $[nf(x)] = nf(x) - p_n(x)$, where $0 \leq p_n(x) < 1$. Hence
$$\sup_{x \in [a,b]} |f_n(x) - f(x)| = \sup_{x \in [a,b]} \left|\frac{p_n(x)}{n}\right| \leq \frac{1}{n},$$
and therefore $f_n \underset{[a,b]}{\rightrightarrows} f$.

3.1.12. Since
$$\sin \sqrt{4\pi^2 n^2 + x^2} = \sin\left(2\pi n \sqrt{1 + \frac{x^2}{4\pi^2 n^2}} + 2n\pi - 2n\pi\right)$$
$$= \sin 2n\pi \left(\sqrt{1 + \frac{x^2}{4\pi^2 n^2}} - 1\right)$$
$$= \sin \frac{x^2}{\sqrt{4n^2\pi^2 + x^2} + 2n\pi},$$

3.1. Sequences of Functions, Uniform Convergence

we see that $\lim\limits_{n\to\infty} n\sin\sqrt{4\pi^2n^2+x^2} = \frac{x^2}{4\pi}$. Moreover, if $x \in [0,a]$, then, using the fact that $\sin x \geq x - \frac{x^3}{3!}$, we get

$$\left| n\sin\sqrt{4\pi^2n^2+x^2} - \frac{x^2}{4\pi} \right| \leq \frac{a^2}{4\pi}\left(1 - \frac{2}{\sqrt{1+\frac{a^2}{4\pi^2n^2}}+1}\right) + \frac{n}{3!}\frac{a^6}{8n^3\pi^3}.$$

This establishes the uniform convergence of the sequence on $[0,a]$.

For $x \in \mathbb{R}$, by the inequality $|\sin x| \leq |x|$, we obtain

$$\left| n\sin\sqrt{4\pi^2n^2+x^2} - \frac{x^2}{4\pi} \right| \geq \frac{x^2}{4\pi}\left(1 - \frac{2}{\sqrt{1+\frac{x^2}{4\pi^2n^2}}+1}\right),$$

which shows that the convergence cannot be uniform on $\mathbb{R}$.

3.1.13. First we show, by induction, that for any positive integer n,

$$0 \leq \sqrt{x} - P_n(x) \leq \frac{2\sqrt{x}}{2+n\sqrt{x}} < \frac{2}{n}, \quad x \in [0,1].$$

For $n = 1$ the inequalities are obvious. Now, assuming the inequalities hold for n, we will prove them for $n+1$. It follows from the induction assumption that

$$0 \leq \sqrt{x} - P_n(x) \leq \sqrt{x}.$$

Hence, by the definition of P_{n+1},

$$\sqrt{x} - P_{n+1}(x) = (\sqrt{x} - P_n(x))\left(1 - \frac{1}{2}(\sqrt{x} + P_n(x))\right).$$

Thus $\sqrt{x} - P_{n+1}(x) \geq 0$. Moreover,

$$\sqrt{x} - P_{n+1}(x) \leq \frac{2\sqrt{x}}{2+n\sqrt{x}}\left(1 - \frac{\sqrt{x}}{2}\right)$$

$$\leq \frac{2\sqrt{x}}{2+n\sqrt{x}}\left(1 - \frac{\sqrt{x}}{2+(n+1)\sqrt{x}}\right)$$

$$= \frac{2\sqrt{x}}{2+(n+1)\sqrt{x}}.$$

Since $|x| = \sqrt{x^2}$, it follows from the proved inequalities that the sequence of polynomials $\{P_n(x^2)\}$ converges uniformly on $[-1,1]$ to the absolute value function $|x|$.

3.1.14. By the mean value theorem,

$$\left| \frac{f\left(x + \frac{1}{n}\right) - f(x)}{\frac{1}{n}} - f'(x) \right| = |f'(\zeta_n) - f'(x)|,$$

where $\zeta_n \in \left(x, x + \frac{1}{n}\right)$. Since the derivative f' is uniformly continuous on $\mathbb{R}$, given $\varepsilon > 0$ there is an n_0 such that if $n \geq n_0$, then

$$|f'(\zeta_n) - f'(x)| < \varepsilon \quad \text{for all } x \in \mathbb{R}.$$

Thus the uniform convergence on $\mathbb{R}$ is proved.

Consider $f(x) = x^3$, $x \in \mathbb{R}$. Then

$$d_n = \sup_{x \in \mathbb{R}} \left| \frac{f\left(x + \frac{1}{n}\right) - f(x)}{\frac{1}{n}} - f'(x) \right| = \sup_{x \in \mathbb{R}} \left| 3x\frac{1}{n} + \frac{1}{n^2} \right| = +\infty,$$

which shows that the convergence is not uniform. So we see that the assumption of the uniform continuity of f' is essential.

3.1.15. Let $\varepsilon > 0$ be arbitrarily chosen. It follows from the uniform convergence of the sequence on $\mathbb{R}$ that there is an $n_0 \in \mathbb{N}$ such that

$$|f_{n_0}(x) - f(x)| < \frac{\varepsilon}{3} \quad \text{for all} \quad x \in \mathbb{R}.$$

Now the uniform continuity of f_{n_0} implies that there exists $\delta > 0$ such that $|f_{n_0}(x) - f_{n_0}(x')| < \frac{\varepsilon}{3}$ whenever $|x - x'| < \delta$. Consequently,

$$|f(x) - f(x')| \leq |f_{n_0}(x) - f(x)| + |f_{n_0}(x) - f_{n_0}(x')| + |f_{n_0}(x') - f(x')| < \varepsilon$$

whenever $|x - x'| < \delta$.

3.1.16. Set $g_n(x) = f_n(x) - f(x)$ for $x \in \mathbf{K}$. We will show that $\{g_n\}$ converges to zero uniformly on $\mathbf{K}$. Let $\varepsilon > 0$ be arbitrarily chosen. Since $\{g_n\}$ is pointwise convergent to zero on $\mathbf{K}$, for $x \in \mathbf{K}$ there is n_x such that

$$0 \leq g_{n_x}(x) < \frac{\varepsilon}{2}.$$

It follows from the continuity of g_{n_x} and from the monotonicity of the sequence $\{g_n\}$ that there is a neighborhood $\mathbf{O}(x)$ of x such that

(1) $\qquad 0 \leq g_n(t) < \varepsilon \quad \text{for} \quad n \geq n_x \text{ and } t \in \mathbf{O}(x).$

3.1. Sequences of Functions, Uniform Convergence

Since $\mathbf{K}$ is compact, there are finitely many points $x_1, \ldots, x_n \in \mathbf{K}$ such that $\mathbf{K} \subset \mathbf{O}(x_1) \cup \mathbf{O}(x_2) \cup \cdots \cup \mathbf{O}(x_n)$. Now if

$$n_0 = \max\{n_{x_1}, n_{x_2}, \ldots, n_{x_n}\},$$

then (1) holds for all $n > n_0$ and all $x \in \mathbf{K}$.

To see that the compactness of $\mathbf{K}$ is essential, consider

$$f_n(x) = \frac{1}{1+nx}, \quad x \in (0,1), \; n = 1, 2, \ldots.$$

Then $d_n = \sup\limits_{x \in (0,1)} |f_n(x) - f(x)| = 1$, and therefore the convergence is not uniform.

The continuity of the limit function is also essential. Indeed, the sequence

$$f_n(x) = x^n, \; x \in [0,1], \; n \in \mathbb{N},$$

fails to converge uniformly on $[0,1]$.

The assumption of continuity of f_n cannot be omitted, as the following example shows. The functions

$$f_n(x) = \begin{cases} 0 & \text{if } x = 0 \text{ or } \frac{1}{n} \le x \le 1, \\ 1 & \text{if } 0 < x < \frac{1}{n} \end{cases}$$

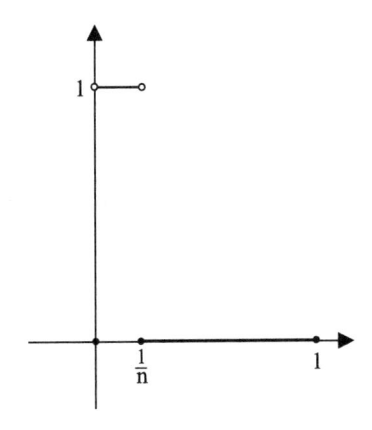

are not continuous. They form a monotonic sequence pointwise convergent to zero on $[0,1]$, but the convergence is not uniform.

Finally, the functions defined by

$$f_n(x) = \begin{cases} 2n^2 x & \text{for } 0 \leq x \leq \frac{1}{2n}, \\ n - 2n^2\left(x - \frac{1}{2n}\right) & \text{for } \frac{1}{2n} < x \leq \frac{1}{n}, \\ 0 & \text{for } \frac{1}{n} < x \leq 1 \end{cases}$$

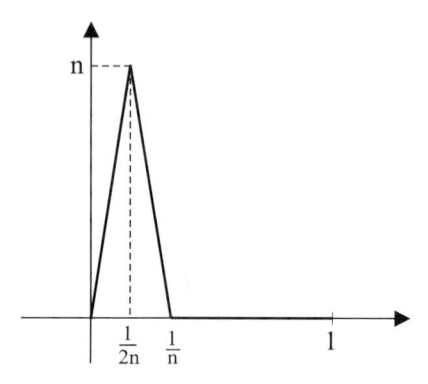

are continuous and form a sequence which is pointwise convergent to the zero function on $[0,1]$. Note that the sequence $\{f_n\}$ is not monotonic and the convergence is not uniform.

3.1.17. Let $\{f_n\}$ be a sequence of continuous functions uniformly convergent on a compact set **K** to the limit function f. Let $\varepsilon > 0$ be given. Choose n_0 such that (see 3.1.7)

$$|f_n(x) - f_{n_0}(x)| < \frac{\varepsilon}{3} \quad \text{for} \quad n > n_0 \quad \text{and all} \quad x \in \mathbf{K}.$$

Next, since each function f_n is uniformly continuous on **K**, one can choose $\delta > 0$ such that if $x, x' \in \mathbf{K}$ and $|x - x'| < \delta$, then

(1) $\qquad |f_k(x) - f_k(x')| < \frac{\varepsilon}{3} \quad \text{for} \quad 1 \leq k \leq n_0.$

Therefore we get

$$|f_n(x) - f_n(x')| \leq |f_n(x) - f_{n_0}(x)| + |f_{n_0}(x) - f_{n_0}(x')|$$
$$+ |f_{n_0}(x') - f_n(x')| < \varepsilon$$

for $|x - x'| < \delta$ and $n > n_0$. This together with (1) proves the equicontinuity of the sequence $\{f_n\}$ on **K**.

3.1. Sequences of Functions, Uniform Convergence 327

3.1.18. Let $\{f_{n_k}\}$ be a subsequence of $\{f_n\}$, and $\{x_n\}$ a sequence of elements of $\mathbf{A}$ converging to $x \in \mathbf{A}$. We define the sequence $\{y_m\}$ by setting

$$y_m = \begin{cases} x_1 & \text{for } 1 \leq m \leq n_1, \\ x_2 & \text{for } n_1 < m \leq n_2, \\ \dots, & \\ x_k & \text{for } n_{k-1} < m \leq n_k, \\ \dots & \end{cases}$$

Then the sequence $\{y_m\}$ converges to x, so $\lim\limits_{m \to \infty} f_m(y_m) = f(x)$. Thus $\lim\limits_{k \to \infty} f_{n_k}(y_{n_k}) = \lim\limits_{k \to \infty} f_{n_k}(x_k) = f(x)$.

3.1.19. Note first that if $\{f_n\}$ converges continuously on $\mathbf{A}$ to f, then $\{f_n\}$ converges pointwise to the same limit function. To see this, it is enough to consider constant sequences all of whose terms are equal to an element of $\mathbf{A}$. Let $x \in \mathbf{A}$ be arbitrarily chosen and let $\{x_n\}$ be a sequence of elements in $\mathbf{A}$ converging to x. Given $\varepsilon > 0$, the pointwise convergence of the sequence implies that there is n_1 (which can depend on x_1) such that

$$|f_{n_1}(x_1) - f(x_1)| < \frac{\varepsilon}{2}.$$

Similarly, there is n_2 (which can depend on x_2), $n_2 > n_1$, such that

$$|f_{n_2}(x_2) - f(x_2)| < \frac{\varepsilon}{2}.$$

Continuing the process, we get the sequence $\{n_k\}$ such that

$$|f_{n_k}(x_k) - f(x_k)| < \frac{\varepsilon}{2}, \quad k \in \mathbb{N}.$$

Moreover, by the result in the foregoing problem,

$$|f_{n_k}(x_k) - f(x)| < \frac{\varepsilon}{2}, \quad k \geq k_0.$$

Consequently, $|f(x_k) - f(x)| \leq |f_{n_k}(x_k) - f(x_k)| + |f_{n_k}(x_k) - f(x)| < \varepsilon$ for $k \geq k_0$.

3.1.20. Let $\{x_n\}$ be a sequence of elements in $\mathbf{A}$ converging to $x \in \mathbf{A}$. Let $\varepsilon > 0$ be given. It follows from the uniform convergence of $\{f_n\}$ that

$$|f(x_n) - f_n(x_n)| \leq \sup_{y \in \mathbf{A}} |f(y) - f_n(y)| < \frac{\varepsilon}{2} \quad \text{for} \quad n \geq n_0.$$

Since f is continuous,

$$|f(x_n) - f(x)| < \frac{\varepsilon}{2} \quad \text{for} \quad n \geq n_1.$$

Hence if $n \geq \max\{n_0, n_1\}$, we have

$$|f_n(x_n) - f(x)| \leq |f_n(x_n) - f(x_n)| + |f(x_n) - f(x)| < \varepsilon.$$

The converse of the statement just proved is not true, as the following example shows. Let $\mathbf{A} = (0,1)$ and $f_n(x) = x^n$. It is easy to see that $\{f_n\}$ fails to converge uniformly to zero on $(0,1)$. But $\{f_n\}$ does converge continuously on $(0,1)$. Indeed, if $\{x_n\}$ is a sequence of points in $(0,1)$ converging to $x \in (0,1)$, then there is $0 < a < 1$ such that $x_n < a$. Therefore $\lim\limits_{n \to \infty} f_n(x_n) = 0$.

3.1.21. The implication (i) $\implies$ (ii) has been proved in the foregoing problem. Our task is to prove (ii) $\implies$ (i). We know (see 3.1.19) that the limit function f is continuous on $\mathbf{K}$. Suppose, contrary to our claim, that $\{f_n\}$ fails to converge uniformly on $\mathbf{K}$. Then there are $\varepsilon_0 > 0$, a sequence $\{n_k\}$ of positive integers, and a sequence $\{x_k\}$ of elements in $\mathbf{K}$ such that

$$|f_{n_k}(x_k) - f(x_k)| > \varepsilon_0.$$

Since $\mathbf{K}$ is compact, we can assume without loss of generality that $\{x_k\}$ converges, say, to $x \in \mathbf{K}$. On the other hand, by 3.1.18,

$$|f_{n_k}(x_k) - f(x)| < \frac{\varepsilon_0}{3} \quad \text{for} \quad k > k_0.$$

Moreover, the continuity of f implies that

$$|f(x_k) - f(x)| < \frac{\varepsilon_0}{3} \quad \text{for} \quad k > k_1.$$

Thus, for sufficiently large k,

$$\varepsilon_0 < |f_{n_k}(x_k) - f(x_k)| \leq |f_{n_k}(x_k) - f(x)| + |f(x) - f(x_k)| < \frac{2}{3}\varepsilon_0,$$

3.1. Sequences of Functions, Uniform Convergence

a contradiction.

3.1.22. Assume, for example, that the functions f_n are increasing on $[a,b]$. Evidently, f is uniformly continuous on $[a,b]$. Let $\varepsilon > 0$ be given. By the uniform continuity of f there is $\delta > 0$ such that

$$|f(x) - f(x')| < \frac{\varepsilon}{2}$$

whenever $|x - x'| < \delta$, $x, x' \in [a,b]$. Now choose $a = x_0 < x_1 < x_2 < \cdots < x_k = b$ so that $|x_i - x_{i-1}| < \delta$, $i = 1, 2, \ldots, k$. Since

$$\lim_{n \to \infty} f_n(x_i) = f(x_i), \quad i = 1, 2, \ldots, k,$$

there exists n_0 such that, if $n > n_0$, then

(1) $$|f_n(x_i) - f(x_i)| < \frac{\varepsilon}{2}, \quad i = 1, 2, \ldots, k.$$

Clearly, for an $x \in [a,b]$ there is an i such that $x_{i-1} \leq x < x_i$. Now the monotonicity of f_n and (1) imply

$$f(x_{i-1}) - \frac{\varepsilon}{2} < f_n(x_{i-1}) \leq f_n(x) \leq f_n(x_i) < f(x_i) + \frac{\varepsilon}{2}$$

for $n > n_0$. Since f must be increasing, we have $f(x_{i-1}) \leq f(x) \leq f(x_i)$, which combined with the uniform continuity of f yields

$$-\varepsilon < f(x_{i-1}) - f(x_i) - \frac{\varepsilon}{2} \leq f_n(x) - f(x) \leq f(x_i) - f(x_{i-1}) + \frac{\varepsilon}{2} < \varepsilon.$$

Thus the uniform convergence of $\{f_n\}$ on $[a,b]$ is proved.

3.1.23. We will first show that there is a subsequence $\{f_{n_k}\}$ convergent on the set of all rationals $\mathbb{Q}$. Since $\mathbb{Q}$ is countable, we can write $\mathbb{Q} = \{r_1, r_2, \ldots\}$. The sequence $\{f_n(r_1)\}$ is bounded, so it contains a convergent subsequence $\{f_{n,1}(r_1)\}$. Next, since $\{f_{n,1}(r_2)\}$ is bounded, there exists a convergent subsequence $\{f_{n,2}(r_2)\}$. Clearly, $\{f_{n,2}(r_1)\}$ is also convergent. Repeating the process, we obtain the sequence of sequences $\{f_{n,1}\}, \{f_{n,2}\}, \ldots$ with the following properties:
- $\{f_{n,k+1}\}$ is a subsequence of $\{f_{n,k}\}$ for $k = 1, 2, \ldots$,
- the sequence $\{f_{n,k}(r_i)\}$ is convergent for $k \in \mathbb{N}$ and $i = 1, 2, \ldots, k$.

So the diagonal sequence $\{f_{n,n}\}$ is convergent on $\mathbb{Q}$. In this way we have constructed the subsequence $\{f_{n_k}\}$ pointwise convergent on $\mathbb{Q}$, say, to f. Clearly, f is increasing on $\mathbb{Q}$. Now we extend f to $\mathbb{R}$ by setting
$$f(x) = \sup\{f(r) : r \in \mathbb{Q},\ r \leq x\}.$$
The extended function f is also increasing on $\mathbb{R}$. Now we show that if f is continuous at x, then $\lim_{k \to \infty} f_{n_k}(x) = f(x)$. To this end, consider two sequences of rationals $\{p_n\}$ and $\{q_n\}$ converging to x and such that $p_n < x < q_n$. The monotonicity of f_{n_k} implies that $f_{n_k}(p_n) \leq f_{n_k}(x) \leq f_{n_k}(q_n)$. Now, letting $k \to \infty$, we get
$$f(p_n) \leq \liminf_{k \to \infty} f_{n_k}(x) \leq \limsup_{k \to \infty} f_{n_k}(x) \leq f(q_n).$$
Next, upon passage to the limit as $n \to \infty$ (see, e. g., 1.1.35), we obtain
$$f(x^-) \leq \liminf_{k \to \infty} f_{n_k}(x) \leq \limsup_{k \to \infty} f_{n_k}(x) \leq f(x^+).$$
It then follows that $f(x) = \lim_{k \to \infty} f_{n_k}(x)$ at each point x of continuity of f. We know that the set $\mathbf{D}$ of points of discontinuity of a monotonic function is countable (see, e. g., 1.2.29). Thus we have $f(x) = \lim_{k \to \infty} f_{n_k}(x)$ on the set $\mathbb{R} \setminus \mathbf{D}$, and since $\{f_{n_k}\}$ is bounded on the countable set $\mathbf{D}$, we can use the diagonal method again to choose a subsequence of $\{f_{n_k}\}$ pointwise convergent on $\mathbf{D}$. Clearly, this subsequence is convergent on all of $\mathbb{R}$.

3.1.24. If $\mathbf{K}$ is a compact subset of $\mathbb{R}$, then there is a closed interval $[a,b]$ such that $\mathbf{K} \subset [a,b]$. Clearly, f is uniformly continuous on $[a,b]$. By the result in 3.1.22 $\{f_{n_k}\}$ converges uniformly on $[a,b]$, and so it also converges uniformly on $\mathbf{K}$.

The following example shows that $\{f_{n_k}\}$ may fail to converge uniformly on $\mathbb{R}$. Put
$$f_n(x) = \left(\frac{1}{\pi}\left(\arctan x + \frac{\pi}{2}\right)\right)^n, \quad x \in \mathbb{R}.$$
Each f_n is strictly increasing on $\mathbb{R}$, and $0 < f_n(x) < 1$. The sequence $\{f_n\}$ is pointwise convergent to $f(x) \equiv 0$. However, the convergence is not uniform.

3.1. Sequences of Functions, Uniform Convergence

3.1.25. We first show that if $\{P_n\}$ is a sequence of polynomials convergent uniformly on $\mathbb{R}$, then, beginning with some value of the index n, all P_n are of the same degree. Indeed, if this were not true, then for every $k \in \mathbb{N}$ there would exist $n_k > k$ such that the degree of P_k would differ from the degree of P_{n_k}. Consequently,

$$\sup_{x \in \mathbb{R}} |P_{n_k}(x) - P_k(x)| = +\infty,$$

contrary to the Cauchy criterion for uniform convergence (see, e.g., 3.1.7). Hence there is $n_0 \in \mathbb{N}$ such that if $n \geq n_0$, then

$$P_n(x) = a_{n,p} x^p + a_{n,p-1} x^{p-1} + \cdots + a_{n,1} x + a_{n,0}.$$

By the Cauchy criterion for uniform convergence again, we see that if $n \geq n_0$, then the coefficients $a_{n,i}$, $i = 1, 2, \ldots, p$, are constant (independent of n), that is,

$$P_n(x) = a_p x^p + a_{p-1} x^{p-1} + \cdots + a_1 x + a_{n,0}.$$

Clearly, such a sequence of polynomials converges uniformly on $\mathbb{R}$ to the polynomial

$$P(x) = a_p x^p + a_{p-1} x^{p-1} + \cdots + a_1 x + a_0,$$

where $a_0 = \lim\limits_{n \to \infty} a_{n,0}$.

3.1.26. Clearly, (i) $\implies$ (ii). We now show that (ii) $\implies$ (iii). Indeed,

(1)
$$\begin{aligned} a_{n,0} + a_{n,1} c_0 + \cdots + a_{n,p} c_0^p &= P_n(c_0), \\ a_{n,0} + a_{n,1} c_1 + \cdots + a_{n,p} c_1^p &= P_n(c_1), \\ &\cdots \\ a_{n,0} + a_{n,1} c_p + \cdots + a_{n,p} c_p^p &= P_n(c_p). \end{aligned}$$

Since the so-called Vandermonde determinant

$$\det \begin{vmatrix} 1 & c_0 & c_0^2 & \cdots & c_0^p \\ 1 & c_1 & c_1^2 & \cdots & c_1^p \\ \vdots & \vdots & \vdots & \ldots & \vdots \\ 1 & c_p & c_p^2 & \cdots & c_p^p \end{vmatrix}$$

is different from zero, the system of linear equations (1) has a unique solution and $a_{n,i}$, $i = 0, 1, 2, \ldots, p$, can be determined using Cramer's rule. Consequently, (ii) implies the convergence of each sequence $\{a_{n,i}\}$, $i = 0, 1, 2, \ldots, p$. The implication (iii) $\implies$ (i) is easy to prove.

3.1.27. Since $\{f_n\}$ is equicontinuous, given $\varepsilon > 0$ one can choose $\delta > 0$ such that for all $n \in \mathbb{N}$

(1) $$|f_n(x) - f_n(y)| < \frac{\varepsilon}{3}$$

whenever $|x - y| < \delta$, $x, y \in \mathbf{K}$. Letting $n \to \infty$, we get

(2) $$|f(x) - f(y)| \leq \frac{\varepsilon}{3}.$$

(Note that this shows that f is uniformly continuous on $\mathbf{K}$.) As $\mathbf{K}$ is compact, there are finitely many open intervals $(x_i - \delta, x_i + \delta)$, $i = 1, 2, \ldots, k$, where $x_i \in \mathbf{K}$, which cover the set $\mathbf{K}$. By pointwise convergence of $\{f_n\}$, there is n_0 such that if $n > n_0$, then

(3) $$|f_n(x_i) - f(x_i)| < \frac{\varepsilon}{3}, \quad i = 1, 2, \ldots, k.$$

Clearly, for $x \in \mathbf{K}$ there is an i such that $|x - x_i| < \delta$. Thus by (1), (2) and (3), if $n > n_0$, then

$$|f_n(x) - f(x)| \leq |f_n(x) - f_n(x_i)| + |f_n(x_i) - f(x_i)| + |f(x_i) - f(x)| < \varepsilon.$$

3.1.28. Observe that $\{f_n\}$ is equicontinuous on $[a, b]$. Indeed, by the mean value theorem,

$$|f_n(x) - f_n(y)| = |f'_n(\zeta)||x - y| \leq M|x - y|$$

for all $x, y \in [a, b]$ and $n \in \mathbb{N}$. Now the desired result follows from the foregoing problem.

3.1.29.

(a) Since $|f_n(x)| \leq \frac{1}{\sqrt{n}}$, the sequence is uniformly convergent on $\mathbb{R}$. We have $f'_n(x) = \sqrt{n} \cos nx$. Hence $\lim\limits_{n \to \infty} f'_n(0) = \lim\limits_{n \to \infty} \sqrt{n} = +\infty$. Moreover, if $x \neq 0$, then the limit $\lim\limits_{n \to \infty} f'_n(x)$ does not exist. Indeed, if $\lim\limits_{n \to \infty} f'_n(x) = l$, then for sufficiently large n we would get

3.1. Sequences of Functions, Uniform Convergence

$|\cos nx| < \frac{1}{2}$. Thus $|\cos 2nx| = 1 - 2\cos^2 nx > \frac{1}{2}$, a contradiction. So we see that $\{f'_n\}$ does not converge at any point.

(b) Since $|f_n(x)| \leq \frac{1}{2n}$, the sequence converges uniformly on $[-1, 1]$. On the other hand,

$$\lim_{n \to \infty} f'_n(x) = \lim_{n \to \infty} \frac{1 - n^2 x^2}{(1 + n^2 x^2)^2} = \begin{cases} 1 & \text{for } x = 0, \\ 0 & \text{for } x \neq 0. \end{cases}$$

The pointwise limit of $\{f'_n\}$ is discontinuous at zero, and therefore the convergence cannot be uniform.

3.1.30. Assume first that $\lim_{x \to x_0} f(x) = l$. Let $\varepsilon > 0$ be given. Then there is $\delta > 0$ such that if $0 < |x - x_0| < \delta$, then $|f(x) - l| < \frac{\varepsilon}{2}$. The uniform convergence of $\{f_n\}$ on **A** implies

$$|f_n(x) - f(x)| < \frac{\varepsilon}{2} \quad \text{for} \quad n \geq n_0, \ x \in \mathbf{A}.$$

Hence
$$|f_n(x) - l| < \varepsilon$$
whenever $0 < |x - x_0| < \delta$ and $n \geq n_0$. Since $\lim_{x \to x_0} f_n(x)$ exists, this implies that $\lim_{n \to \infty} \lim_{x \to x_0} f_n(x) = l$.

Assume now that $\lim_{n \to \infty} \lim_{x \to x_0} f_n(x) = l$. Set $\lim_{x \to x_0} f_n(x) = g_n(x_0)$. So we have $\lim_{n \to \infty} g_n(x_0) = l$. Let $\varepsilon > 0$ be given. By the uniform convergence of $\{f_n\}$ there is n_1 such that $n > n_1$ implies

(1) $$|f_n(x) - f(x)| < \frac{\varepsilon}{3}, \quad x \in \mathbf{A}.$$

By the above there is n_2 such that if $n > n_2$, then

(2) $$|g_n(x_0) - l| < \frac{\varepsilon}{3}.$$

Fix $n_0 > \max\{n_1, n_2\}$. Since $\lim_{x \to x_0} f_{n_0}(x) = g_{n_0}(x_0)$, we have

(3) $$|f_{n_0}(x) - g_{n_0}(x_0)| < \frac{\varepsilon}{3}$$

if $|x - x_0| < \delta_{n_0}$. By (1), (2) and (3), we see that $\lim_{x \to x_0} f(x) = l$.

The equality $\lim_{n \to \infty} \lim_{x \to \infty} f_n(x) = \lim_{x \to \infty} f(x)$ can be established in much the same way.

3.1.31. Let $\varepsilon > 0$ be given. Choose n_0 such that if $n, m \geq n_0$, then

(1) $$|f_n(x_0) - f_m(x_0)| < \frac{\varepsilon}{2}$$

and

(2) $$|f'_n(t) - f'_m(t)| < \frac{\varepsilon}{2(b-a)}, \quad t \in [a,b].$$

This combined with the mean value theorem applied to the function $f_n - f_m$ gives

(3) $$|f_n(x) - f_m(x) - f_n(t) + f_m(t)| < \frac{\varepsilon|x-t|}{2(b-a)} \leq \frac{\varepsilon}{2}$$

for $n, m \geq n_0$ and $x, t \in [a, b]$. Now, by (3) and (1),

$$|f_n(x) - f_m(x)| \leq |f_n(x) - f_m(x) - f_n(x_0) + f_m(x_0)|$$
$$+ |f_n(x_0) - f_m(x_0)| < \varepsilon.$$

Thus the Cauchy criterion for uniform convergence is satisfied (see, e.g., 3.1.7). Let $x \in [a,b]$ be arbitrarily chosen. Define the functions h and h_n by

$$h(t) = \frac{f(t) - f(x)}{t - x}, \quad h_n(t) = \frac{f_n(t) - f_n(x)}{t - x}, \quad t \in [a,b], \ t \neq x.$$

Then $\lim_{t \to x} h_n(t) = f'_n(x)$, $n = 1, 2, \ldots$. By (3),

$$|h_n(t) - h_m(t)| < \frac{\varepsilon}{2(b-a)}, \quad n, m \geq n_0,$$

which means that $\{h_n\}$ is uniformly convergent (evidently to h) on $[a,b] \setminus \{x\}$. Applying the result in the foregoing problem to the sequence $\{h_n\}$ and the set $[a,b] \setminus \{x\}$, we get $\lim_{n \to \infty} f'_n(x) = \lim_{t \to x} h(t) = f'(x)$.

3.1.32. The equality

$$1 = (x + (1-x))^n = \sum_{k=0}^{n} \binom{n}{k} x^k (1-x)^{n-k}$$

gives

$$f(x) = \sum_{k=0}^{n} f(x) \binom{n}{k} x^k (1-x)^{n-k}.$$

3.1. Sequences of Functions, Uniform Convergence

Consequently,

$$(1) \quad |B_n(f,x) - f(x)| \leq \sum_{k=0}^{n} \left| f\left(\frac{k}{n}\right) - f(x) \right| \binom{n}{k} x^k (1-x)^{n-k}.$$

By the uniform continuity of f on $[0,1]$, given $\varepsilon > 0$ there is $\delta > 0$ such that

$$|f(x) - f(x')| < \varepsilon$$

whenever $|x - x'| < \delta$, $x, x' \in [0,1]$. Clearly, there is $M > 0$ such that $|f(x)| \leq M$ for $x \in [0,1]$. Let x be arbitrarily chosen in $[0,1]$. Then the set $\{0, 1, 2, \ldots, n\}$ can be decomposed into the two sets

$$\mathbf{A} = \left\{ k : \left|\frac{k}{n} - x\right| < \delta \right\} \quad \text{and} \quad \mathbf{B} = \left\{ k : \left|\frac{k}{n} - x\right| \geq \delta \right\}.$$

If $k \in \mathbf{A}$, then

$$\left| f\left(\frac{k}{n}\right) - f(x) \right| < \varepsilon,$$

and so

$$(2) \quad \sum_{k \in \mathbf{A}} \left| f\left(\frac{k}{n}\right) - f(x) \right| < \varepsilon \sum_{k \in \mathbf{A}} \binom{n}{k} x^k (1-x)^{n-k} \leq \varepsilon.$$

If $k \in \mathbf{B}$, then

$$\frac{(k - nx)^2}{n^2 \delta^2} \geq 1,$$

and by the inequality given in 2.5.52 we get

$$\sum_{k \in \mathbf{B}} \left| f\left(\frac{k}{n}\right) - f(x) \right| \binom{n}{k} x^k (1-x)^{n-k}$$

$$\leq \frac{2M}{n^2 \delta^2} \sum_{k \in \mathbf{B}} (k - nx)^2 \binom{n}{k} x^k (1-x)^{n-k} \leq \frac{M}{2n\delta^2}.$$

This combined with (1) and (2) yields

$$|B_n(f,x) - f(x)| \leq \varepsilon + \frac{M}{2n\delta^2}, \quad x \in [0,1].$$

3.1.33. If $[a,b] = [0,1]$, then we take $P(x) = B_n(f,x)$. If $[a,b] \neq [0,1]$, then we can apply the result in the foregoing problem to the function $g(y) = f(a+y(b-a))$, $y \in [0,1]$. So, given $\varepsilon > 0$, there is a Bernstein's polynomial $B_n(g,y)$ such that

$$|g(y) - B_n(g,y)| < \varepsilon, \quad y \in [0,1].$$

Putting $x = a + y(b-a)$, we obtain

$$\left| f(x) - B_n\left(g, \frac{x-a}{b-a}\right) \right| < \varepsilon.$$

3.2. Series of Functions, Uniform Convergence

3.2.1.

(a) If $x \in (-1,1]$, then $\lim_{n\to\infty} \frac{1}{1+x^n} \neq 0$. So the series diverges by the nth term test for divergence. If $|x| > 1$, then $|x|^n \geq 2$ for sufficiently large n. Hence

$$\left| \frac{1}{1+x^n} \right| \leq \frac{1}{|x|^n - 1} \leq \frac{2}{|x|^n},$$

and by the comparison test the series converges.

(b) Clearly, the series converges if $x = 0$. If $x \neq 0$, then

$$\frac{x^n}{1+x^n} = \frac{1}{1+\frac{1}{x^n}}.$$

Therefore by (a) the series converges for $-1 < x < 1$.

(c) If $x = 0$, the series diverges. If $x \neq 0$, then

$$\frac{2^n + x^n}{1 + 3^n x^n} = \frac{\left(\frac{2}{3x}\right)^n + \frac{1}{3^n}}{1 + \frac{1}{3^n x^n}}.$$

So the nth term of the series converges to zero if and only if $\left|\frac{2}{3x}\right| < 1$, that is, if $|x| > \frac{2}{3}$. The comparison test shows that the series converges if $x \in (-\infty, -2/3) \cup (2/3, \infty)$.

(d) We have

$$\frac{x^{n-1}}{(1-x^n)(1-x^{n+1})} = \frac{1}{x(1-x)} \left(\frac{1}{1-x^n} - \frac{1}{1-x^{n+1}} \right).$$

3.2. Series of Functions, Uniform Convergence

Hence

$$S_N(x) = \sum_{n=1}^{N} \frac{x^{n-1}}{(1-x^n)(1-x^{n+1})}$$

$$= \frac{1}{x(1-x)} \left(\frac{1}{1-x} - \frac{1}{1-x^{N+1}} \right).$$

Consequently,

$$\lim_{N \to \infty} S_N(x) = \begin{cases} \frac{1}{(1-x)^2} & \text{if } |x| < 1, \\ \frac{1}{x(1-x)^2} & \text{if } |x| > 1. \end{cases}$$

So the series converges on $\mathbb{R} \setminus \{-1, 1\}$.

(e) We have

$$\frac{x^{2^n-1}}{1-x^{2^n}} = \frac{1}{1-x^{2^{n-1}}} - \frac{1}{1-x^{2^n}}.$$

Hence

$$\lim_{N \to \infty} S_N(x) = \begin{cases} \frac{x}{1-x} & \text{if } |x| < 1, \\ \frac{1}{1-x} & \text{if } |x| > 1. \end{cases}$$

So the series converges on $\mathbb{R} \setminus \{-1, 1\}$.

(f) If $x \leq 0$, then the series diverges by the nth term test for divergence. For $x > 0$, by the Cauchy condensation test (see, e.g., I, 3.2.28) the given series converges if and only if $\sum_{n=2}^{\infty} \frac{n^x}{2^{n(x-1)}}$ does. The root test shows that the latter converges if $x > 1$ and diverges if $x < 1$. If $x = 1$, then the series diverges. Summing up, we see that the domain of convergence is $(1, \infty)$.

(g) Since $x^{\ln n} = n^{\ln x}$, the series converges if $\ln x < -1$ and diverges if $\ln x \geq -1$. Thus the domain of convergence is $\left(0, \frac{1}{e}\right)$.

(h) We have

$$\sin^2 \left(2\pi \sqrt{n^2 + x^2} \right) = \sin^2 \left(2n\pi \frac{\frac{x^2}{n^2}}{\sqrt{1 + \frac{x^2}{n^2}} + 1} \right) \leq \frac{\pi^2 x^4}{n^2}.$$

The comparison test shows that the series converges for all x.

3.2.2.

(a) Since $\arctan x + \arctan \frac{1}{x} = \frac{\pi}{2}$ for $x > 0$, we see that

$$\frac{\pi}{2} - \arctan(n^2(1+x^2)) = \arctan \frac{1}{n^2(1+x^2)} < \frac{1}{n^2(1+x^2)} \leq \frac{1}{n^2}.$$

By the M-test of Weierstrass (dominated convergence test) the series is uniformly convergent on $\mathbb{R}$.

(b) For $x \in [2, \infty)$,

$$\frac{\ln(1+nx)}{nx^n} \leq \frac{1}{x^{n-1}} \leq \frac{1}{2^{n-1}},$$

and consequently, the uniform convergence of the series follows from the M-test of Weierstrass.

(c) Since $\sup\{n^2 x^2 e^{-n^2|x|} : x \in \mathbb{R}\} = \frac{4}{n^2 e^2}$, the M-test of Weierstrass shows that the series converges uniformly on $\mathbb{R}$.

(d) The series converges pointwise to

$$S(x) = \begin{cases} 1 & \text{if } x \in [-1,1] \setminus \{0\}, \\ 0 & \text{if } x = 0. \end{cases}$$

Since S fails to be continuous, the convergence cannot be uniform on $[-1, 1]$.

(e) Note that

$$\sup_{1/2 \leq |x| \leq 2} \left| \frac{n^2}{\sqrt{n!}} (x^n + x^{-n}) \right| \leq \frac{n^2}{\sqrt{n!}} (2^n + 2^n) = \frac{n^2}{\sqrt{n!}} 2^{n+1}.$$

Since $\sum_{n=1}^{\infty} \frac{n^2}{\sqrt{n!}} 2^{n+1}$ converges, for example by the ratio test, the M-test of Weierstrass shows that the series converges uniformly on **A**.

(f) The series does not converge uniformly on **A** because the Cauchy criterion for uniform convergence fails to hold. Indeed, if $0 <$

3.2. Series of Functions, Uniform Convergence 339

$\frac{1}{3^n x} \le \frac{\pi}{2}$, then

$$|S_{n+m}(x) - S_n(x)| = 2^{n+1} \sin \frac{1}{3^{n+1}x} + \cdots + 2^{n+m} \sin \frac{1}{3^{n+m}x}$$

$$\ge 2^{n+1} \frac{2}{\pi} \frac{1}{3^{n+1}x} + \cdots + 2^{n+m} \frac{2}{\pi} \frac{1}{3^{n+m}x}$$

$$\ge 2^{n+1} \frac{2}{\pi 3^{n+1}x}.$$

Putting $x = \frac{1}{3^n}$, we obtain

$$\left| S_{n+m}\left(\frac{1}{3^n}\right) - S_n\left(\frac{1}{3^n}\right) \right| \ge \frac{2^{n+2}}{3\pi} \ge \frac{2^3}{3\pi}.$$

(g) The uniform convergence of the series follows from the M-test of Weierstrass. We have

$$\ln\left(1 + \frac{x^2}{n \ln^2 n}\right) \le \frac{x^2}{n \ln^2 n} < \frac{a^2}{n \ln^2 n},$$

and the Cauchy condensation test shows that $\sum_{n=2}^{\infty} \frac{a^2}{n \ln^2 n}$ converges.

3.2.3. Let $S(x) = \sum\limits_{n=1}^{\infty} f_n(x)$ and $S_n(x) = \sum\limits_{k=1}^{n} f_k(x)$. Then

$$\sup\{S(x) - S_n(x) : x \in [0,1]\} = 1/(n+1),$$

which shows that the series converges uniformly on $[0,1]$. Since $\sup\{f_n(x) : x \in [0,1]\} = 1/n$, the M-test of Weierstrass fails.

3.2.4. We have

$$S_n(x) = \sum_{k=1}^{n} \frac{x}{((k-1)x+1)(kx+1)}$$

$$= \sum_{k=1}^{n} \left(\frac{1}{(k-1)x+1} - \frac{1}{kx+1}\right) = 1 - \frac{1}{nx+1}.$$

Hence

$$f(x) = \lim_{n\to\infty} S_n(x) = \begin{cases} 0 & \text{if } x = 0, \\ 1 & \text{if } x > 0. \end{cases}$$

Clearly, f is not continuous at zero.

3.2.5.

(a) The series converges absolutely on $\mathbb{R}$, since

$$\sum_{n=0}^{\infty}\left|\frac{x^n \sin(nx)}{n!}\right| \leq \sum_{n=0}^{\infty}\frac{|x|^n}{n!} = e^{|x|}.$$

Clearly, the convergence is uniform on each bounded interval. So the continuity of the sum follows from the result in 1.2.34.

(b) Since

$$\sum_{n=0}^{\infty}|x|^{n^2} \leq \sum_{n=0}^{\infty}|x|^n = \frac{1}{1-|x|},$$

the series converges absolutely on $(-1,1)$. Moreover, the convergence is uniform on each compact subset of $(-1,1)$. Thus the sum is continuous on $(-1,1)$.

(c) The series converges absolutely for $-1/2 < x < 1/2$, and, as in (a), one can show that its sum is continuous on $(-1/2, 1/2)$.

(d) The series converges absolutely for $1/e - 1 < x < e - 1$, and its sum is continuous on $(1/e - 1, e - 1)$.

3.2.6. Clearly, the series converges for $x = 0$. Using, for example, the result in I, 3.2.16, we see that the series converges if $0 < |x| < 1$. If $|x| \geq 1$, the series diverges. Reasoning similar to that used in the solution of the preceding problem shows that the sum is continuous on the domain of convergence.

3.2.7. Note first that the series

$$\sum_{n=1}^{\infty}\frac{\sin(n^2 x)}{n^2}$$

is uniformly convergent on $\mathbb{R}$, so its sum $\tilde{S}$ is continuous on $\mathbb{R}$. Moreover, if $S_n(x) = \frac{x\sin(k^2 x)}{k^2}$, then $\lim\limits_{n\to\infty} S_n(x) = x\tilde{S}(x)$. Consequently, the sum of the given series is also continuous on $\mathbb{R}$.

3.2. Series of Functions, Uniform Convergence 341

3.2.8. Suppose that $\sum_{n=1}^{\infty} f_n(x)$ converges uniformly on **A** to S. This means that
$$d_n = \sup_{x \in \mathbf{A}} |S_n(x) - S(x)| \xrightarrow[n \to \infty]{} 0,$$
where $S_n(x) = \sum_{k=1}^{n} f_k(x)$. Since f is bounded, we also have
$$d'_n = \sup_{x \in \mathbf{A}} |f(x)S_n(x) - f(x)S(x)| \xrightarrow[n \to \infty]{} 0.$$

To see that boundedness of f is essential, take $\mathbf{A} = (0, 1]$, $f(x) = \frac{1}{x}$, and $f_n(x) = \frac{1}{2^{n-1}}$. Then the series $\sum_{n=1}^{\infty} f_n(x)$ converges uniformly on **A**, but $\sum_{n=1}^{\infty} \frac{1}{x} f_n(x)$ fails to converge uniformly on **A**, because
$$\sup_{x \in (0,1]} \left| \sum_{k=n+1}^{\infty} \frac{1}{x} f_k(x) \right| = \sup_{x \in (0,1]} \frac{2}{x 2^n} = +\infty.$$

It is easy to see that if $\frac{1}{f}$ is bounded on **A**, then the converse holds.

3.2.9. For $x \in \mathbf{A}$ the series $\sum_{n=1}^{\infty} (-1)^n f_n(x)$ converges by the Leibniz theorem. Moreover, by the result in I, 3.4.14,
$$\sup_{x \in \mathbf{A}} |r_n(x)| = \sup_{x \in \mathbf{A}} \left| \sum_{k=n+1}^{\infty} (-1)^{k+1} f_k(x) \right| \leq \sup_{x \in \mathbf{A}} f_{n+1}(x).$$

This combined with condition (3) proves the uniform convergence of the given series on **A**.

3.2.10. The three series (a), (b) and (c) satisfy the assumptions of the assertion in the foregoing problem.

3.2.11. By the Cauchy inequality,
$$\sup_{x \in \mathbf{A}} \left| \sum_{k=n}^{n+m} c_k f_k(x) \right| \leq \left(\sum_{k=n}^{n+m} c_k^2 \right)^{1/2} \sup_{x \in \mathbf{A}} \left(\sum_{k=n}^{n+m} f_k^2(x) \right)^{1/2}.$$

So it suffices to apply the Cauchy criterion for uniform convergence.

3.2.12.

(a) $\mathbf{A} = \left[\frac{1}{6}, \frac{1}{2}\right)$ and $\mathbf{B} = \left(\frac{1}{6}, \frac{1}{2}\right)$. The series converges uniformly on $\left[\frac{1}{6}, \frac{1}{3}\right]$, because

$$\sup_{x \in \left[\frac{1}{6}, \frac{1}{3}\right]} \left| \sum_{k=n+1}^{\infty} \frac{1}{k} 2^k (3x-1)^k \right| \leq \sup_{x \in \left[\frac{1}{6}, \frac{1}{3}\right]} \frac{|6x-2|^{n+1}}{n+1} = \frac{1}{n+1}.$$

(b) $\mathbf{A} = \left(-\infty, -\frac{1}{2}\right]$ and $\mathbf{B} = \left(-\infty, -\frac{1}{2}\right)$. The series converges uniformly on $[-2, -1]$, because

$$\sup_{x \in [-2,-1]} \left| \sum_{k=n+1}^{\infty} \frac{1}{k} \left(\frac{x+1}{x}\right)^k \right| \leq \sum_{k=n+1}^{\infty} \frac{1}{k 2^k}.$$

3.2.13. Summation by parts gives

$$S_n(x) = \sum_{k=1}^{n} f_k(x) g_k(x) = \sum_{k=1}^{n-1} G_k(x)(f_k(x) - f_{k+1}(x)) + G_n(x) f_n(x).$$

This together with assumption (3) implies

$$|S_{n+m}(x) - S_n(x)|$$
$$= \left| \sum_{k=n}^{n+m-1} G_k(x)(f_k(x) - f_{k+1}(x)) + G_{n+m}(x) f_{n+m}(x) - G_n(x) f_n(x) \right|$$
$$\leq M \left(\sum_{k=n}^{n+m-1} |f_k(x) - f_{k+1}(x)| + |f_{n+m}(x)| + |f_n(x)| \right).$$

Now let $\varepsilon > 0$ be given. Then it follows from (1) and (2) that, for $m \in \mathbb{N}$ and for sufficiently large n,

$$\sup_{x \in \mathbf{A}} |S_{n+m}(x) - S_n(x)|$$
$$\leq M \sup_{x \in \mathbf{A}} \left(\sum_{k=n}^{n+m-1} |f_k(x) - f_{k+1}(x)| + |f_{n+m}(x)| + |f_n(x)| \right) < \varepsilon.$$

Thus the Cauchy criterion for uniform convergence can be applied to $\sum_{n=1}^{\infty} f_n(x) g_n(x)$.

3.2. Series of Functions, Uniform Convergence 343

To prove the Dirichlet test for uniform convergence, note that the monotonicity and the uniform convergence to zero of $\{f_n(x)\}$ imply (1) and (2). Moreover, since the sequence of partial sums of $\sum_{n=1}^{\infty} g_n(x)$ is uniformly bounded on **A**, we see that condition (3) is also satisfied. Consequently, the series $\sum_{n=1}^{\infty} f_n(x)g_n(x)$ converges uniformly on **A**.

3.2.14. The Dirichlet test for uniform convergence will be applied.

(a) Take
$$f_n(x) = \frac{1}{n} \quad \text{and} \quad g_n(x) = (-1)^{n+1}x^n.$$

(b) Here we take
$$f_n(x) = \frac{1}{n} \quad \text{and} \quad g_n(x) = \sin(nx)$$
and note that
$$\left|\sum_{k=1}^{n} \sin(kx)\right| \leq \frac{1}{\sin\frac{x}{2}} \leq \frac{1}{\sin\frac{\delta}{2}}.$$

(c) Since
$$2\left|\sum_{k=1}^{n} \sin(k^2 x) \sin(kx)\right| = \left|\sum_{k=1}^{n} (\cos(k(k-1)x) - \cos(k(k+1)x))\right|$$
$$= |1 - \cos(n(n+1)x)| \leq 2$$
and $\{\frac{1}{n+x^2}\}$ is decreasing and uniformly convergent to zero, the Dirichlet test shows that the series converges uniformly on $\mathbb{R}$.

(d) We have
$$(*) \quad \sum_{n=1}^{\infty} \frac{\sin(nx)\arctan(nx)}{n}$$
$$= \sum_{n=1}^{\infty} \left(\frac{\sin(nx)\left(\arctan(nx) - \frac{\pi}{2}\right)}{n} + \frac{\frac{\pi}{2}\sin(nx)}{n}\right).$$

Since $\sum_{n=1}^{\infty} \frac{\frac{\pi}{2}\sin(nx)}{n}$ converges uniformly on $[\delta, 2\pi - \delta]$ (see (b)), the sequence of its partial sums is uniformly bounded. Moreover, the sequence $\{\arctan(nx) - \pi/2\}$ is increasing and satisfies the Cauchy criterion for uniform convergence on $[\delta, 2\pi - \delta]$, because

$$\arctan((m+n)x) - \arctan(nx) = \arctan\frac{mx}{1+(m+n)nx^2}$$
$$\leq \arctan\frac{mx}{(m+n)nx^2}$$
$$\leq \arctan\frac{1}{n\delta}.$$

So $\{\arctan(nx) - \pi/2\}$ converges uniformly to zero. It then follows, by (*), that the given series converges uniformly on **A**.

(e) We have

$$\sum_{n=1}^{\infty}(-1)^{n+1}\frac{1}{n^x} = \sum_{n=1}^{\infty}(-1)^{n+1}\frac{1}{n^{x-\frac{a}{2}}}\frac{1}{n^{\frac{a}{2}}}.$$

Since $\sum_{n=1}^{\infty}(-1)^{n+1}\frac{1}{n^{\frac{a}{2}}}$ converges, the sequence of its partial sums is bounded. Moreover, the sequence $\left\{\frac{1}{n^{x-\frac{a}{2}}}\right\}$ decreases and converges uniformly to zero on $[a, \infty)$.

(f) Note that for $x \in [0, \infty)$,

$$\left|\sum_{k=1}^{n}(-1)^{k+1}\frac{1}{e^{kx}}\right| = \left|\frac{1 - \frac{(-1)^n}{e^{nx}}}{e^x + 1}\right| \leq 1.$$

Moreover, the sequence $\left\{\frac{1}{\sqrt{n+x^2}}\right\}$ decreases and converges uniformly to zero on $[0, \infty)$.

3.2.15. Summation by parts yields

$$S_n(x) = \sum_{k=1}^{n} f_k(x)g_k(x) = \sum_{k=1}^{n-1} G_k(x)(f_k(x) - f_{k+1}(x)) + G_n(x)f_n(x),$$

where $G_n(x) = \sum_{k=1}^{n} g_k(x)$. Since f_1 is bounded on **A**, condition (2) implies that there is $M > 0$ such that $|f_n(x)| \leq M$ for all $x \in$ **A** and

3.2. Series of Functions, Uniform Convergence

all $n \in \mathbb{N}$. Since $\{G_n\}$ converges uniformly on $\mathbf{A}$, say, to G, we obtain

$$S_{n+m}(x) - S_n(x)$$
$$= \sum_{k=n}^{n+m-1} G_k(x)(f_k(x) - f_{k+1}(x)) + G_{n+m}(x)f_{n+m}(x) - G_n(x)f_n(x)$$
$$= \sum_{k=n}^{n+m-1} (f_k(x) - f_{k+1}(x))(G_k(x) - G(x))$$
$$+ (G_{n+m}(x) - G(x))f_{n+m}(x) - (G_n(x) - G(x))f_n(x).$$

This combined with (2) and the uniform boundedness of $\{f_n(x)\}$ shows that $\{S_n\}$ satisfies the Cauchy criterion for uniform convergence.

To prove the Abel test for uniform convergence, it suffices to note that the monotonicity and the uniform boundedness of $\{f_n\}$ imply the pointwise convergence to a bounded function, and so conditions (1) and (2) are satisfied.

3.2.16.

(a) The sequence $\{\arctan(nx)\}$ satisfies conditions (1') and (2') in the Abel test for uniform convergence. Moreover, $\sum_{n=1}^{\infty} \frac{(-1)^{n+1}}{n+x^2}$ converges uniformly on $\mathbb{R}$ (see 3.2.10(a)).

(b) The Abel test for uniform convergence can be applied, because the series

$$\sum_{n=2}^{\infty} \frac{(-1)^{n+1}}{\sqrt{n} + \cos x}$$

is uniformly convergent on $\mathbf{A}$ (see 3.2.10(c)) and the sequence $\{\cos \frac{x}{n}\}$ is bounded and monotonic for $n > \frac{2R}{\pi}$.

(c) The series

$$\sum_{n=1}^{\infty} \frac{(-1)^{[\sqrt{n}]}}{n}$$

converges (see, e.g., I, 3.4.8) and the sequence $\{\frac{\sqrt{n}}{\sqrt{n+x}}\}$ is monotonic and bounded on $[0, \infty)$. Thus the Abel test for uniform convergence can be applied.

3.2.17. The result follows immediately from 3.1.30.

3.2.18. To prove (a) and (b), one can use the results in 3.2.14, 3.2.17, and in I, 3.1.32(a).

(c) Since
$$\sum_{n=1}^{\infty}(x^n - x^{n+1}) = \begin{cases} x & \text{for } x \in [0,1), \\ 0 & \text{for } x = 1, \end{cases}$$
we get
$$\lim_{x \to 1^-} \sum_{n=1}^{\infty}(x^n - x^{n+1}) = 1.$$

(d) Note first that $\sum_{n=1}^{\infty} \frac{1}{2^n n^x}$ is uniformly convergent on $[0, \infty)$, by the M-test of Weierstrass. Thus, by the foregoing problem,
$$\lim_{x \to 0^+} \sum_{n=1}^{\infty} \frac{1}{2^n n^x} = \sum_{n=1}^{\infty} \frac{1}{2^n} = 1.$$

(e) Since
$$\sup_{x \in \mathbb{R}} \frac{x^2}{1+n^2 x^2} = \frac{1}{n^2},$$
the series $\sum_{n=1}^{\infty} \frac{x^2}{1+n^2 x^2}$ converges uniformly on $\mathbb{R}$. Now using the result in 3.1.30 we obtain
$$\lim_{x \to \infty} \sum_{n=1}^{\infty} \frac{x^2}{1+n^2 x^2} = \sum_{n=1}^{\infty} \frac{1}{n^2} = \frac{\pi^2}{6}.$$

3.2.19. Observe first that $\sum_{n=1}^{\infty} a_n x^n$ converges uniformly on $[0,1]$. This follows immediately from the Abel test for uniform convergence stated in 3.2.15, with $f_n(x) = x^n$ and $g_n(x) = a_n$. Now by 3.2.17 we see that the limit is $\sum_{n=1}^{\infty} a_n$.

3.2.20. Since the f_n are continuous on $[0,1]$, we see that
$$\sup_{x \in [0,1)} \sum_{k=n}^{n+m} f_k(x) = \sup_{x \in [0,1]} \sum_{k=n}^{n+m} f_k(x).$$

3.2. Series of Functions, Uniform Convergence   347

Thus by the Cauchy criterion, the uniform convergence of $\sum_{n=1}^{\infty} f_n(x)$ on $[0,1)$ implies the uniform convergence of $\sum_{n=1}^{\infty} f_n(x)$ on $[0,1]$.

3.2.21. $\mathbf{A} = (0,\infty)$. The convergence is not uniform. Indeed, if the series were uniformly convergent on $\mathbf{A}$, then by the result in the foregoing problem it would converge for $x=0$, a contradiction.

3.2.22. Note that

$$r_n(x) = \sum_{k=n+1}^{\infty} f_k(x) = f(x) - S_n(x),$$

where $S_n(x)$ denotes the nth partial sum of $\sum_{n=1}^{\infty} f_n(x)$. By assumption, the sequence $\{r_n(x)\}$ is monotonic and convergent to zero at each fixed x in $[a,b]$. Hence the Dini theorem (see, e.g., 3.1.16) implies the uniform convergence of $\{r_n(x)\}$, and consequently, the uniform convergence of the series on $[a,b]$.

3.2.23. No. Consider

$$\sum_{n=0}^{\infty} (-1)^n (1-x)x^n, \quad \mathbf{A} = [0,1].$$

By the result stated in 3.2.9 this series converges uniformly on $\mathbf{A}$. On the other hand, the sum of the series $\sum_{n=0}^{\infty} (1-x)x^n$ is

$$S(x) = \begin{cases} 1 & \text{for } x \in [0,1), \\ 0 & \text{for } x = 1. \end{cases}$$

Since S is not continuous, the convergence cannot be uniform.

3.2.24. Since the f_n are monotonic on $[a,b]$,

$$|r_n(x)| = \left| \sum_{k=n+1}^{\infty} f_k(x) \right| \leq \sum_{k=n+1}^{\infty} |f_k(x)| \leq \sum_{k=n+1}^{\infty} \max\{|f_k(a)|, |f_k(b)|\}.$$

This shows that if the series converges absolutely at the endpoints of the interval $[a,b]$, then it converges absolutely and uniformly on the whole $[a,b]$.

3.2.25. Let **A** be a bounded set disjoint with the elements of $\{a_n\}$. Since $\sum\limits_{n=1}^{\infty} \frac{1}{|a_n|}$ converges, we have $\lim\limits_{n\to\infty} |a_n| = +\infty$. Consequently, one can choose n_0 such that if $n \geq n_0$, then $|x - a_n| \geq 1$ for $x \in \mathbf{A}$. Hence, for sufficiently large n,

$$\frac{1}{|x-a_n|} = \frac{1}{|a_n|} \cdot \frac{1}{\left|\frac{x}{a_n} - 1\right|} \leq \frac{1}{|a_n|} \cdot \frac{1}{1 - \frac{M}{|a_n|}},$$

where $M = \sup\limits_{x \in \mathbf{A}} |x|$. Finally, observe that if $\sum\limits_{n=1}^{\infty} \frac{1}{|a_n|}$ converges, then $\sum\limits_{n=1}^{\infty} \left(\frac{1}{|a_n|} \cdot \frac{1}{1 - \frac{M}{|a_n|}}\right)$ also converges.

3.2.26. Write

$$\sum_{n=1}^{\infty} \frac{a_n}{n^x} = \sum_{n=1}^{\infty} \frac{a_n}{n^{x_0}} \cdot \frac{1}{n^{x-x_0}}$$

and apply the Abel test for uniform convergence (see, e.g., 3.2.15).

3.2.27. It has been shown in the solution of 3.2.7 that the given series converges to a continuous function on $\mathbb{R}$. We now show that the convergence is not uniform on $\mathbb{R}$.

Observe first that if n_0 is odd, then the sum

$$\sum_{n=n_0}^{\infty} \frac{\sin(n^2 x)}{n^2}$$

is different from zero at each $x_k = \frac{\pi}{2} + 2k\pi$, $k \in \mathbb{N}$. Moreover,

(1) $$\sum_{n=n_0}^{\infty} \frac{\sin(n^2 x_k)}{n^2} = \sum_{l=0}^{\infty} \frac{\sin(n_0^2 \frac{\pi}{2})}{(n_0 + 2l)^2}.$$

If the series $\sum\limits_{n=1}^{\infty} x \frac{\sin(n^2 x)}{n^2}$ were convergent to f uniformly on $\mathbb{R}$, then, given $\varepsilon > 0$, there would exist an odd n_0 such that

$$\left| f(x) - \sum_{n=1}^{n_0-1} x \frac{\sin(n^2 x)}{n^2} \right| < \varepsilon \quad \text{for all} \quad x \in \mathbb{R}.$$

3.2. Series of Functions, Uniform Convergence 349

In particular, we would get

$$\left| \frac{f(x_k)}{x_k} - \sum_{n=1}^{n_0-1} \frac{\sin(n^2 \frac{\pi}{2})}{n^2} \right| < \frac{\varepsilon}{\frac{\pi}{2} + 2k\pi},$$

and consequently,

$$\lim_{k \to \infty} \frac{f(x_k)}{x_k} = \sum_{n=1}^{n_0-1} \frac{\sin(n^2 \frac{\pi}{2})}{n^2}.$$

On the other hand, by (1),

$$\frac{f(x_k)}{x_k} = \sum_{n=1}^{\infty} \frac{\sin(n^2 x_k)}{n^2} = \sum_{n=1}^{n_0-1} \frac{\sin(n^2 \frac{\pi}{2})}{n^2} + \sin\left(n_0^2 \frac{\pi}{2}\right) \sum_{l=0}^{\infty} \frac{1}{(n_0+2l)^2},$$

contrary to

$$\sin\left(n_0^2 \frac{\pi}{2}\right) \sum_{l=0}^{\infty} \frac{1}{(n_0+2l)^2} \neq 0.$$

3.2.28. The assertion follows immediately from the result in 3.1.31.

3.2.29. By the M-test of Weierstrass the series $\sum\limits_{n=1}^{\infty} \frac{1}{n^2+x^2}$ converges uniformly on $\mathbb{R}$. Moreover, since

$$\left| \left(\frac{1}{n^2+x^2} \right)' \right| = \left| \frac{-2x}{(n^2+x^2)^2} \right| \leq \frac{1}{n^3},$$

$\sum\limits_{n=1}^{\infty} \left(\frac{1}{n^2+x^2} \right)'$ also converges uniformly on $\mathbb{R}$. Hence by the result in the foregoing problem f is differentiable on each compact interval and consequently on $\mathbb{R}$.

3.2.30. Note first that $\sum\limits_{n=1}^{\infty} \frac{\cos(nx)}{1+n^2}$ converges uniformly on $\mathbb{R}$. The series

$$\sum_{n=1}^{\infty} \left(\frac{\cos(nx)}{1+n^2} \right)' = \sum_{n=1}^{\infty} \frac{-n \sin(nx)}{1+n^2}$$

converges uniformly on the indicated interval by the Dirichlet test for uniform convergence stated in 3.2.13. Therefore the differentiability of f follows from 3.2.28.

3.2.31. The series $\sum\limits_{n=1}^{\infty}(-1)^{n+1}\ln\left(1+\frac{x}{n}\right)$ converges, e.g., for $x = 0$. The series

$$\sum_{n=1}^{\infty}\left((-1)^{n+1}\ln\left(1+\frac{x}{n}\right)\right)' = \sum_{n=1}^{\infty}(-1)^{n+1}\frac{1}{n+x}$$

converges uniformly on $[0,\infty)$ by the result stated in 3.2.9. So the result in 3.2.28 shows that f is differentiable on $[0,\infty)$ and

$$f'(0) = \sum_{n=1}^{\infty}(-1)^{n+1}\frac{1}{n} = \ln 2, \quad f'(1) = \sum_{n=1}^{\infty}(-1)^{n+1}\frac{1}{n+1} = 1 - \ln 2.$$

Finally, applying 3.1.30 we find that $\lim\limits_{x\to\infty} f'(x) = 0$.

3.2.32. By the Abel test for uniform convergence (see, e.g., 3.2.15), $\sum\limits_{n=1}^{\infty}(-1)^{n+1}\frac{1}{\sqrt{n}}\arctan\frac{x}{\sqrt{n}}$ converges uniformly on $\mathbb{R}$. The derived series $\sum\limits_{n=1}^{\infty}\frac{(-1)^{n+1}}{n+x^2}$ is also uniformly convergent on $\mathbb{R}$ (see 3.2.10(a)). So one can apply 3.2.28.

3.2.33. Clearly, the series $\sum\limits_{n=1}^{\infty}\frac{\sin(nx^2)}{1+n^3}$ converges uniformly on $\mathbb{R}$. The derived series $\sum\limits_{n=1}^{\infty}\frac{2xn\cos(nx^2)}{1+n^3}$ converges uniformly on each bounded interval. Therefore by 3.2.28 f' is continuous on each bounded interval, and thus f' is continuous on $\mathbb{R}$.

3.2.34. The M-test of Weierstrass shows that the series and the derived series

$$\sum_{n=1}^{\infty}n\sqrt{n}(\tan x)^{n-1}\frac{1}{\cos^2 x}$$

are uniformly convergent on each compact subinterval of $(-\pi/4, \pi/4)$. Therefore by 3.2.28 f' is continuous on $(-\pi/4, \pi/4)$.

3.2.35. The M-test of Weierstrass shows that the given series converges uniformly on $[0,\infty)$. By the M-test again, we see that the derived series

$$\sum_{n=0}^{\infty}\frac{-ne^{-nx}}{1+n^2}$$

3.2. Series of Functions, Uniform Convergence

is uniformly convergent on each interval $[a, \infty)$, $a > 0$. Thus f is in $C^1(0, \infty)$. Repeating the above process k times, we conclude that $\sum_{n=0}^{\infty} \frac{(-1)^k n^k e^{-nx}}{1+n^2}$ converges uniformly on each $[a, \infty)$, $a > 0$. This shows that $f \in C^\infty(0, \infty)$.

If $f'(0)$ existed, then since

$$\frac{f(x) - f(0)}{x} = \sum_{n=0}^{\infty} \frac{e^{-nx} - 1}{x(1+n^2)} \leq \sum_{n=0}^{N} \frac{e^{-nx} - 1}{x(1+n^2)}$$

for $x > 0$ and $N \geq 1$, we would get

$$\lim_{x \to 0^+} \frac{f(x) - f(0)}{x} \leq \sum_{n=0}^{N} \frac{-n}{1+n^2}.$$

Upon passage to the limit as $N \to \infty$, we would obtain $f'(0) \leq -\infty$, a contradiction.

3.2.36. Clearly, the series converges uniformly on each bounded interval. Thus f is continuous on $\mathbb{R}$. Moreover, for $x \neq 0$,

$$\sum_{n=1}^{\infty} \left(\frac{|x|}{x^2 + n^2}\right)' = \sum_{n=1}^{\infty} \frac{n^2 \operatorname{sgn}(x) - x|x|}{(x^2 + n^2)^2}.$$

Thus the derived series is uniformly convergent on each bounded interval that does not contain zero. Consequently, f' is continuous at each $x \neq 0$. Now we show that $f'(0)$ does not exist. Since

$$\frac{f(h) - f(0)}{h} = \left(\frac{|h|}{h} \sum_{n=1}^{\infty} \frac{1}{h^2 + n^2}\right)$$

and (see, e.g., 3.2.17)

$$\lim_{h \to 0} \sum_{n=1}^{\infty} \frac{1}{h^2 + n^2} = \sum_{n=1}^{\infty} \frac{1}{n^2} = \frac{\pi^2}{6},$$

the limit $\lim_{h \to 0} \frac{f(h) - f(0)}{h}$ does not exist.

3.2.37. Observe first that the series $\sum_{n=1}^{\infty} \frac{1}{n^x}$ converges uniformly on each interval $[x_0, \infty)$, $x_0 > 1$ (see, e.g., 3.2.26). Thus the Riemann ζ-function is continuous on $(1, \infty)$. For a $k \in \mathbb{N}$, the series

(1) $$\sum_{n=1}^{\infty} (-1)^k \frac{\ln^k n}{n^x}$$

is also uniformly convergent on each $[x_0, \infty)$, $x_0 > 1$, because

$$\frac{\ln^k n}{n^x} \leq \frac{n^{\frac{x_0-1}{2}}}{n^{x_0}} = \frac{1}{n^{\frac{x_0+1}{2}}}$$

for sufficiently large n. Consequently, each kth derivative of the Riemann ζ-function is continuous on $(1, \infty)$.

3.2.38. By (1) there is an $x_0 \in (0, 1]$ such that $f(x_0) \neq 0$. Now, by (2) and by Taylor's formula with the Lagrange form for the remainder, we get

$$f(x_0) = \frac{f^{(n)}(\theta_n) x_0^n}{n!},$$

where $\theta_n \in (0, 1)$. Hence

(∗) $$f^{(n)}(\theta_n) = \frac{n! f(x_0)}{x_0^n}.$$

Now (3) implies that $\sup_{x \in [0,1]} |a_n f^{(n)}(x)| \xrightarrow[n \to \infty]{} 0$. This means that, given $\varepsilon > 0$, there is n_0 such that if $n > n_0$, then $|a_n f^{(n)}(\theta_n)| < \varepsilon$. It then follows by (∗) that

$$|n! a_n| < \frac{\varepsilon x_0^n}{|f(x_0)|}.$$

3.2.39. Clearly, for $x \in \mathbb{Z}$ we have $f_n(x) = 0$. So $\sum_{n=1}^{\infty} f_n(x) = 0$. Now let $x = \frac{r}{s}$, where r and s and co-prime integers and $s > 1$. If p is a prime number different from s, then $f_p(x) \geq \frac{1}{ps}$. Indeed, for any $a \in \mathbb{Z}$,

$$\left| \frac{r}{s} - \frac{a}{p} \right| = \frac{|rp - as|}{sp} \geq \frac{1}{sp}.$$

3.2. Series of Functions, Uniform Convergence

Consequently,
$$\sum_{n=1}^{\infty} f_n(x) \geq \sum_{p \in \mathbf{P}} \frac{1}{sp},$$
where $\mathbf{P}$ denotes the set of all prime numbers different from s. So (see, e.g., I, 3.2.72) the series $\sum_{n=1}^{\infty} f_n(x)$ is divergent for all $x \in \mathbb{Q} \setminus \mathbb{Z}$. For an irrational x, set
$$\mathbf{A} = \left\{ n \in \mathbb{N} : \frac{1}{4} < nx - [nx] < \frac{1}{2} \right\},$$
$$A(m) = \sharp \{ n \in \mathbf{A} : n < m \},$$
where $\sharp \mathbf{B}$ denotes the number of elements of the set $\mathbf{B}$. It follows from the fact that for an irrational x the numbers $nx - [nx]$ are uniformly distributed modulo 1 (see, e.g., Theorem 25.1 in P. Billingsley, Probability and Measure, Wiley, New York, 1979, pp. 282-283) that $\lim_{m \to \infty} \frac{A(m)}{m} = \frac{1}{4}$. Consequently, $\sum_{n \in \mathbf{A}} \frac{1}{4n} = +\infty$. Note that for $n \in \mathbf{A}$,
$$f_n(x) = x - \frac{[nx]}{n} \geq \frac{1}{4n}.$$

It then follows that $\sum_{n=1}^{\infty} f_n(x)$ diverges for $x \in \mathbb{R} \setminus \mathbb{Q}$.

3.2.40. Since g is bounded, the series converges uniformly on $\mathbb{R}$ to f. Hence f is continuous on $\mathbb{R}$. Our task is now to show that f is nowhere differentiable. Let a real number x and a positive integer m be arbitrarily chosen. If there is an integer in $\left(4^m x, 4^m x + \frac{1}{2}\right)$, then there is no integer in $\left(4^m x - \frac{1}{2}, 4^m x\right)$. So we can always find $\delta_m = \pm \frac{1}{2} 4^{-m}$ such that there is no integer in the open interval with the endpoints $4^m x$ and $4^m(x + \delta_m)$. By the definition of g,
$$\left| \frac{g(4^n(x + \delta_m)) - g(4^n x)}{\delta_m} \right| = \begin{cases} 0 & \text{if } n > m, \\ 4^n & \text{if } 0 \leq n \leq m. \end{cases}$$

Note here that, for a fixed m,
$$\frac{g(4^n(x + \delta_m)) - g(4^n x)}{\delta_m}$$

have the same sign for $n = 0, 1, \ldots, m$. Hence

$$\left| \frac{f(x+\delta_m) - f(x)}{\delta_m} \right| = \left| \sum_{n=0}^{\infty} \left(\frac{3}{4} \right)^n \frac{g(4^n(x+\delta_m)) - g(4^n x)}{\delta_m} \right|$$

$$= \left| \sum_{n=0}^{m} \left(\frac{3}{4} \right)^n \frac{g(4^n(x+\delta_m)) - g(4^n x)}{\delta_m} \right|$$

$$= \sum_{n=0}^{m} \left(\frac{3}{4} \right)^n 4^n$$

$$= \frac{3^{m+1} - 1}{2}.$$

Since $\lim_{m \to \infty} \delta_m = 0$, it follows from the above that $\lim_{h \to 0} \frac{f(x+h) - f(x)}{h}$ does not exist. This shows that f is nowhere differentiable. The graphs of three first partial sums $S_0(x), S_1(x)$ and $S_2(x)$ of the series defining f are sketched below.

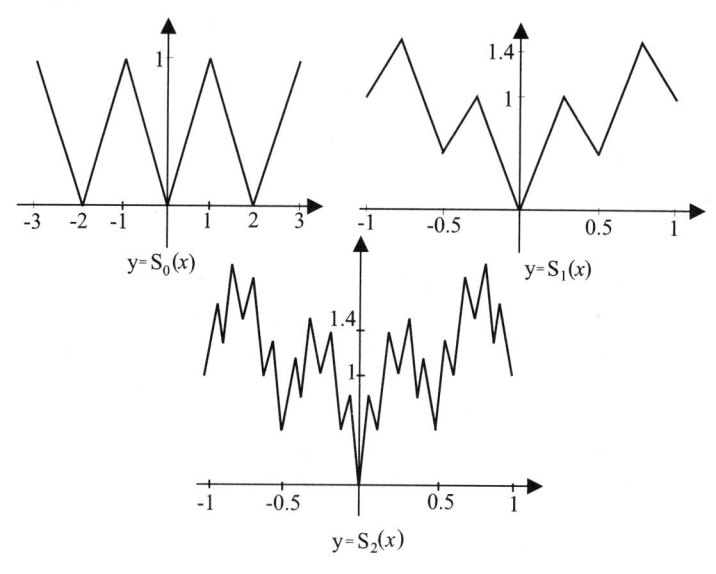

3.3. Power Series

3.3.1. Define R to be the supremum of the set of the $r \in [0, \infty)$ for which $\{|a_n|r^n\}$ is a bounded sequence. If R is positive, then for $0 \leq \rho < R$ there is a positive constant, say C_ρ, such that $|a_n|\rho^n \leq C_\rho$. Hence $\varlimsup_{n \to \infty} \sqrt[n]{|a_n|} \leq \frac{1}{\rho}$. Since the last inequality holds for each $\rho \in [0, R)$, we get

(i) $$\varlimsup_{n \to \infty} \sqrt[n]{|a_n|} \leq \frac{1}{R}.$$

Note that inequality (i) holds also for $R = 0$. To show that the inverse inequality also holds, suppose that $R < \infty$; then for $\rho > R$ the sequence $\{|a_n|\rho^n\}$ is unbounded. Consequently, it contains a subsequence such that $|a_{n_k}|\rho^{n_k} \geq 1$. So

$$\varlimsup_{n \to \infty} \sqrt[n]{|a_n|} \geq \varlimsup_{k \to \infty} \sqrt[n_k]{|a_{n_k}|} \geq \frac{1}{\rho}.$$

Since $\rho > R$ can be arbitrarily chosen, we get

(ii) $$\varlimsup_{n \to \infty} \sqrt[n]{|a_n|} \geq \frac{1}{R}.$$

Note that (ii) obviously holds for $R = \infty$. Combining (i) and (ii), we see that $\frac{1}{R} = \varlimsup_{n \to \infty} \sqrt[n]{|a_n|}$. Now the root test shows that the series $\sum_{n=0}^{\infty} a_n(x - x_0)^n$ converges absolutely for $|x - x_0| < R$ and diverges for $|x - x_0| > R$.

3.3.2.

(a) The radius of convergence of the series is 1, and it therefore converges for $|x| < 1$ and diverges for $|x| > 1$. For $x = 1, -1$ the series diverges. Thus the open interval $(-1, 1)$ is the interval of convergence.

(b) The radius of convergence is $+\infty$, and therefore the series converges for all $x \in \mathbb{R}$.

(c) The domain of convergence is the closed interval $[-1/2, 1/2]$.

(d) We have
$$\frac{1}{R} = \varlimsup_{n \to \infty} \sqrt[n]{(2 + (-1)^n)^n} = 3.$$

Thus the series converges on $(-1/3, 1/3)$. Clearly, the series diverges at the endpoints of the interval of convergence.

(e) Since
$$\frac{1}{R} = \varlimsup_{n\to\infty} \frac{2+(-1)^n}{5+(-1)^{n+1}} = \frac{3}{4},$$
the series converges on $(-4/3, 4/3)$. At the endpoints the series diverges.

(f) Since
$$\frac{1}{R} = \varlimsup_{n\to\infty} \sqrt[n]{|a_n|} = \lim_{n\to\infty} \sqrt[n]{2^n} = 1,$$
one can easily find that the interval of convergence is $(-1, 1)$.

(g) Since
$$\frac{1}{R} = \varlimsup_{n\to\infty} \sqrt[n]{|a_n|} = \lim_{n\to\infty} \sqrt[n!]{2^{n^2}} = 1,$$
one can easily find that the interval of convergence is $(-1, 1)$.

(h) We have
$$\frac{1}{R} = \varlimsup_{n\to\infty} \sqrt[n]{|a_n|} = \varlimsup_{n\to\infty} \left(1 + \frac{1}{n}\right)^{(-1)^n n} = e.$$

Therefore the series converges on $(-1/e, 1/e)$. At the endpoints the series diverges by the nth term test for divergence. Indeed, if $x = 1/e$, then
$$\lim_{n\to\infty} a_{2n} = \lim_{n\to\infty} \frac{\left(1 + \frac{1}{2n}\right)^{4n^2}}{e^{2n}} = e^{-1/2}$$
and if $x = -1/e$, then $\lim_{n\to\infty} |a_{2n}| = e^{-1/2}$.

3.3.3.

(a) The radius of convergence is $\sqrt{2}$ and the interval of convergence is $[1 - \sqrt{2}, 1 + \sqrt{2}]$.

(b) The radius of convergence of $\sum_{n=1}^{\infty} \frac{n}{n+1} y^n$ is 1. Thus the series $\sum_{n=1}^{\infty} \frac{n}{n+1} \left(\frac{2x+1}{x}\right)^n$ converges on $(-1, -1/3)$. Clearly it diverges at $x = -1$ and $x = -1/3$.

(c) The radius of convergence of $\sum_{n=1}^{\infty} \frac{n4^n}{3^n} y^n$ is $3/4$. Consequently, the series $\sum_{n=1}^{\infty} \frac{n4^n}{3^n} x^n (1-x)^n$ converges on $(-1/2, 3/2)$. One can easily see that it diverges at the endpoints.

(d) Since the radius of convergence is 4, the series converges on $(-3, 5)$. For $x = 5$ the series diverges because the sequence of its terms $\{\frac{(n!)^2}{(2n)!} 4^n\}$ monotonically increases. For $x = -3$ we get the series $\sum_{n=1}^{\infty} (-1)^n \frac{(n!)^2}{(2n)!} 4^n$, which diverges by the nth term test for divergence.

(e) The radius of convergence of $\sum_{n=1}^{\infty} \sqrt{n} y^n$ is 1. Therefore the series $\sum_{n=1}^{\infty} \sqrt{n} (\tan x)^n$ converges on the set

$$\bigcup_{n \in \mathbb{Z}} \left(-\frac{\pi}{4} + n\pi, \frac{\pi}{4} + n\pi \right).$$

If $x = -\frac{\pi}{4} + n\pi$ or $x = \frac{\pi}{4} + n\pi$, the series diverges.

(f) The domain of convergence is

$$(-\infty, -\tan 1) \cup (\tan 1, \infty).$$

3.3.4.

(a) Suppose that, for example, $R_1 < R_2$. Then for $|x| < R_1$ the series $\sum_{n=0}^{\infty} (a_n + b_n) x^n$ converges as the sum of two convergent series. For $R_1 < |x| < R_2$, the series diverges as the sum of a divergent and a convergent series. Thus $R = R_1 = \min\{R_1, R_2\}$. If $R_1 = R_2$, then clearly $R \geq R_1$. To show that the inequality can be strict, set $a_n = -1$, $b_n = 1$ for $n = 0, 1, 2, \ldots$. Then $R_1 = R_2 = 1$ and $R = \infty$.

(b) Since (see, e.g., I, 2.4.16)

$$\frac{1}{R} = \varlimsup_{n \to \infty} \sqrt[n]{|a_n b_n|} \leq \varlimsup_{n \to \infty} \sqrt[n]{|a_n|} \cdot \varlimsup_{n \to \infty} \sqrt[n]{|b_n|} = \frac{1}{R_1} \cdot \frac{1}{R_2},$$

we obtain $R \geq R_1 R_2$. The following example shows that the inequality can be strict. Set

$$a_{2n} = 0, \ a_{2n+1} = 1, \quad b_{2n} = 1, \ b_{2n+1} = 0, \quad n = 0, 1, 2, \ldots.$$

Then $R_1 = R_2 = 1$ and $R = \infty$.

3.3.5.

(a) It follows from
$$a_n = \frac{a_n}{b_n} \cdot b_n$$
and from (b) in the foregoing problem that $R_1 \geq RR_2$. To see that the inequality may be strict, consider, for example, the series $\sum_{n=0}^{\infty} a_n x^n$ and $\sum_{n=0}^{\infty} b_n x^n$, where

$$a_n = \begin{cases} 1 & \text{for even } n, \\ 2^n & \text{for odd } n \end{cases}$$

and
$$b_n = \begin{cases} 2^n & \text{for even } n, \\ 1 & \text{for odd } n. \end{cases}$$

Then $R_1 = R_2 = R = 1/2$.

(b) It suffices to observe that if $|x| < \min\{R_1, R_2\}$, then by the Mertens theorem (see, e.g., I, 3.6.1) the Cauchy product of the series $\sum_{n=0}^{\infty} a_n x^n$ and $\sum_{n=0}^{\infty} b_n x^n$ converges. The following example shows that the inequality $R \geq \min\{R_1, R_2\}$ can be strict. The Cauchy product of $\sum_{n=0}^{\infty} a_n x^n$ and $\sum_{n=0}^{\infty} b_n x^n$, where

$$a_0 = 1, \ a_n = -\left(\frac{3}{2}\right)^n, \quad b_0 = 1, \ b_n = \left(\frac{3}{2}\right)^{n-1}\left(2^n + \frac{1}{2^{n+1}}\right)$$

is $\sum_{n=0}^{\infty} \left(\frac{3}{4}\right)^n x^n$ (see, e.g., I, 3.6.11). Here $R_1 = 2/3$, $R_2 = 1/3$ and $R = 4/3$. The next example shows that R can be infinite even while both R_1 and R_2 are finite. If

$$a_n = \begin{cases} 2 & \text{for } n = 0, \\ 2^n & \text{for } n = 1, 2, \ldots \end{cases}$$

3.3. Power Series

and
$$b_n = \begin{cases} -1 & \text{for } n = 0, \\ 1 & \text{for } n = 1, 2, \ldots, \end{cases}$$
then $R_1 = 1/2$, $R_2 = 1$ and $R = +\infty$.

3.3.6. We will use 3.3.1(2).

(a) For $0 < \varepsilon < L$ there is n_0 such that if $n \geq n_0$, then
$$\sqrt[n]{\frac{L-\varepsilon}{n^\alpha}} \leq \sqrt[n]{|a_n|} \leq \sqrt[n]{\frac{L+\varepsilon}{n^\alpha}}.$$
Hence $\lim\limits_{n\to\infty} \sqrt[n]{|a_n|} = 1$ and $R = 1$.

(b) One can show, as in (a), that $R = \alpha$.

(c) $R = \infty$.

3.3.7.

(a) Since $\varlimsup\limits_{n\to\infty} \sqrt[n]{|2^n a_n|} = \frac{2}{R}$, the radius of convergence is equal to $\frac{1}{2}R$.

(b) If $\varepsilon > 0$ is so small that $\frac{1}{R} - \varepsilon > 0$, then for infinitely many n,
$$n\sqrt[n]{|a_n|} > n\left(\frac{1}{R} - \varepsilon\right).$$
Consequently, $\varlimsup\limits_{n\to\infty} n\sqrt[n]{|a_n|} = +\infty$ and $R = 0$.

(c) Since $\lim\limits_{n\to\infty} \frac{n}{\sqrt[n]{n!}} = e$, we see that the radius of convergence is R/e (see, e.g., I, 2.4.20).

(d) Since there is a sequence of positive integers $\{n_k\}$ such that
$$\frac{1}{R} = \lim_{k\to\infty} \sqrt[n_k]{|a_{n_k}|},$$
we conclude that the radius of convergence is R^2.

3.3.8. It follows immediately from the result in 3.1.25 that the only such power series are polynomials.

3.3.9. The radius of convergence of the series is $+\infty$. Termwise differentiation gives

$$f'(x) = \left(\sum_{n=0}^{\infty} \frac{x^{2n+1}}{(2n+1)!!}\right)' = 1 + \sum_{n=1}^{\infty} \frac{x^{2n}}{(2n-1)!!} = 1 + xf(x).$$

3.3.10. As in the solution of the foregoing problem, for $x \in \mathbb{R}$ we get

$$f''(x) + f'(x) + f(x) = \sum_{n=0}^{\infty} \frac{x^n}{n!} = e^x.$$

3.3.11. For $x \in (-1, 1)$, set

$$g(x) = \frac{f(xx_0) - f(x_0)}{x - 1}.$$

Then $\lim\limits_{x \to 1^-} g(x) = x_0 f'(x_0)$. Moreover (see, e.g., I, 3.6.4),

$$g(x) = \frac{1}{1-x} f(x_0) - \frac{1}{1-x} f(x_0 x) = \sum_{n=0}^{\infty} (f(x_0) - S_n(x_0)) x^n.$$

So if $0 < x < 1$ and $m = 0, 1, 2, \ldots$, we get

$$g(x) = \sum_{n=0}^{\infty} (f(x_0) - S_n(x_0)) x^n > (f(x_0) - S_m(x_0)) x^m.$$

Consequently, $x_0 f'(x_0) = \lim\limits_{x \to 1^-} g(x) \geq f(x_0) - S_m(x_0) > 0$.

3.3.12. We first show that $\sum\limits_{n=0}^{\infty} a_n x^n$, $\sum\limits_{n=0}^{\infty} S_n x^n$ and $\sum\limits_{n=0}^{\infty} (n+1) T_n x^n$ converge for $|x| < 1$. Since $\{T_n\}$ is bounded, there is $C > 0$ such that $|T_n| \leq C$ for all n. Then, for $|x| < 1$,

$$\sum_{n=0}^{\infty} (n+1)|T_n x^n| \leq \sum_{n=0}^{\infty} (n+1) C|x|^n = \frac{C}{(1-|x|)^2}.$$

Convergence of $\sum\limits_{n=0}^{\infty} S_n x^n$ for $|x| < 1$ follows from the equality

$$\sum_{n=0}^{N} S_n x^n = S_0 + \sum_{n=1}^{N} ((n+1) T_n - n T_{n-1}) x^n.$$

3.3. Power Series

Similarly, since $\sum_{n=0}^{N} a_n x^n = a_0 + \sum_{n=1}^{N}(S_n - S_{n-1})x^n$, the convergence of $\sum_{n=0}^{\infty} S_n x^n$ implies the convergence of $\sum_{n=0}^{\infty} a_n x^n$ for $|x| < 1$.

The stated equalities follow from the Mertens theorem (see, e.g., I, 3.6.1).

3.3.13. We have

$$\frac{|x|}{1-|x|}|f'(x)| \leq \sum_{n=0}^{\infty} |x|^n \sum_{k=0}^{\infty} 2^k x^{2^k} = \sum_{n=1}^{\infty} \left(\sum_{2^k \leq n} 2^k \right) |x|^n$$

$$= \sum_{n=1}^{\infty} \left(\sum_{k=0}^{[\log_2 n]} 2^k \right) |x|^n$$

$$\leq 2 \sum_{n=1}^{\infty} n|x|^n = 2\frac{|x|}{(1-|x|)^2}.$$

Thus the desired inequality is satisfied with $M = 2$.

3.3.14. The uniform convergence of $\sum_{n=0}^{\infty} a_n x^n$ on $[0,1]$ follows from the Abel test for uniform convergence (see the solution of 3.2.19). To prove (2) it suffices to apply 3.1.30 (see also the solution of 3.2.19).

3.3.15. We first show that

(1) $$\varlimsup_{x \to 1^-} f(x) \leq \varlimsup_{n \to \infty} S_n.$$

We have (see 3.3.12)

(2) $$f(x) = (1-x) \sum_{n=0}^{\infty} S_n x^n \quad \text{for} \quad |x| < 1.$$

If $\varlimsup_{n \to \infty} S_n = +\infty$, then (1) is obvious. If $\varlimsup_{n \to \infty} S_n = S \in \mathbb{R}$, then by (2) we get

(3) $$S - f(x) = (1-x) \sum_{n=0}^{\infty} (S - S_n) x^n.$$

362 Solutions. 3: Sequences and Series of Functions

Let $\varepsilon > 0$ be given. Then there is n_0 such that $S_n < S + \varepsilon$ whenever $n > n_0$. So, by (3), for $x \in (0,1)$,

$$S - f(x) \geq (1-x) \sum_{n=0}^{n_0} (S - S_n)x^n - \varepsilon(1-x) \sum_{n=n_0+1}^{\infty} x^n$$

$$= (1-x) \sum_{n=0}^{n_0} (S - S_n)x^n - \varepsilon x^{n_0+1}$$

$$\geq (1-x) \sum_{n=0}^{n_0} (S - S_n)x^n - \varepsilon.$$

Consequently,

$$f(x) \leq S + \varepsilon - (1-x) \sum_{n=0}^{n_0} (S - S_n)x^n.$$

Since there is $\delta > 0$ such that if $x \in (1-\delta, 1)$, then

$$\left| (1-x) \sum_{n=0}^{n_0} (S - S_n)x^n \right| < \varepsilon,$$

we see that $f(x) \leq S + 2\varepsilon$. So (1) is proved in the case of finite $\varlimsup_{n\to\infty} S_n$. Now if $\varlimsup_{n\to\infty} S_n = -\infty$, then clearly, $\lim_{n\to\infty} S_n = -\infty$. Thus for an $M \in \mathbb{R}$ one can choose n_1 such that if $n > n_1$, then $S_n < M$. Consequently, for $x \in (0,1)$ we get

$$M - f(x) = (1-x) \sum_{n=0}^{n_1} (M - S_n)x^n + (1-x) \sum_{n=n_1+1}^{\infty} (M - S_n)x^n$$

$$\geq (1-x) \sum_{n=0}^{n_1} (M - S_n)x^n.$$

So $f(x) \leq M - (1-x) \sum_{n=0}^{n_1} (M - S_n)x^n$. Since there is $\delta > 0$ such that if $x \in (1-\delta, 1)$, then,

$$\left| (1-x) \sum_{n=0}^{n_0} (M - S_n)x^n \right| < \varepsilon,$$

3.3. Power Series

we obtain $f(x) \leq M + \varepsilon$. Hence

$$\varliminf_{x \to 1^-} f(x) \leq \varlimsup_{x \to 1^-} f(x) \leq M.$$

Since M can be arbitrarily chosen, this shows that $\lim\limits_{x \to 1^-} f(x) = -\infty$. This ends the proof of (1). The inequality

$$\lim_{n \to \infty} S_n \leq \varliminf_{x \to 1^-} f(x)$$

can be established analogously.

3.3.16. Set

$$A_n = \frac{\sum\limits_{k=0}^{n} k|a_k|}{n}.$$

Then $\lim\limits_{n \to \infty} A_n = 0$ (see e.g., I, 2.3.2). By assumption, if $x_n = 1 - \frac{1}{n}$, then $\lim\limits_{n \to \infty} f(x_n) = L$. Thus, given $\varepsilon > 0$, there is n_0 such that if $n \geq n_0$, then

$$|f(x_n) - L| < \frac{\varepsilon}{3}, \quad A_n < \frac{\varepsilon}{3} \quad \text{and} \quad n|a_n| < \frac{\varepsilon}{3}.$$

Putting $S_n = \sum\limits_{k=0}^{n} a_k$, we get

$$S_n - L = f(x) - L + \sum_{k=0}^{n} a_k(1 - x^k) - \sum_{k=n+1}^{\infty} a_k x^k, \quad |x| < 1.$$

Now note that if $x \in (0, 1)$, then

$$(1 - x^k) = (1 - x)(1 + x + \cdots + x^{k-1}) \leq k(1 - x).$$

Consequently,

$$|S_n - L| \leq |f(x) - L| + (1 - x)\sum_{k=0}^{n} k|a_k| + \frac{\varepsilon}{3n(1 - x)}.$$

Finally, taking $x = x_n$ gives

$$|S_n - L| \leq \frac{\varepsilon}{3} + \frac{\varepsilon}{3} + \frac{\varepsilon}{3} = \varepsilon.$$

3.3.17. Consider, for example, the series $\sum_{n=0}^{\infty}(-1)^n x^n$.

3.3.18. It follows from the Abel theorem (see 3.3.14) that if the series $\sum_{n=1}^{\infty} a_n$ converges, then the limit $\lim_{x\to 1^-} f(x)$ exists. To show that the other implication holds, assume that $\lim_{x\to 1^-} f(x) = g \in \mathbb{R}$. Then, by assumption, for $0 < x < 1$ we get

$$\sum_{n=1}^{k} a_n x^n \leq f(x) \leq g, \quad k \in \mathbb{N}.$$

Hence $\sum_{n=1}^{k} a_n = \lim_{x\to 1^-} \sum_{n=1}^{k} a_n x^n \leq g$, which implies the convergence of $\sum_{n=1}^{\infty} a_n$.

3.3.19. Define

$$b_0 = 0, \quad b_n = a_1 + 2a_2 + \cdots + na_n, \quad n \in \mathbb{N}.$$

Then

$$f(x) = a_0 + \sum_{n=1}^{\infty} \frac{b_n - b_{n-1}}{n} x^n$$

$$= a_0 + \sum_{n=1}^{\infty} b_n \left(\frac{x^n}{n} - \frac{x^{n+1}}{n+1} \right)$$

$$= a_0 + \sum_{n=1}^{\infty} b_n \left(\frac{x^n - x^{n+1}}{n+1} + \frac{x^n}{n(n+1)} \right)$$

$$= a_0 + (1-x) \sum_{n=1}^{\infty} \frac{b_n}{n+1} x^n + \sum_{n=1}^{\infty} \frac{b_n}{n(n+1)} x^n.$$

Since $\lim_{n\to\infty} \frac{b_n}{n+1} = 0$, one can show that

$$\lim_{x\to 1^-} (1-x) \sum_{n=1}^{\infty} \frac{b_n}{n+1} x^n = 0.$$

3.3. Power Series

Now applying the Tauber theorem, we get

$$\sum_{n=1}^{\infty} \frac{b_n}{n(n+1)} = L - a_0.$$

Moreover,

$$\lim_{N\to\infty} \sum_{n=1}^{N} \frac{b_n}{n(n+1)} = \lim_{N\to\infty} \sum_{n=1}^{N} b_n \left(\frac{1}{n} - \frac{1}{n+1} \right)$$

$$= \lim_{N\to\infty} \left(\sum_{n=1}^{N} \frac{b_n - b_{n-1}}{n} - \frac{b_N}{N+1} \right)$$

$$= \lim_{N\to\infty} \sum_{n=1}^{N} a_n.$$

Thus $\sum_{n=0}^{\infty} a_n = L$.

3.3.20. It follows from the convergence of the series $\sum_{n=1}^{\infty} na_n^2$ and the result in I, 3.5.9(b) that

$$\lim_{n\to\infty} \frac{a_1^2 + 2^2 a_2^2 + \cdots + n^2 a_n^2}{n} = 0.$$

By the Cauchy inequality,

$$\left(\sum_{k=1}^{n} ka_k \right)^2 \leq n \left(\sum_{k=1}^{n} k^2 a_k^2 \right).$$

Consequently,

$$\lim_{n\to\infty} \frac{\left(\sum_{k=1}^{n} ka_k \right)^2}{n} \leq \lim_{n\to\infty} \frac{\sum_{k=1}^{n} k^2 a_k^2}{n} = 0.$$

The desired result follows from the foregoing problem.

3.3.21. Let $\varepsilon > 0$ be given. By assumption there is $n_0 \in \mathbb{N}$ such that if $n > n_0$, then $|a_n - Ab_n| < \varepsilon b_n$. Thus, for $x \in (0,1)$,

$$|f(x) - Ag(x)| = \left|\sum_{n=0}^{\infty}(a_n - Ab_n)x^n\right|$$

$$\leq \left|\sum_{n=0}^{n_0}(a_n - Ab_n)x^n\right| + \left|\sum_{n=n_0+1}^{\infty}(a_n - Ab_n)x^n\right|$$

$$\leq \sum_{n=0}^{n_0}|a_n - Ab_n| + \varepsilon \sum_{n=n_0+1}^{\infty} b_n x^n$$

$$\leq \sum_{n=0}^{n_0}|a_n - Ab_n| + \varepsilon g(x).$$

Since $\lim\limits_{x \to 1^-} g(x) = +\infty$, for x sufficiently close to 1 we get

$$\sum_{n=0}^{n_0}|a_n - Ab_n| < \varepsilon g(x).$$

Hence $|f(x) - Ag(x)| < 2\varepsilon g(x)$ for x sufficiently close to 1.

3.3.22. Note that by the Mertens theorem (see, e.g., I, 3.6.1),

$$f(x) = (1-x)\sum_{n=0}^{\infty} S_n x^n \quad \text{and} \quad g(x) = (1-x)\sum_{n=0}^{\infty} T_n x^n$$

for $|x| < 1$. Thus the result stated in the foregoing problem gives

$$\lim_{x \to 1^-} \frac{f(x)}{g(x)} = \lim_{x \to 1^-} \frac{\frac{f(x)}{1-x}}{\frac{g(x)}{1-x}} = A.$$

3.3.23. Consider

$$f(x) = \frac{1}{(1+x)^2(1-x)} = (1-x)\sum_{n=0}^{\infty}(n+1)x^{2n}$$

$$= \sum_{n=0}^{\infty}(n+1)(x^{2n} - x^{2n+1})$$

3.3. Power Series

and
$$g(x) = \frac{1}{1-x} = \sum_{n=0}^{\infty} x^n.$$

Then $\lim_{x\to 1^-} \frac{f(x)}{g(x)} = \frac{1}{4}$. On the other hand, since $S_{2n+1} = 0$, $S_{2n} = n+1$ and $T_n = n$, the limit $\lim_{n\to\infty} \frac{S_n}{T_n}$ does not exist.

3.3.24. For $x \in (0,1)$,

(1) $$f(x) \geq \sum_{k=0}^{n} a_k x^k \geq x^n S_n,$$

because all the coefficients a_n are nonnegative. Putting $x = e^{-\frac{1}{n}}$, we get
$$e^{-1} S_n \leq f\left(e^{-\frac{1}{n}}\right).$$

Thus, by assumption, given $\varepsilon > 0$, there is n_0 such that if $n \geq n_0$, then
$$e^{-1} S_n \leq \frac{A+\varepsilon}{1-e^{-\frac{1}{n}}} < 2(A+\varepsilon)n.$$

The last inequality follows from the fact that $\lim_{n\to\infty} \ln\left(1 - \frac{1}{2n}\right)^n = -\frac{1}{2} > -1$. So we have

(2) $$S_n \leq A_2 n \quad \text{with some} \quad A_2 \geq 2(A+\varepsilon)e.$$

Now by (2) we get
$$f(x) = (1-x) \sum_{n=0}^{\infty} S_n x^n$$
$$< (1-x) S_n \sum_{k=0}^{n-1} x^k + A_2(1-x) \sum_{k=n}^{\infty} k x^k$$
$$< S_n + A_2 n x^n + \frac{A_2 x^{n+1}}{1-x}.$$

If in (1) we put $x = e^{-\alpha/n}$, $\alpha > 0$, as above we obtain
$$f\left(e^{-\frac{\alpha}{n}}\right) > \frac{A-\varepsilon}{1-e^{-\frac{\alpha}{n}}} > (A-\varepsilon)\frac{n}{\alpha}.$$

The last inequality follows from $e^{-\frac{\alpha}{n}} > 1 - \frac{\alpha}{n}$. Consequently,

$$(A-\varepsilon)\frac{n}{\alpha} < S_n + A_2 n e^{-\alpha} + \frac{2A_2 n e^{-\alpha}}{\alpha},$$

or in other words,

$$S_n > n\frac{A - \varepsilon - 2A_2 e^{-\alpha} - A_2 \alpha e^{-\alpha}}{\alpha}.$$

If we take α sufficiently large, we get $S_n > A_1 n$ with some positive constant A_1.

3.3.25. We start with some considerations which we will need in the proof of the theorem. Assume that φ is continuous on $[0,1]$ except for one point $c \in (0,1)$ at which one-sided limits $\varphi(c^+)$ and $\varphi(c^-)$ exist and $\varphi(c) = \varphi(c^+)$ or $\varphi(c) = \varphi(c^-)$. Our aim is now to show that, given $\varepsilon > 0$, there are polynomials P_1 and P_2 such that

$$\int_0^1 (P_2(x) - \varphi(x))dx < \varepsilon \quad \text{and} \quad \int_0^1 (\varphi(x) - P_1(x))dx < \varepsilon.$$

To this end suppose, for example, that $\varphi(c^-) < \varphi(c^+)$ and $\varphi(c) = \varphi(c^+)$. Clearly, one can choose $\delta_1 > 0$ so small that the inequality $|\varphi(c - \delta_1) - \varphi(x)| < \varepsilon/4$ holds for $x \in (c - \delta_1, c)$. Set

$$M = \sup\{|\varphi(x) - \varphi(c)| : x \in (c - \delta_1, c)\}$$

and take $\delta < \min\{\delta_1, \varepsilon/(4M), c, 1-c\}$. Now define

$$g(x) = \begin{cases} \varphi(x) & \text{if } x \in [0, c-\delta] \cup [c,1], \\ \max\{l(x), \varphi(x)\} & \text{if } x \in (c-\delta, c), \end{cases}$$

where $l(x)$ is the linear function such that $l(c - \delta) = \varphi(c - \delta)$ and $l(c) = \varphi(c)$. Then g is continuous and $\varphi \leq g$ on $[0,1]$. By the approximation theorem of Weierstrass (see, e.g., 3.1.33) there is a polynomial P_2 such that

$$|g(x) - P_2(x)| < \frac{\varepsilon}{2} \quad \text{for} \quad x \in [0,1].$$

Likewise, we define

$$h(x) = \begin{cases} \varphi(x) & \text{if } x \in [0, c) \cup [c+\delta, 1], \\ \min\{l_1(x), \varphi(x)\} & \text{if } x \in [c, c+\delta), \end{cases}$$

3.3. Power Series

where $l_1(x)$ is the linear function such that $l_1(c) = \varphi(c^-)$ and $l_1(c+\delta) = \varphi(c+\delta)$. Clearly, h is continuous and $h \le \varphi$ on $[0,1]$. By the approximation theorem of Weierstrass there is a polynomial P_1 such that
$$|h(x) - P_1(x)| < \frac{\varepsilon}{2} \quad \text{for} \quad x \in [0,1].$$
Moreover, we have
$$\int_0^1 (g(x) - \varphi(x))dx = \int_{(c-\delta,c)} (g(x) - \varphi(x))dx.$$
If we put
$$\mathbf{A} = \{x \in (c-\delta, c) : g(x) = l(x)\} \quad \text{and} \quad \mathbf{B} = (c-\delta, c) \setminus \mathbf{A},$$
then we get
$$\int_{(c-\delta,c)} (g(x) - \varphi(x))dx = \int_{\mathbf{A}} (g(x) - \varphi(x))dx$$
$$\le \int_{(c-\delta,c)} |l(x) - \varphi(x)|dx$$
$$\le \int_{(c-\delta,c)} (|\varphi(x) - \varphi(c-\delta)| + |\varphi(c-\delta) - l(x)|dx$$
$$< \frac{\varepsilon}{4} + M\frac{\delta}{2} < \frac{\varepsilon}{2}.$$
It follows from the above that
$$\int_0^1 (P_2(x) - \varphi(x))dx$$
$$= \int_0^1 (P_2(x) - g(x))dx + \int_0^1 (g(x) - \varphi(x))dx < \varepsilon.$$
In an entirely similar manner one can show that
$$\int_0^1 (\varphi(x) - P_1(x))dx < \varepsilon.$$
We now turn to the proof of the theorem of Hardy and Littlewood. Without loss of generality we can assume that $A = 1$. We first show that
$$\lim_{x \to 1^-} (1-x) \sum_{n=0}^{\infty} a_n x^n P(x^n) = \int_0^1 P(t)dt$$

for any polynomial P. Clearly, it is enough to prove the equality for $P(x) = x^k$. We have

$$\lim_{x \to 1^-} (1-x) \sum_{n=0}^{\infty} a_n x^{n+kn} = \lim_{x \to 1^-} \frac{1-x}{1-x^{k+1}}(1-x^{k+1}) \sum_{n=0}^{\infty} a_n x^{(k+1)n}$$

$$= \frac{1}{k+1} = \int_0^1 t^k dt.$$

Now we define φ by

$$\varphi(x) = \begin{cases} 0 & \text{for } 0 \leq x < e^{-1}, \\ \frac{1}{x} & \text{for } e^{-1} \leq x \leq 1. \end{cases}$$

Our task is to show that

(1) $$\lim_{x \to 1^-} (1-x) \sum_{n=0}^{\infty} a_n x^n \varphi(x^n) = \int_0^1 \varphi(t) dt = 1.$$

It follows from the considerations presented at the beginning of the solution that, given $\varepsilon > 0$, there exist polynomials P_1 and P_2 such that

$$P_1(x) - \frac{\varepsilon}{2} \leq h(x) \leq \varphi(x) \leq g(x) \leq P_2(x) + \frac{\varepsilon}{2}$$

and

$$\int_0^1 (P_2(x) - \varphi(x)) dx < \varepsilon, \quad \int_0^1 (\varphi(x) - P_1(x)) dx < \varepsilon.$$

Since $a_n \geq 0$, we get

$$\varlimsup_{x \to 1^-} (1-x) \sum_{n=0}^{\infty} a_n x^n \varphi(x^n) \leq \varlimsup_{x \to 1^-} (1-x) \sum_{n=0}^{\infty} a_n x^n P_2(x^n) + \frac{\varepsilon}{2}$$

$$= \int_0^1 P_2(t) dt + \frac{\varepsilon}{2} < \int_0^1 \varphi(t) dt + \frac{3\varepsilon}{2}.$$

Consequently,

$$\varlimsup_{x \to 1^-} (1-x) \sum_{n=0}^{\infty} a_n x^n \varphi(x^n) \leq \int_0^1 \varphi(t) dt.$$

In much the same way one can show that

$$\varliminf_{x \to 1^-} (1-x) \sum_{n=0}^{\infty} a_n x^n \varphi(x^n) \geq \int_0^1 \varphi(t) dt.$$

3.3. Power Series

So (1) is proved. Therefore

$$1 = \lim_{N\to\infty} (1-e^{-1/N}) \sum_{n=0}^{\infty} a_n e^{-n/N} \varphi(e^{-n/N}) = \lim_{N\to\infty} (1-e^{-1/N}) \sum_{n=0}^{N} a_n.$$

Since $\lim_{N\to\infty} (1 - e^{-1/N})N = 1$, we obtain

$$\lim_{N\to\infty} \frac{\sum_{n=0}^{N} a_n}{N} = 1.$$

3.3.26. If $|na_n| \leq C$, then for $x \in (0,1)$,

$$|f''(x)| \leq \sum_{n=2}^{\infty} n(n-1)|a_n|x^{n-2} \leq C \sum_{n=2}^{\infty} (n-1)x^{n-2} = C\frac{1}{(1-x)^2}.$$

It then follows by 2.3.23 that

$$\lim_{x\to 1^-} (1-x)f'(x) = 0 = \lim_{x\to 1^-} (1-x) \sum_{n=1}^{\infty} na_n x^{n-1}.$$

Now since

$$F(x) = \sum_{n=1}^{\infty} \left(1 - \frac{na_n}{C}\right) x^{n-1} = \frac{1}{1-x} - \frac{f'(x)}{C},$$

we obtain $\lim_{x\to 1^-} (1-x)F(x) = 1$, which combined with the above theorem of Hardy and Littlewood yields

$$\lim_{n\to\infty} \frac{\sum_{k=1}^{n} \left(1 - \frac{ka_k}{C}\right)}{n} = 1.$$

Hence

$$\lim_{n\to\infty} \frac{\sum_{k=1}^{n} ka_k}{n} = 0.$$

To end the proof it suffices to apply the result given in 3.3.19.

372   Solutions. 3: Sequences and Series of Functions

3.3.27. Suppose, contrary to our claim, that $\lim_{n\to\infty} a_n = 0$. Then, given $\varepsilon > 0$, there is n_0 such that if $n > n_0$, then $|a_n| < \varepsilon/2$. So,

$$|(1-x)f(x)| \leq \left|(1-x)\sum_{k=0}^{n_0} a_k x^k\right| + \frac{\varepsilon}{2},$$

which implies

$$\lim_{x\to 1^-} |(1-x)f(x)| = 0,$$

contrary to the assumption.

3.4. Taylor Series

3.4.1. Suppose that $|f^{(n)}(x)| \leq M$ for $n \in \mathbb{N}$ and $x \in [a,b]$. By Taylor's formula with the Lagrange form for the remainder (see, e.g., 2.3.3 (a)) we have

$$f(x) = \sum_{k=0}^{n} \frac{f^{(k)}(x_0)}{k!}(x-x_0)^k + r_n(x),$$

where

$$|r_n(x)| = \left|\frac{f^{(n+1)}(x_0 + \theta(x-x_0))}{(n+1)!}(x-x_0)^{n+1}\right| \leq M\frac{(b-a)^{n+1}}{(n+1)!}.$$

Hence $\lim_{n\to\infty} r_n(x) = 0$. Consequently,

$$f(x) = \lim_{n\to\infty} \sum_{k=0}^{n} \frac{f^{(k)}(x_0)}{k!}(x-x_0)^k = \sum_{k=0}^{\infty} \frac{f^{(k)}(x_0)}{k!}(x-x_0)^k.$$

3.4.2. No, because $f^{(n)}(0) = 0$ for $n = 0, 1, 2, \ldots$, and $f(x) \neq 0$ for $x \neq 0$.

3.4.3. By the M-test of Weierstrass the series $\sum_{n=0}^{\infty} \frac{\cos(n^2 x)}{e^n}$ and the derived series $\sum_{n=0}^{\infty} \frac{-n^2 \sin(n^2 x)}{e^n}$ converge absolutely and uniformly on $\mathbb{R}$. So f' is continuous on $\mathbb{R}$. Repeating the reasoning, we see that f is

3.4. Taylor Series

in $C^\infty(\mathbb{R})$. Moreover, one can find that $f^{(2k-1)}(0) = 0$ and $f^{(2k)}(0) = (-1)^k \sum_{n=0}^{\infty} \frac{n^{4k}}{e^n}$. Thus

$$\frac{|f^{(2k)}(0)|x^{2k}}{(2k)!} > \left(\frac{n^2 x}{2k}\right)^{2k} e^{-n}, \quad x \neq 0, \quad n = 0, 1, 2, \dots.$$

If we take $n = 2k$, we get

$$\frac{|f^{(2k)}(0)|x^{2k}}{(2k)!} > \left(\frac{2kx}{e}\right)^{2k} > 1 \quad \text{for} \quad x \neq 0 \quad \text{and} \quad k > \left|\frac{e}{2x}\right|.$$

Therefore the Taylor series of f about zero diverges for $x \neq 0$ and the equality cannot hold if $x \neq 0$.

3.4.4. Suppose first that $x > 0$. The Lagrange form for the remainder in Taylor's formula of $f(x) = (1+x)^\alpha$ is

$$r_n(x) = \frac{\alpha(\alpha-1)\cdots(\alpha-n)}{(n+1)!} x^{n+1} (1+\theta x)^{\alpha-n-1}.$$

For $|x| < 1$ we have

$$\lim_{n \to \infty} \frac{\alpha(\alpha-1)\cdots(\alpha-n)}{(n+1)!} x^{n+1} = 0.$$

To see this one can apply, for example, I, 2.2.31. Consequently, to prove that $\lim_{n \to \infty} r_n(x) = 0$ it suffices to show that $\{(1+\theta x)^{\alpha-n-1}\}$ is a bounded sequence. This follows from the following obvious inequalities:

$$1 \leq (1+\theta x)^\alpha \leq (1+x)^\alpha \leq 2^\alpha \quad \text{for} \quad \alpha \geq 0$$

and

$$2^\alpha \leq (1+x)^\alpha \leq (1+\theta x)^\alpha \leq 1 \quad \text{for} \quad \alpha < 0,$$

and $(1+\theta x)^{-n} \leq 1$. Thus we have proved the given equality for $0 < x < 1$. Now we turn to the case $x < 0$. The Cauchy form for the remainder in Taylor's formula of $f(x) = (1+x)^\alpha$ (see, e.g., 2.3.3(b)) is

$$r_n(x) = \frac{\alpha(\alpha-1)\cdots(\alpha-n)}{(n+1)!} x^{n+1} (1-\theta)^n (1+\theta x)^{\alpha-n-1}.$$

As above, it suffices to show that $\{(1-\theta)^n(1+\theta x)^{\alpha-n-1}\}$ is a bounded sequence. Since $x \in (-1,0)$, we see that

$$(1-\theta)^n \leq \left(\frac{1-\theta}{1+\theta x}\right)^n < 1.$$

Moreover,

$$1 \leq (1+\theta x)^{\alpha-1} \leq (1+x)^{\alpha-1} \quad \text{if} \quad \alpha \leq 1$$

and

$$(1+x)^{\alpha-1} \leq (1+\theta x)^{\alpha-1} \leq 1 \quad \text{if} \quad \alpha \geq 1.$$

This ends the proof of the equality for $x \in (-1, 0)$.

3.4.5. Suppose first that $x \neq 0$. Then the equality

$$|x| = \sqrt{1 - (1-x^2)}$$

and Newton's binomial formula with $\alpha = 1/2$ (see the foregoing problem) give

$$|x| = 1 - \frac{1}{2}(1-x^2) - \sum_{n=2}^{\infty} \frac{1 \cdot 3 \cdots (2n-3)}{2^n n!}(1-x^2)^n$$

$$= 1 - \frac{1}{2}(1-x^2) - \sum_{n=2}^{\infty} \frac{(2n-3)!!}{(2n)!!}(1-x^2)^n.$$

Moreover, note that the series $\sum_{n=2}^{\infty} \frac{(2n-3)!!}{(2n)!!}$ converges because, by the Wallis formula (see, e.g., I, 3.8.38),

$$\lim_{n \to \infty} \frac{\frac{(2n-3)!!}{(2n)!!}}{\frac{1}{(2n-1)\sqrt{n}}} = \frac{1}{\sqrt{\pi}}.$$

Therefore the Abel theorem (see, e.g., 3.3.14) shows that the equality holds also for $x = 0$.

3.4.6. Termwise differentiation shows that f is in $C^\infty(-R, R)$. Moreover,

$$f^{(k)}(x) = \sum_{n=k}^{\infty} n(n-1)(n-2) \cdots (n-k+1) a_n x^{n-k}.$$

3.4. Taylor Series

Hence $f^{(k)}(0) = k!a_k$ for $k = 0, 1, 2, \ldots$.

3.4.7. Observe that
$$f(x) = \sum_{n=0}^{\infty} a_n((x - x_0) + x_0)^n$$
$$= \sum_{n=0}^{\infty} a_n \sum_{k=0}^{n} \binom{n}{k} (x - x_0)^k x_0^{n-k}$$
$$= \sum_{k=0}^{\infty} \left(\sum_{n=k}^{\infty} \binom{n}{k} a_n x_0^{n-k} \right) (x - x_0)^k.$$

To see that the last equality holds, note that
$$\sum_{n=0}^{\infty} \sum_{k=0}^{n} \left| a_n \binom{n}{k} (x - x_0)^k x_0^{n-k} \right| = \sum_{n=0}^{\infty} |a_n|(|x - x_0| + |x_0|)^n.$$

Consequently, the double series on the left side of this equality converges absolutely for $|x - x_0| + |x_0| < R$, and therefore the result in I, 3.7.23 can be applied. Now differentiating term by term we get
$$f^{(k)}(x_0) = \sum_{n=k}^{\infty} \binom{n}{k} a_n x_0^{n-k} k! \quad \text{for} \quad k = 0, 1, 2, \ldots.$$

Thus
$$\frac{f^{(k)}(x_0)}{k!} = \sum_{n=k}^{\infty} \binom{n}{k} a_n x_0^{n-k} \quad \text{for} \quad k = 0, 1, 2, \ldots.$$

3.4.8. Set $c_n = a_n - b_n$ and

(1) $$f(x) = \sum_{n=0}^{\infty} c_n x^n, \quad x \in (-R, R).$$

Then $f(x) = 0$ for $x \in \mathbf{A}$. Now let $\mathbf{B}$ be the set of all limit points of $\mathbf{A}$ that are in $(-R, R)$, and put $\mathbf{C} = (-R, R) \setminus \mathbf{B}$. Then $\mathbf{C}$ is open. By assumption $\mathbf{B}$ is nonempty. Clearly, $(-R, R) = \mathbf{B} \cup \mathbf{C}$. Our task is now to prove that $\mathbf{B}$ is also open. To this end take $x_0 \in \mathbf{B}$. By (1) and the result in the foregoing problem,

(2) $$f(x) = \sum_{n=0}^{\infty} d_n (x - x_0)^n, \quad |x - x_0| < R - |x_0|.$$

Now we show that $d_n = 0$ for $n = 0, 1, 2, \ldots$. If this were not the case, then we would find the least nonnegative integer k for which $d_k \neq 0$ and we would get

$$f(x) = (x - x_0)^k g(x),$$

where

$$g(x) = \sum_{n=0}^{\infty} d_{k+n}(x - x_0)^n, \quad |x - x_0| < R - |x_0|.$$

Since g is continuous at x_0 and $g(x_0) = d_k \neq 0$, there would exist $\delta > 0$ such that $g(x) \neq 0$ for $|x - x_0| < \delta$, contrary to the fact that x_0 is in **B**. Thus $d_n = 0$ for $n = 0, 1, 2, \ldots$, and consequently, $f(x) = 0$ for $|x - x_0| < R - |x_0|$. So we have proved that **B** is open. Since $(-R, R)$ is a connected set, we see that $\mathbf{C} = \emptyset$ and $\mathbf{B} = (-R, R)$.

3.4.9. We will apply the result stated in 3.4.6.

(a) Since

$$\sin x = \sum_{n=0}^{\infty} (-1)^n \frac{x^{2n+1}}{(2n+1)!}, \quad x \in \mathbb{R},$$

we get

$$\sin x^3 = \sum_{n=0}^{\infty} (-1)^n \frac{x^{6n+3}}{(2n+1)!}, \quad x \in \mathbb{R}.$$

(b) In view of the identity $\sin^3 x = \frac{3}{4} \sin x - \frac{1}{4} \sin 3x$, $x \in \mathbb{R}$, we get

$$\sin^3 x = \frac{3}{4} \sum_{n=0}^{\infty} (-1)^{n+1}(3^{2n} - 1) \frac{x^{2n+1}}{(2n+1)!}, \quad x \in \mathbb{R}.$$

(c) We have $\sin x \cos 3x = \frac{1}{2}(\sin 4x - \sin 2x)$, $x \in \mathbb{R}$. Thus

$$\sin x \cos 3x = \frac{1}{2} \sum_{n=0}^{\infty} (-1)^n (4^{2n+1} - 2^{2n+1}) \frac{x^{2n+1}}{(2n+1)!}, \quad x \in \mathbb{R}.$$

(d) We have $\sin^6 x + \cos^6 x = \frac{5}{8} + \frac{3}{8} \cos 4x$, $x \in \mathbb{R}$, and

$$\cos x = \sum_{n=0}^{\infty} (-1)^n \frac{x^{2n}}{(2n)!}, \quad x \in \mathbb{R}.$$

3.4. Taylor Series

Consequently,

$$\sin^6 x + \cos^6 x = \frac{5}{8} + \frac{3}{8}\sum_{n=0}^{\infty}(-1)^n 4^{2n}\frac{x^{2n}}{(2n)!}, \quad x \in \mathbb{R}.$$

(e) Since

$$\ln(1+x) = \sum_{n=1}^{\infty}(-1)^{n+1}\frac{x^n}{n}, \quad x \in (-1,1),$$

we get

$$\frac{1}{2}\ln\frac{1+x}{1-x} = \frac{1}{2}(\ln(1+x) - \ln(1-x)) = \sum_{n=0}^{\infty}\frac{x^{2n+1}}{2n+1}, \quad x \in (-1,1).$$

(f) Clearly, $\ln(1+x+x^2) = \ln\frac{1-x^3}{1-x}$, $x \in (-1,1)$. Therefore, as in (e) we have

$$\ln(1+x+x^2) = \sum_{n=1}^{\infty}a_n x^n, \quad x \in (-1,1),$$

where

$$a_n = \begin{cases} -\frac{2}{n} & \text{for } n = 3k, \ k = 1, 2, 3, \ldots, \\ \frac{1}{n} & \text{for } n \neq 3k, \ k = 1, 2, 3, \ldots. \end{cases}$$

(g) Since $\frac{1}{1-5x+6x^2} = \frac{3}{1-3x} - \frac{2}{1-2x}$, we obtain

$$\frac{1}{1-5x+6x^2} = \sum_{n=0}^{\infty}(3^{n+1} - 2^{n+1})x^n, \quad x \in (-1/3, 1/3).$$

(h) We know that

$$e^x = \sum_{n=0}^{\infty}\frac{x^n}{n!}, \quad x \in \mathbb{R},$$

and

$$\frac{1}{1-x} = \sum_{n=0}^{\infty}x^n, \quad x \in (-1,1).$$

By the Mertens theorem (see, e.g., I, 3.6.1) the Cauchy product of these two series converges for $|x| < 1$, and

$$\frac{e^x}{1-x} = \sum_{n=0}^{\infty}\left(1 + \frac{1}{1!} + \frac{1}{2!} + \cdots + \frac{1}{n!}\right)x^n.$$

3.4.10.

(a) We have
$$f(x+1) = (x+2)e^{x+1} = \sum_{n=0}^{\infty} \frac{e(n+2)}{n!}x^n, \quad x \in \mathbb{R}.$$

Hence
$$f(x) = \sum_{n=0}^{\infty} \frac{e(n+2)}{n!}(x-1)^n, \quad x \in \mathbb{R}.$$

(b) As in 3.4.9(h), one can show that
$$f(x+1) = e\sum_{n=0}^{\infty}\left((-1)^n \sum_{k=0}^{n}\frac{(-1)^k}{k!}\right)x^n, \quad x \in (-1,1,).$$

Thus
$$f(x) = e\sum_{n=0}^{\infty}\left((-1)^n \sum_{k=0}^{n}\frac{(-1)^k}{k!}\right)(x-1)^n, \quad x \in (0,2).$$

(c) Apply the identity
$$\frac{\cos x}{x} = \frac{\cos 1 \cos(x-1) - \sin 1 \sin(x-1)}{1+(x-1)}.$$

(d) Reasoning similar to that presented in the solution of 3.4.9(h) yields
$$\frac{\ln x}{x} = \sum_{n=1}^{\infty}\left((-1)^{n+1}\sum_{k=1}^{n}\frac{1}{k}\right)(x-1)^n, \quad x \in (0,2).$$

3.4.11.

(a) By 3.4.4,
$$\frac{1}{\sqrt{1+x}} = 1 + \sum_{n=1}^{\infty}\frac{(-1)^n(2n-1)!!}{(2n)!!}x^n$$
for $|x| < 1$. Hence
$$\frac{1}{\sqrt{1-x^2}} = 1 + \sum_{n=1}^{\infty}\frac{(2n-1)!!}{(2n)!!}x^{2n}.$$

3.4. Taylor Series

Set
$$S(x) = x + \sum_{n=1}^{\infty} \frac{(2n-1)!!}{(2n)!!(2n+1)} x^{2n+1}$$
and note that $(\arcsin x)' = \frac{1}{\sqrt{1-x^2}} = S'(x)$. So $\arcsin x = S(x) + C$. Moreover, since $S(0) = 0 = \arcsin 0$, we obtain $S(x) = \arcsin x$.

(b) Set
$$S(x) = \sum_{n=0}^{\infty} (-1)^n \frac{1}{2n+1} x^{2n+1}.$$
In view of the well known identity
$$\frac{1}{1+x^2} = \sum_{n=0}^{\infty} (-1)^n x^{2n}, \quad |x| < 1,$$
we get $(\arctan x)' = \frac{1}{1+x^2} = S'(x)$. Thus $S(x) = \arctan x + C$. Since $\arctan 0 = S(0) = 0$, we see that $C = 0$.

To obtain the first equality it is enough to put $x = \frac{1}{2}$ in (a). To obtain the second one, observe that $\sum_{n=0}^{\infty} (-1)^n \frac{1}{2n+1}$ converges and apply the Abel theorem (see, e.g, 3.3.14) to the power series (b).

3.4.12.

(a) Applying the Taylor series expansion for $\arctan x$ (given in the foregoing problem) and for $\ln(1+x^2)$, we obtain

$$x \arctan x - \frac{1}{2} \ln(1+x^2) = \sum_{n=1}^{\infty} \frac{(-1)^{n-1} x^{2n}}{2n(2n-1)}, \quad x \in (-1, 1).$$

(b) Applying the Taylor series expansion for $\arcsin x$ (given in the foregoing problem) and Newton's binomial formula (see 3.4.4), we get

$$x \arcsin x + \sqrt{1-x^2} = 1 + \frac{x^2}{2} + \sum_{n=2}^{\infty} \frac{(2n-3)!!}{(2n)!!(2n-1)} x^{2n}, \quad x \in (-1, 1).$$

3.4.13.

(a) Let
$$f(x) = \sum_{n=1}^{\infty} \frac{1}{n(n+1)} x^{n+1}, \quad |x| \leq 1.$$
Then
$$f'(x) = \sum_{n=1}^{\infty} \frac{1}{n} x^n = -\ln(1-x), \quad |x| < 1.$$
So
$$f(x) = (1-x)\ln(1-x) + x \quad \text{for} \quad |x| < 1.$$
Now the Abel theorem gives
$$\sum_{n=1}^{\infty} \frac{(-1)^{n+1}}{n(n+1)} = 2\ln 2 - 1.$$

(b) For $x \in \mathbb{R}$, we have
$$\sum_{n=0}^{\infty} \frac{(-1)^n n}{(2n+1)!} x^{2n+1} = \frac{1}{2} x \sum_{n=0}^{\infty} \frac{(-1)^n}{(2n)!} x^{2n} - \frac{1}{2} \sum_{n=0}^{\infty} \frac{(-1)^n}{(2n+1)!} x^{2n+1}$$
$$= \frac{1}{2}(x \cos x - \sin x).$$
Putting $x = 1$, we get
$$\sum_{n=0}^{\infty} \frac{(-1)^n n}{(2n+1)!} = \frac{1}{2}(\cos 1 - \sin 1).$$

(c) It follows from the equality
$$\frac{1}{n^2+n-2} = \frac{1}{3}\left(\frac{1}{n-1} - \frac{1}{n+2}\right)$$
that if $0 < |x| < 1$, then
$$\sum_{n=2}^{\infty} \frac{(-1)^n}{n^2+n-2} x^{n-1} = \frac{1}{3} \sum_{n=2}^{\infty} \frac{(-1)^n}{n-1} x^{n-1} - \frac{1}{3} \sum_{n=2}^{\infty} \frac{(-1)^n}{n+2} x^{n-1}$$
$$= \frac{1}{3} \sum_{n=1}^{\infty} (-1)^{n-1} \frac{x^n}{n} + \frac{1}{3x^3} \sum_{n=4}^{\infty} (-1)^{n-1} \frac{x^n}{n}$$
$$= \frac{1}{3}\ln(1+x) + \frac{1}{3x^3}\left(\ln(1+x) - x + \frac{x^2}{2} - \frac{x^3}{3}\right).$$

3.4. Taylor Series

This combined with the Abel theorem gives

$$\sum_{n=2}^{\infty} \frac{(-1)^n}{n^2+n-2} = \frac{2}{3}\ln 2 - \frac{5}{18}.$$

(d) The sum is $\pi/2 - \ln 2$. To see this, apply 3.4.12(a) and the Abel theorem.

(e) By Newton's binomial formula (see 3.4.4),

$$\frac{1}{\sqrt{1+x^2}} = 1 + \sum_{n=1}^{\infty} \frac{(-1)^n(2n-1)!!}{(2n)!!} x^{2n} \quad \text{for } |x|<1,$$

and hence, by the Abel theorem,

$$\sum_{n=1}^{\infty} \frac{(-1)^n(2n-1)!!}{(2n)!!} = \frac{1}{\sqrt{2}}.$$

(f) Clearly,

$$\sum_{n=0}^{\infty} \frac{(3x)^{n+1}}{n!} = 3xe^{3x}, \quad x \in \mathbb{R}.$$

So

$$3\sum_{n=0}^{\infty} \frac{(3x)^n(n+1)}{n!} = (3xe^{3x})' = e^{3x}(3+9x).$$

Putting $x = 1$ gives

$$\sum_{n=0}^{\infty} \frac{3^n(n+1)}{n!} = 4e^3.$$

3.4.14. The interval of convergence of the series is $(-1, 1)$. Let $S(x)$ denote its sum in that interval. Then

$$S'(x) = 2\sum_{n=1}^{\infty} \frac{((n-1)!)^2}{(2n-1)!}(2x)^{2n-1}$$

and

$$S''(x) = 4\sum_{n=1}^{\infty} \frac{((n-1)!)^2}{(2n-2)!}(2x)^{2n-2}.$$

It then follows that

$$(1-x^2)S''(x) - xS'(x) = 4, \quad |x|<1.$$

Multiplying both sides of this equality by $(1-x^2)^{-\frac{1}{2}}$ produces

$$\left(\sqrt{1-x^2}S'(x)\right)' = \frac{4}{\sqrt{1-x^2}}.$$

Consequently,

$$S'(x) = \frac{4}{\sqrt{1-x^2}}\arcsin x + \frac{C}{\sqrt{1-x^2}},$$

and therefore $S(x) = 2(\arcsin x)^2 + C\arcsin x + D$. Since $S'(0) = S(0) = 0$, we obtain $S(x) = 2(\arcsin x)^2$.

If $x = \pm 1$ we get the series

$$\sum_{n=1}^{\infty} \frac{((n-1)!)^2}{(2n)!} 4^n,$$

which converges by the Gauss criterion (see, e.g., I, 3.2.25). Indeed, we have

$$\frac{a_{n+1}}{a_n} = 1 - \frac{6}{4n} + O\left(\frac{1}{n^2}\right).$$

So the Abel theorem gives

$$\sum_{n=1}^{\infty} \frac{((n-1)!)^2}{(2n)!} 4^n = \frac{\pi^2}{2}.$$

3.4.15. For $a \in \mathbf{I}$,

$$f(x) = f(a) + \frac{f'(a)}{1!}(x-a) + \cdots + \frac{f^{(n)}(a)}{n!}(x-a)^n + R_n(x),$$

where

$$R_n(x) = \frac{1}{n!} \int_a^x f^{(n+1)}(s)(x-s)^n ds.$$

Applying the change of variable formula twice, we obtain

$$R_n(x) = \frac{1}{n!} \int_0^{x-a} f^{(n+1)}(u+a)(x-u-a)^n du$$

$$= \frac{(x-a)^{n+1}}{n!} \int_0^1 f^{(n+1)}((x-a)t+a)(1-t)^n dt.$$

3.4. Taylor Series 383

The monotonicity of $f^{(n+1)}$ implies that if $a < x < b$, $b \in \mathbf{I}$, then

$$0 \leq R_n(x) \leq \frac{(x-a)^{n+1}}{n!} \int_0^1 f^{(n+1)}((b-a)t + a)(1-t)^n dt$$

$$= \left(\frac{x-a}{b-a}\right)^{n+1} R_n(b).$$

Clearly, $R_n(b) \leq f(b)$. Thus

$$0 \leq R_n(x) \leq \left(\frac{x-a}{b-a}\right)^n f(b) \quad \text{for} \quad a < x < b, \ a, b \in \mathbf{I},$$

and therefore, $\lim\limits_{n \to \infty} R_n(x) = 0$. This shows that Taylor's series converges to f uniformly on each compact subinterval of $\mathbf{I}$. Since $a < b$ can be arbitrarily chosen in $\mathbf{I}$, the analyticity of f follows from 3.4.7.

3.4.16. The proof is similar to that of 3.4.1.

3.4.17 [18]. Let x_0 be arbitrarily chosen in $\mathbf{I}$. By assumption, there is an $r > 0$ such that

$$f(x) = \sum_{n=0}^{\infty} \frac{f^{(n)}(x_0)}{n!}(x-x_0)^n \quad \text{for} \quad |x - x_0| < r.$$

Differentiating m times yields

$$f^{(m)}(x) = \sum_{n=m}^{\infty} \frac{f^{(n)}(x_0)}{n!} n(n-1)\cdots(n-m+1)(x-x_0)^{n-m}.$$

Hence

$$|f^{(m)}(x)| \leq \sum_{n=m}^{\infty} \frac{|f^{(n)}(x_0)|}{n!} n(n-1)\cdots(n-m+1)|x-x_0|^{n-m}.$$

It follows from the definition of the radius of convergence of the power series (see, e.g., 3.3.1) that for $0 < \rho < r$ there is a positive C such that

$$\frac{|f^{(n)}(x_0)|}{n!} \leq \frac{C}{\rho^n}.$$

Consequently,

$$|f^{(m)}(x)| \leq \sum_{n=m}^{\infty} \frac{C}{\rho^n} n(n-1)\cdots(n-m+1)|x-x_0|^{n-m}.$$

Therefore, in view of the identity

$$\sum_{n=m}^{\infty} n(n-1)\cdots(n-m+1)x^{n-m} = \frac{m!}{(1-x)^{m+1}}, \quad |x| < 1,$$

we arrive at

$$|f^{(m)}(x)| \leq \rho^{-m} \sum_{n=m}^{\infty} \frac{C}{\rho^{n-m}} n(n-1)\cdots(n-m+1)|x-x_0|^{n-m}$$

$$= \frac{Cm!}{\rho^m \left(1 - \frac{|x-x_0|}{\rho}\right)^{m+1}} \leq \frac{C\rho m!}{(\rho-\rho_1)^m}$$

for $|x - x_0| < \rho_1 < \rho$. So we can take $\mathbf{J} = (x_0 - \rho_1, x_0 + \rho_1)$, $A = C\rho$ and $B = \rho - \rho_1$.

3.4.18 [18]. Set

$$f(x) = \frac{1}{1 - A(x-1)} \quad \text{for} \quad |x-1| < \frac{1}{A}$$

and

$$g(t) = \frac{1}{1-t} \quad \text{for} \quad |t| < 1.$$

Then

$$h(t) = (f \circ g)(t) = \frac{1-t}{1-(A+1)t}.$$

Clearly,

$$f(x) = \sum_{n=0}^{\infty} A^n(x-1)^n, \quad g(t) = \sum_{n=0}^{\infty} t^n.$$

Moreover,

$$h(t) = \frac{1}{1-(A+1)t} - \frac{t}{1-(A+1)t}$$

$$= \sum_{n=0}^{\infty} (1+A)^n t^n - \sum_{n=0}^{\infty} (1+A)^n t^{n+1}$$

$$= 1 + \sum_{n=1}^{\infty} A(1+A)^{n-1} t^n.$$

Since $g^{(n)}(0) = n!$, $f^{(n)}(g(0)) = f^{(n)}(1) = n!A^n$ and $h^{(n)}(0) = n!A(1+A)^{n-1}$, application of the Faà di Bruno formula gives the desired equality.

3.4. Taylor Series

3.4.19 [18]. Let x_0 be arbitrarily chosen in $\mathbf{I}$ and let $y_0 = f(x_0)$. It follows from 3.4.17 that there are intervals $\mathbf{I}_1 \subset \mathbf{I}$ and $\mathbf{J}_1 \subset \mathbf{J}$ (containing x_0 and y_0, respectively) and positive constants A, B, C and D such that

$$|f^{(n)}(x)| \leq A \frac{n!}{B^n} \quad \text{for} \quad x \in \mathbf{I}_1$$

and

$$|g^{(n)}(y)| \leq C \frac{n!}{D^n} \quad \text{for} \quad y \in \mathbf{J}_1.$$

By the formula of Faà di Bruno,

$$h^{(n)}(x) = \sum \frac{n!}{k_1! k_2! \cdots k_n!} g^{(k)}(f(x)) \Big(\frac{f^{(1)}(x)}{1!}\Big)^{k_1} \Big(\frac{f^{(2)}(x)}{2!}\Big)^{k_2} \cdots \Big(\frac{f^{(n)}(x)}{n!}\Big)^{k_n},$$

where $k = k_1 + k_2 + \cdots + k_n$ and the sum is taken over all $k_1, k_2, \ldots, k_n$ such that $k_1 + 2k_2 + \cdots + nk_n = n$. This combined with the result in the preceding problem gives

$$|h^{(n)}(x)| \leq \sum \frac{n!}{k_1! k_2! \cdots k_n!} \frac{Ck!}{D^k} \Big(\frac{A}{B^1}\Big)^{k_1} \Big(\frac{A}{B^2}\Big)^{k_2} \cdots \Big(\frac{A}{B^n}\Big)^{k_n}$$

$$= \sum \frac{n!}{k_1! k_2! \cdots k_n!} \frac{Ck!}{D^k} \frac{A^k}{B^n} = \frac{n!C}{B^n} \sum \frac{k!}{k_1! k_2! \cdots k_n!} \frac{A^k}{D^k}$$

$$= \frac{n!C}{B^n} \frac{A}{D} \Big(1 + \frac{A}{D}\Big)^{n-1}.$$

It now follows from the result in 3.4.16 that h is real analytic on $\mathbf{I}$.

3.4.20. It follows from 3.4.15 that $g(x) = f(-x)$ is real analytic on the interval $-\mathbf{I} = \{x : -x \in \mathbf{I}\}$. Since $x \mapsto -x$ is real analytic, the result follows from the foregoing problem.

3.4.21 [18]. Consider $g(t) = 1 - \sqrt{1 - 2t}$, $|t| < 1/2$, and $f(x) = \frac{1}{1-x}$, $|x| < 1$. Then

$$h(t) = f(g(t)) = \frac{1}{\sqrt{1-2t}} = g'(t).$$

So $g^{(n+1)}(t) = h^{(n)}(t)$. Moreover, by the Newton binomial formula (see 3.4.4),

$$g(t) = -\sum_{n=1}^{\infty} \binom{\frac{1}{2}}{n} (-2t)^n.$$

Clearly, $f(x) = \sum_{n=0}^{\infty} x^n$. Consequently, $g^{(n)}(0) = -n!\binom{\frac{1}{2}}{n}(-2)^n$ and $f^{(n)}(g(0)) = n!$. Finally, by the formula of Faà di Bruno,

$$-(n+1)!\binom{\frac{1}{2}}{n+1}(-2)^{n+1} = g^{(n+1)}(0) = h^{(n)}(0)$$

$$= n! \sum \frac{k!}{k_1! k_2! \cdots k_n!} \left(-\binom{\frac{1}{2}}{1}(-2)\right)^{k_1} \cdots \left(-\binom{\frac{1}{2}}{n}(-2)^n\right)^{k_n}$$

$$= (-2)^n n! \sum \frac{(-1)^k k!}{k_1! k_2! \cdots k_n!} \binom{\frac{1}{2}}{1}^{k_1} \cdots \binom{\frac{1}{2}}{n}^{k_n},$$

where $k = k_1 + k_2 + \cdots + k_n$ and the sum is taken over all $k_1, k_2, \ldots, k_n$ such that $k_1 + 2k_2 + \cdots + nk_n = n$.

3.4.22 [18]. Observe first that if f satisfies the assumptions stated in the problem, then its inverse g exists in an open interval containing $f(x_0)$. Moreover,

$$g'(y) = h(g(y)), \quad \text{where} \quad h(x) = \frac{1}{f'(x)}.$$

It is also clear that since f is in C^∞, so is g. Our task is now to prove that g satisfies the assumptions of 3.4.16. We know, by 3.4.19, that h is analytic in some open interval containing x_0 (as a composition of two analytic functions). Therefore, by 3.4.17, there are positive constants A and B such that

(1) $$|h^{(n)}(x)| \leq A \frac{n!}{B^n}$$

in some open interval $\mathbf{I}_0 \subset \mathbf{I}$ containing x_0. Now induction will be used to show that there is an open interval $\mathbf{K}$ containing $f(x_0)$ such that

(2) $$|g^{(n)}(y)| \leq n!(-1)^{n-1}\binom{\frac{1}{2}}{n}\frac{(2A)^n}{B^{n-1}} \quad \text{for} \quad y \in \mathbf{K}.$$

We choose $\mathbf{K}$ so that $g(\mathbf{K})$ is contained in $\mathbf{I}_0$. Then, by (1), we have $|g'(y)| = |h(g(y))| \leq A$, which proves (2) for $n = 1$. Assuming (2) to hold for $k = 1, 2, \ldots, n$, we will prove it for $n + 1$. By the foregoing

3.4. Taylor Series

problem we get

$$|g^{(n+1)}(y)| = |(h \circ g)^{(n)}(y)|$$

$$\leq n! \sum \frac{k!}{k_1!k_2!\cdots k_n!} \frac{A}{B^k} \left(\binom{\frac{1}{2}}{1}(2A)\right)^{k_1} \cdots \left((-1)^{n-1}\binom{\frac{1}{2}}{n}\frac{(2A)^n}{B^{n-1}}\right)^{k_n}$$

$$= (-1)^n n! \frac{(2A)^n}{B^n} A \sum \frac{(-1)^k k!}{k_1!k_2!\cdots k_n!} \binom{\frac{1}{2}}{1}^{k_1} \cdots \binom{\frac{1}{2}}{n}^{k_n}$$

$$= (-1)^n n! \frac{(2A)^n}{B^n} A2(n+1) \binom{\frac{1}{2}}{n+1}$$

$$= (-1)^n (n+1)! \frac{(2A)^{n+1}}{B^n} \binom{\frac{1}{2}}{n+1}.$$

This completes the proof of (2). Thus the analyticity of g on **K** follows from 3.4.16.

3.4.23. It follows from $f^{-1}(x) = f'(x)$ that f maps the interval $(0, \infty)$ onto itself and that f is in C^∞ on that interval. Hence $f'(x) > 0$, and f is strictly increasing on $(0, \infty)$. Differentiating the equality $f(f'(x)) = x$, we see that $f''(x) > 0$ for $x \in (0, \infty)$. We claim that $(-1)^n f^{(n)}(x) > 0$ for $x \in (0, \infty)$ and $n \geq 2$. We will prove this by induction, using the formula of Faà di Bruno (see 2.1.38). Suppose that $(-1)^m f^{(m)}(x) > 0$ for $m = 2, 3, \ldots, n$. Then

$$0 = \sum \frac{n!}{k_1!k_2!\cdots k_{n-1}!} f^{(k)}(f'(x)) \left(\frac{f''(x)}{1!}\right)^{k_1} \left(\frac{f^{(3)}(x)}{2!}\right)^{k_2}$$

$$\cdots \left(\frac{f^{(n)}(x)}{(n-1)!}\right)^{k_{n-1}} + f'(f'(x))f^{(n+1)}(x),$$

where $k = k_1 + \cdots + k_{n-1}$ and the sum is taken over all $k_1, \ldots, k_{n-1}$ such that $k_1 + 2k_2 + \cdots + (n-1)k_{n-1} = n$. Since the sign of each term under $\sum$ is

$$\text{sgn}\left((-1)^k(-1)^{2k_1}(-1)^{3k_2}\cdots(-1)^{nk_{n-1}}\right) = (-1)^n,$$

we get

$$\text{sgn}\left(f'(f'(x))f^{(n+1)}(x)\right) = \text{sgn} f^{(n+1)}(x) = -(-1)^n.$$

Now the result in 3.4.20 shows that f is analytic on $(0,\infty)$.

3.4.24. We know, by the foregoing problem, that each f satisfying the assumption is analytic on $(0,\infty)$. We first show that there is exactly one number $a > 0$ such that $f(x) < x$ if $x \in (0, a)$, and $f(x) > x$ if $x > a$. To this end, observe that by the monotonicity of f we have $\lim\limits_{x \to 0^+} f(x) = 0$, which together with the equality $f'(f(x))f'(x) = xf'(x)$ gives

(1) $$f(f(x)) = \int_0^x tf'(t)dt.$$

Now if $f(x)$ were greater than x for $0 < x < 1$, then (1) would imply

$$\int_0^x f'(t)(t-1) > 0,$$

contrary to the fact that $f'(x) > 0$ for $x > 0$. On the other hand, if $f(x) < x$ for all $x \in (0, \infty)$, then (1) would imply

$$f(x) > f(f(x)) = \int_0^x tf'(t)dt > \int_0^x f(t)f'(t)dt = \frac{1}{2}(f(x))^2,$$

which in turn would give $f(x) < 2$ for $x > 0$, contrary to the fact that $f((0,\infty)) = (0,\infty)$. Consequently, by the intermediate value property there is a fixed point a of f. Since $f(x) < x$ for $x \in (0, a)$, we see that $f'(y) = f^{-1}(y) > y$ for $y \in (0, a)$. Likewise, $f'(y) < y$ for $y > a$.

Now we turn to the proof of the uniqueness. Suppose, contrary to our claim, that there are two such functions f_1 and f_2. Let a_1 and a_2 be the fixed points of f_1 and f_2, respectively. Clearly, we can assume that $a_1 \geq a_2$. Set $g = f_1 - f_2$. If $a_1 = a_2 = a$, then $g(a) = 0$ and $f^{-1} = f'$ implies that $g^{(n)}(a) = 0$ for $n \in \mathbb{N}$. So since g is analytic, g is a constant function (equal to zero) on $(0,\infty)$. If $a_1 > a_2$, then $f_1(x) < x \leq f_2(x)$ and $f_1'(x) > x \geq f_2'(x)$ for $[a_2, a_1)$. Therefore $g(x) < 0$ and $g'(x) > 0$ for $x \in [a_2, a_1)$. Since $\lim\limits_{x \to 0^+} g(x) = 0$, there is $b \in (0, a_2)$ such that $g'(b) = 0$ and $g'(x) > 0$ for $x \in (b, a_1)$, and $g(x) < 0$ for $x \in [b, a_1)$. Set $f_1'(b) = f_2'(b) = b'$. Then $b' \in (b, a_2)$, because $b < f_2'(b) = b' < f_2'(a_2) = a_2$. Hence $g(b') < 0$. On the other hand, $f_1(b') = f_1(f_1'(b)) = b$ and $f_2(b') = f_2(f_2'(b)) = b$, a contradiction.

3.4. Taylor Series 389

3.4.25. If $f(x) = ax^c$, then $f'(x) = acx^{c-1}$ and $f^{-1}(x) = a^{-\frac{1}{c}}x^{\frac{1}{c}}$. This gives $c = \frac{1+\sqrt{5}}{2}$ and $a = c^{1-c}$.

3.4.26. By Taylor's formula proved in 2.3.10,

$$\ln(1+x) = 2\sum_{n=0}^{N} \frac{1}{2n+1}\left(\frac{x}{2+x}\right)^{2n+1} + R_N(x),$$

where

$$R_N(x) = \frac{2}{(2N+1)(1+\theta x)^{2N+3}}\left(\frac{x}{2}\right)^{2N+3}.$$

Clearly, $\lim\limits_{N\to\infty} R_N(x) = 0$ for $x \in (0,2)$. Consequently,

$$\ln(1+x) = 2\sum_{n=0}^{\infty} \frac{1}{2n+1}\left(\frac{x}{2+x}\right)^{2n+1}.$$

3.4.27 [Tung-Po Lin, Amer. Math. Monthly 81 (1974), 879-883]. By definition,

$$\frac{L(x,y)}{M_p(x,y)} = \frac{\frac{x-y}{\ln x - \ln y}}{\left(\frac{x^p+y^p}{2}\right)^{1/p}} = \frac{2^{1/p}(x-y)}{(x^p+y^p)^{1/p}\ln\frac{x}{y}}$$

for distinct positive x and y and for $p \neq 0$. Dividing the numerator and denominator by y and putting $z = \left(\frac{x}{y}\right)^p$, we obtain

$$\frac{L(x,y)}{M_p(x,y)} = \frac{2^{1/p}\left(z^{1/p}-1\right)}{(z+1)^{1/p}\ln z^{1/p}}.$$

Now writing

$$z = \frac{1+w}{1-w} \qquad \left(w = \frac{z-1}{z+1},\ 0 < |w| < 1\right)$$

and multiplying the numerator and denominator by $\frac{(1-w)^{1/p}}{2w}$, we arrive at

$$\frac{L(x,y)}{M_p(x,y)} = \frac{p2^{1/p}\left(\left(\frac{1+w}{1-w}\right)^{1/p} - 1\right)}{\left(\frac{1+w}{1-w}+1\right)^{1/p}\ln\frac{1+w}{1-w}}$$

$$= \frac{\frac{p\left((1+w)^{1/p}-(1-w)^{1/p}\right)}{2w}}{\frac{\ln(1+w)-\ln(1-w)}{2w}} = \frac{f(w,p)}{g(w)}.$$

Clearly,
$$g(w) = \sum_{n=0}^{\infty} \frac{1}{2n+1} w^{2n},$$

and by 3.4.4,

$$f(w,p) = 1 + \sum_{n=1}^{\infty} \frac{1}{2n+1}\left(\left(\frac{1}{p}-1\right)\left(\frac{1}{p}-2\right)\cdots\left(\frac{1}{p}-2n\right)\cdot\frac{1}{(2n)!}\right)w^{2n}.$$

Consequently, to prove that $f(w,p) < g(w)$ it suffices to show that for any positive integer n,

$$\left(\frac{1}{p}-1\right)\left(\frac{1}{p}-2\right)\cdots\left(\frac{1}{p}-2n\right)\frac{1}{(2n)!} \leq 1$$

and the strict inequality holds for at lest one n. For $n = 1$, we have

$$\left(\frac{1}{p}-1\right)\left(\frac{1}{p}-2\right)\cdot\frac{1}{2} \leq 1 \quad \text{for} \quad p \geq \frac{1}{3},$$

because
$$\frac{1}{2p^2} - \frac{3}{2p} + 1 = 1 - \frac{1}{2p}\left(3-\frac{1}{p}\right) < 1, \quad \text{if} \quad 0 < \frac{1}{p} < 3.$$

Hence
$$Q_n = \frac{\left(\frac{1}{p}-1\right)\left(\frac{1}{p}-2\right)\left(\frac{1}{p}-3\right)\cdots\left(\frac{1}{p}-2n\right)}{(2n)!}$$

$$= \underbrace{\left(1-\frac{1}{p}\right)\left(1-\frac{1}{2p}\right)}_{Q_1}\underbrace{\left(1-\frac{1}{3p}\right)\cdots\left(1-\frac{1}{2np}\right)}_{<1}$$

for $p \geq \frac{1}{3}$. So $Q_1 \leq 1$ for $p \geq \frac{1}{3}$, and the last formula shows that $Q_n < 1$ for $n = 2, 3, \ldots$.

3.4. Taylor Series

3.4.28 [Tung-Po Lin, Amer. Math. Monthly 81 (1974), 879-883]. We adopt notation from the solution of the foregoing problem. We have $Q_1 > 1$ for $p < \frac{1}{3}$. Thus there is $0 < h < 1$ such that if $0 < w < h$, then $f(w, p) > g(w)$. Now observe that the inequality $0 < w < h$ can be rewritten in the form

$$1 < z < r^p, \quad \text{where} \quad r = \left(\frac{1+h}{1-h}\right)^{1/p} \quad \text{and} \quad z = \left(\frac{x}{y}\right)^p.$$

This means that there is an $r > 1$ such that $L(x, y) > M_p(x, y)$ if $1 < \frac{x}{y} < r$.

3.4.29 [Tung-Po Lin, Amer. Math. Monthly 81 (1974), 879-883]. Putting $\frac{x}{y} = \frac{(1+w)^2}{(1-w)^2}$, $0 < |w| < 1$, we get

$$\frac{L(x,y)}{M_0(x,y)} = \frac{\frac{x-y}{\ln x - \ln y}}{(xy)^{1/2}} = \frac{\frac{\frac{x}{y}-1}{\ln \frac{x}{y}}}{\left(\frac{x}{y}\right)^{1/2}} = \frac{\frac{(1+w)^2}{(1-w)^2}-1}{4(w+\frac{1}{3}w^3+\frac{1}{5}w^5+\cdots)}$$
$$= \frac{1}{1-w^2} \cdot \frac{1}{1+\frac{1}{3}w^2+\frac{1}{5}w^4+\cdots}$$
$$= \frac{1+w^2+w^4+w^6+\cdots}{1+\frac{1}{3}w^2+\frac{1}{5}w^4+\frac{1}{7}w^6+\cdots} > 1,$$

which combined with 2.5.42 and 2.5.43 implies the desired result.

3.4.30 [Tung-Po Lin, Amer. Math. Monthly 81 (1974), 879-883]. Under the notation introduced in the solution of 3.4.27, we have

$$\frac{L(x,y)}{M_p(x,y)} = \frac{p\left((1+w)^{1/p} - (1-w)^{1/p}\right)}{\ln \frac{1+w}{1-w}} \xrightarrow[w\to 1]{} 0.$$

Since $z = \left(\frac{x}{y}\right)^p = \frac{1+w}{1-w}$, we get $L(x,y) < M_p(x,y)$ for sufficiently large z.

Bibliography - Books

References

[1] J. Banaś, S. Wędrychowicz, *Zbiór zadań z analizy matematycznej*, Wydawnictwa Naukowo-Techniczne, Warszawa, 1994.

[2] V. I. Bernik, O.V. Melnikov, I. K. Žuk,, *Sbornik olimpiadnych zadač po matematike*, Narodnaja Asveta, Minsk, 1980.

[3] P. Biler, A. Witkowski, *Problems in Mathematical Analysis*, Marcel Dekker, Inc, New York and Basel, 1990.

[4] T. J. Bromwich, *An Introduction to the Theory of Infinite Series*, Macmillan and Co., Limited, London, 1949.

[5] R. P. Boas, *A Primer of Real Analytic Functions*, Birkhäuser Verlag, Basel Boston Berlin, 1992.

[6] L. Carleson, T. W. Gamelin, *Complex Dynamics*, Springer-Verlag, New York Berlin Heidelberg, 1993.

[7] B. P. Demidovič, *Sbornik zadač i upražnenij po matematičeskomu analizu*, Nauka, Moskva, 1969.

[8] J. Dieudonné, *Foundations of Modern Analysis*, Academic Press, New York San Francisco London, 1969.

[9] A. J. Dorogovcev, *Matematičeskij analiz. Spravočnoe posobe*, Vyščaja Škola, Kiev, 1985.

[10] A. J. Dorogovcev, *Matematičeskij analiz. Sbornik zadač*, Vyščaja Škola, Kiev, 1987.

[11] G. M. Fichtenholz, *Differential-und Integralrechnung, I, II, III*, V.E.B. Deutscher Verlag Wiss., Berlin, 1966-1968.

[12] B. R. Gelbaum, J. M. H. Olmsted, *Theorems and Counterexamples in Mathematics*, Springer-Verlag, New York Berlin Heidelberg, 1990.

[13] E. Hille, *Analysis Vol.I*, Blaisdell Publishing Company, New York Toronto London, 1964.

[14] W. J. Kaczor, M. T. Nowak, *Problems in Mathematical Analysis I. Real Numbers, Sequences and Series*, American Mathematical Society, Providence, RI, 2000.

[15] G. Klambauer, *Mathematical Analysis*, Marcel Dekker, Inc., New York, 1975.

[16] G. Klambauer, *Problems and Propositions in Analysis*, Marcel Dekker, Inc., New York and Basel, 1979.

[17] K. Knopp, *Theorie und Anwendung der Unendlichen Reihen*, Springer-Verlag, Berlin and Heidelberg, 1947.

[18] S. G. Krantz, H. R. Parks, *A Primer of Real Analytic Functions*, Birkhäuser Verlag, 1992.

[19] L. D. Kudriavtsev, A. D. Kutasov, V. I. Chejlov, M. I. Shabunin, *Problemas de Análisis Matemático. Límite, Continuidad, Derivabilidad (Spanish)*, Mir, Moskva, 1989.

[20] K. Kuratowski, *Introduction to Calculus*, Pergamon Press, Oxford-Edinburgh-New York; Polish Scientific Publishers, Warsaw, 1969.

[21] S. Łojasiewicz, *An Introduction to the Theory of Real Functions*, A Wiley-Interscience Publication, John Wiley & Sons, Ltd., Chichester, 1988.

[22] D. S. Mitrinović, *Elementary Inequalities*, P. Noordhoff Ltd., Groningen, 1964.

[23] G. Pólya, G. Szegö, *Problems and theorems in analysis I*, Spriger-Verlag, Berlin Heidelberg New York, 1978.

[24] R. Remmert, *Theory of Complex Functions*, Springer-Verlag, New York Berlin Heidelberg, 1991.

[25] J.I. Rivkind, *Zadači po matematičeskomu analizu*, Vyšejšaja Škola, Mińsk, 1973.

[26] W. I. Rozhov, G. D. Kurdevanidze, N. G. Panfilov, *Sbornik zadač matematičeskich olimpiad*, Izdat. Univ. Druzhby Narodov, Moskva, 1987.

[27] W. Rudin, *Principles of Mathematical Analysis*, McGraw-Hill Book Company, New York, 1964.

[28] W. Rzymowski, *Convex Functions*, preprint.

[29] W. A. Sadovničij, A. S. Podkolzin, *Zadači studenčeskich olimpiad po matematike*, Nauka, Moskva, 1978.

[30] R. Sikorski, *Funkcje rzeczywiste*, PWN, Warszawa, 1958.

[31] H. Silverman, *Complex variables*, Houghton Mifflin Company, Boston, 1975.

[32] E.C. Titchmarsh, *The Theory of Functions*, Oxford University Press, London, 1944.

[33] G. A. Tonojan, W. N. Sergeev, *Studenčeskije matematičeskije olimpiady*, Izdatelstwo Erevanskogo Universiteta, Erevan, 1985.

Index

Abel test for uniform convergence, 92
Abel theorem, 99
approximation theorem of Weierstrass, 87

Baire property, 79
Baire theorem, 23
Bernstein polynomial, 87
Bernstein theorem, 105

Cantor set, 146
Cauchy criterion for uniform convergence, 82
Cauchy functional equation, 27
Cauchy theorem, 8
continuous convergence, 84
convergence in the Cesàro sense, 30

deleted neighborhood, 3
Dini derivative, 281
Dini theorem, 84
Dirichlet series, 94
Dirichlet test for uniform convergence, 91

equicontinuity, 84
extended real number system, 18

Faà di Bruno formula, 44
fixed point, 15
function
  concave, 61
  continuous in the Cesàro sense, 30
  convex, 14, 61
  decreasing, 3
  first Baire class, 34

increasing, 3
midpoint-convex, 66
monotone, 3
piecewise strictly monotone, 17
real analytic, 105
semicontinuous, 19
strictly convex, 61
strictly decreasing, 3
strictly increasing, 3
strictly monotone, 3
subadditive, 67
uniformly continuous, 24
uniformly differentiable, 49
fundamental period, 13

Hamel basis, 197
Hardy and Littlewood theorem, 101
Hölder inequality, 65

incommensurate, 13
intermediate value property, 14
iterate, 9

Jensen equation, 28
Jensen inequality, 62

logarithmic mean, 74

Minkowski inequality, 65
  generalized, 66
modulus of continuity, 27

Newton binomial formula, 102
Newton method, 76

397

oscillation, 23, 33

power mean, 74

radius of convergence, 97
remainder term of Taylor series
    Cauchy form, 53
    integral form, 53
    Lagrange form, 53
    Peano form, 52
    Schlömilch-Roche form, 52
residual set, 78
Riemann zeta-function, 96

Schwarz derivative, 77
    lower, 77
    uniform, 79
    upper, 77
strong derivative, 77
    lower, 77
    upper, 77
symmetric derivative, 77

Tauber theorem, 100

uniform convergence, 81

Weierstrass function, 96